Multiplicity Yurs

Cloning, Stem Cell Research, and
Regenerative Medicine

Multiplicity Y urs

Cloning, Stem Cell Research, and Regenerative Medicine

Dr. Hwa A. Lim *Ph.D., MBA*
Silicon Valley, California, USA

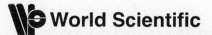

World Scientific

NEW JERSEY • LONDON • SINGAPORE • BEIJING • SHANGHAI • HONG KONG • TAIPEI • CHENNAI

Published by

World Scientific Publishing Co. Pte. Ltd.

5 Toh Tuck Link, Singapore 596224

USA office: 27 Warren Street, Suite 401-402, Hackensack, NJ 07601

UK office: 57 Shelton Street, Covent Garden, London WC2H 9HE

British Library Cataloguing-in-Publication Data
A catalogue record for this book is available from the British Library.

MULTIPLICITY YOURS
Cloning, Stem Cell Research, and Regenerative Medicine

ISBN 981-256-865-4
ISBN 981-256-866-2 (pbk)

Printed by FulIsland Offset Printing (S) Pte Ltd, Singapore

Hwa A. Lim and his "clone," HAL.[α]

Dr. Hwa A. Lim is an internationally respected authority on bioinformatics and biotechnology. Currently he is active in both the academic and the private sectors. He is on the boards of several biotech companies; an adjunct professor of Molecular & Cell Biology; and Mathematical Sciences of the University of Texas at Dallas; a visiting professor of the Institute of Genetics of the Chinese Academy of Sciences.

Besides his many appointments as scholar, technologist and entrepreneur, Dr. Lim is an articulate and well sought-after speaker at international meetings. He travels extensively to do business, to lecture, and to mingle with locals to experience and learn firsthand – something that he enjoys as a part-time writer writing on diverse topics. He is a Kingstone Best-Seller author and the author of fourteen books in English.

Dr. Lim is credited with coining the neologism "Bioinformatics," establishing and shaping the field, and initiating the world's very first bioinformatics conference series. These credits earn him the title "The Father of Bioinformatics."

As a bioinformaticist, he has served as a bioinformatics expert for the United Nations to help set up biotech research parks, and as a review panelist for United States federal agencies (including National Cancer Institute, National Science Foundation), and as a consultant for prominent consulting firms (VAXA, McKinsey), financial firms (Robertson Stephens, Prudential), biotech, pharmaceutical and healthcare companies (Eli Lilly and Company, Monsanto and Company), organizations, and governments (China, France, India, Korea, Malaysia, and Taiwan). He has the distinction of being a key member of two separate teams that took two distinct companies IPO in the United States.

In 1997, he founded a professional biotech and nanotech consulting company. Prior to this venture, he was director and vice president of two biotech companies (1995-1997), Program Director and tenured state-line faculty of the Supercomputer Computations Research Institute, Florida State University (1987-1995).

Dr. Lim obtained his Ph.D. (science), M.A. (science), and MBA (strategy and business laws) from the United States, his B.Sc. (honours) and ARCS from Imperial College of Sc. Tech. & Medicine, the University of London, United Kingdom. Hal resides in Silicon Valley, California, USA. He can be reached at hal_lim@yahoo.com.

[α] HAL are the initials of the author, Hwa Aun Lim. HAL is also the name of a supercomputer, HAL9000, in the flick "Odyssey 2001." Go to hal_lim@yahoo.com for comments.

Selected Titles by Dr. Hwa A. Lim

1. Nikolay A. Kolchanov and Hwa A. Lim, *Computer Analysis of Genetic Macromolecules: Structure, function and evolution*, (World Scientific Publishing Co., New Jersey, 1994), 556 pages.
2. Ralf Hofestädt and Hwa A. Lim (Hrsg.), *Molecular Bioinformatics – Sequence analysis*, (Shaker Verlag, Aachen, Germany, 1997), 60 pages.
3. John R. McCarrey, John L. VandeBerg, and Hwa A. Lim (eds.), *Genes, Gene Families, and Isozymes*, The Journal of Experimental Zoology, Vol. 282, No. 1/2, (Wiley-Liss, New York, 1998), 283 pages.
4. Hwa A. Lim (Guest editor), *Pathways of Bioinformatics: From data to diseases*, Briefings in Bioinformatics, Vol. 3(1), (Henry-Stewart Publishers, London, 2002), 109 pages.
5. Hwa A. Lim, *Genetically Yours: Bioinforming, biopharming, and biofarming*, (World Scientific Publishing Co., New Jersey, 2002), 417 pages.
6. Hwa A. Lim, *Change: In business, corporate governance, education, scandals, technology, and warfare*, (EN Publishing, Inc., Santa Clara, California, 2003), 488 pages.

Praises for Selected Titles by Dr. Hwa A. Lim

Genetically Yours:

"Genetically Yours which I enjoyed reading. You were kind to share your work with me, and I appreciate your thoughtfulness."

Honorable William Jefferson Clinton
42nd President, United States

"Thoughtful and provocative book comes at the perfect time."

Malorye A. Branca
Vice President, Cambridge Healthtech Institute Publications
Cambridge, USA

"I can only say BRAVO!"

Da Hsuan Feng, Ph.D.
Vice President, Research and Economic Development
The University of Texas at Dallas, Richardson, Texas, USA

"Genetically Yours is a true 'Bio-Bible'…"

Olivier Griperay
Vice President, Biotech Projects, Invest in France Agency North America
The French Agency for International Investment, USA

"Well timed and beautifully written."

Salah Mandil, Ph.D.
Director, Health Informatics & Telematics
World Health Organization, Geneva, Switzerland

"A must read for anyone interested in a fresh perspective."

Dalia Cohen, Ph.D.
Vice President, Global Head of Functional Genomics
Novartis Pharmaceuticals, New Jersey, USA

"An informative and comprehensive book."

T.V. Venkatesh, Ph.D.
Project Leader, Bioinformatics
Monsanto & Company, Missouri, USA

Change:

[Change] addresses a number of issues I care deeply about, and I appreciate your continued thoughtfulness.

Honorable William Jefferson Clinton
42nd President, United States

"This is a book for anyone who wants to understand and master change."

Honorable Congressman Curt Weldon
Vice Chairman, House Armed Services and Homeland Security Committees
United States House of Representatives, Washington, DC, USA

"The book delivers important insights each of us can use in business and in our personal, professional, and family lives."

Joseph R. Whaley, Esq.
Attorney at law
Rockville, Maryland, USA

"Precisely a book on rate of change."

Da Hsuan Feng, Ph.D.
Vice President for Research and Economic Development
The University of Texas at Dallas, Texas, USA

"This book dissects the world of change into main chunks of vibrant and varied changes, and placing before the reader a series of snapshots of the factors that comprise the changes in each of these chunks."

Salah H. Mandil, Ph.D.
Vice President of eStrategies, WiseKey SA
Geneva, Switzerland

"Once read, you will never look at the nature of things in quite the same way, but then again, this is the nature of change."

Gary N. Keller, MSB
Chief Executive Officer
Xomix Ltd., Chicago, Illinois, USA

Sex is so Good, why Clone?

"The importance of human intimacy in procreation is beautifully presented from an ethical and factual point of view, reminding us of the true value of human relationships."

Betty Pace, M.D.
Director, Sickle Cell Research Center
The University of Texas at Dallas, Texas, USA

"I highly recommend this great book for general readership, life science or legal ethics class reading."

Teruhiko Wakayama, Ph.D.
Inventor of the Honolulu Cloning Technique, and Cloner of the first adult mouse
Head of Laboratory, Laboratory for Genomic Programming
Center for Developmental Biology, RIKEN, Kobe, Japan

"This book reviews current life science knowledge, and the social, legal and ethical pros and cons of this complex subject."

Olivier Griperay
Vice President, Life Sciences, Invest in France Agency, North America
The French Government Agency for International Investment, New York, USA

"This book is a refreshing way to present the question of sex."

Professor Antoine Danchin
Head of the Genetics of Bacterial Genomes Unit
Centre National de la Recherché Scientifique URA 2171, France

"With his signature light-hearted clarity, the author describes and analyses the scientific achievements in cloning thus far."

Salah H. Mandil, Ph.D.
Senior Expert Consultant on eHealth & eStrategies
The International Telecommunications Union, Switzerland

"The author has painstakingly put together a tome full of information."

Nor Azman Nordin
Editor/Asst Manager – Publication, Symbiosis

To

My Parents
From whom I inherit good genes

My Brothers and Sisters
With whom I share common genes

My Teachers, Friends and Colleagues
Among whom we have varying views on the issues of genes

Acknowledgments

This book represents a cumulative experience of more than ten years as faculty member and later concurrent appointments in the corporate world. It also represents the past ten years of lectures to experts, non-technical professionals, university and high school students, and laypersons. And very importantly, it reflects months of research and writing, including following a pair of twins for a few years.

Prior to becoming very actively involved in the corporate world and in advising governments, I was a tenured university faculty member and researcher. One of the things I learned during those academically prolific years was the importance of evidence and the necessity to document sources. It chagrins me to have to admit it would be impossible for me to provide an exhaustive literature citation.

From the hoards of interesting articles and reports that friends have forwarded electronically, informative conversations I have had and interesting lectures I have attended, I have pillaged cheerfully whenever it was appropriate. Some of the original sources of these articles and conversations have since been lost.

Writing a book is never an easy task, particularly when between the covers is a mixed bag of topics, some of which are very popular and some are controversial with a bag of ethical ramifications. Sex and reproduction are very popular topics; cloning, human cloning in particular, is a very popular but hotly debated issue; stem cell research and its applications in regenerative medicine always bring rays of hope to many people awaiting breakthroughs. The task of writing, however, was made easier by the heavy media coverage: *CNN News*, *The New York Times*, *The San Francisco Chronicle*, *The Wall Street Journal*, *The Washington Post*, *Time Magazine*, *Newsweek*; which are the first portals for me to go into the subjects of interest by reading up credible and peer-reviewed scholarly reports: *Scientific America*, *Science*, *Nature*, many other online or offline publications, journals, books too numerous to mention each by name. Whenever appropriate, I acknowledge them in the text as footnotes. The task now has shifted to providing an unbiased account – I have to make sure the sources are credible and my understanding of the subject matter is not swayed or influenced by any particular interest groups.

The editors of the publishing house spent a considerable amount of effort and time to ensure the book is error-free. Thus while taking pride for keying in every single word in this book myself, I also assume full responsibility – any remaining errors are entirely mine. The same may not be said of translated variants of this original version in English.

I acknowledge the patience of many friends, colleagues and mentors, not mutually exclusive, with whom I have business ventures and collaborations, for being so understanding and so supportive. Their varying views widen my perspective on many of the issues touched upon in this book. Their inputs keep my view from becoming too skewed so that I can present an unbiased view for intelligent readers to make judgments for themselves.

Writing this book takes time. I applaud the great flexibility D'Trends has been granting me. I do most of my reading, thinking and observations while I am on business trips, in flights, between flights, in traffic jams, and very importantly, during the question-and-answer session after each of my lectures.

Late evenings are some of my most productive times. I enjoy watching Tom and Jerry cartoons in the wee hours to break up my writing task into more manageable chunks. During the day and early evenings, on a regular basis, I also watch cartoons with young friends. Their inquisitive and naive questions have also wended their way into this book. My references to computer animated cartoon characters throughout the book are also results of these enjoyable and relaxing sessions.

Writing while ballroom, Latin and hustle dancing is an impossible task even for professional dancers, let alone for an amateur dancer like myself.[β] This once-a-week activity helps me unwind and increases my overall productivity. I thank all the dancers for not stepping on my toes.

Lest I forget, I must mention my loyal two-year-old pet dog. Daily and without fail, she would wait for me, wag her tail to welcome me home and then take me running in a nearby trail every other day if I still behave at the end of a long workday. Otherwise, she would normally lie next to me while I am tapping away on my laptop computer keyboard writing this book. I am seriously thinking about having her genetic material cryogenically preserved so that I can have her cloned when the sad day comes. As you will see, the cat has been cloned, so has the dog, though more recently than the cat (2005).

[β] Ballroom dancing is now an event recognized by the International Olympics Committee (IOC). To qualify as an Olympics event, it has been renamed dancesport. On September 8, 1997, IOC announced that it had granted outright recognition to the International Dancesport Federation (IDSF) as a full member of IOC. Still dancesport will not be a program in the Olympics until the 2008 Olympics in Beijing, China.

There is one more group of people to acknowledge – those who have been the center of almost everything I have undertaken – my family. I am particularly indebted to my mother, brothers, sisters, and immediate relatives for reasons that go beyond this book. They have provided unending support in whatever I undertake, made countless sacrifices and provided the emotional support that makes writing this book possible. I am also grateful to them for being the persons who inspire me to undertake the challenge of writing this book.

I hope you will enjoy this odyssey into the biosphere as you flip through the pages in this book as much as I enjoy writing this book.

Hwa A. Lim
hal_lim@yahoo.com
Spring of a prolific year
Silicon Valley
California, USA

Preface

According to many scholars, the biblical character Noah built Noah's Ark. At the command of God, Noah built the craft Noah's Ark to accommodate 50,000 species of animals and one million species of insects to save them from a divinely planned universal flood. The animals and insects, pairs of each – male and female – constituted a core of breeding stock, that is, to propagate by coital reproduction.

When I finally completed the first draft of this book, yet another Noah, a cloned gaur bull, a nearly extinct wild ox, had come and gone for more than thirty months. Noah was born at 7:30 p.m. on Monday, January 8, 2001. While healthy at birth, Noah died within 48 hours of a common type of dysentery believed unrelated to cloning. The birth of Noah was significant in various ways: First, it was the first endangered species cloned and born alive; Second, the gaur embryo was created by fusing a skin cell from a deceased gaur with a cow's egg; Third, the embryo was implanted into the womb of a cow serving as a surrogate mother and brought to term. Thus Noah was a trans-species clone – the gaur's genetic material nourished in a cow's egg to develop into an embryo.[x]

Though saddened by the news of the untimely passing of Noah, scientists expressed optimism that science has advanced to the point of being able to successfully create a healthy trans-species gaur clone. Scientists are learning to improve and perfect the process and have continued hope for its inevitable role in the conservation of endangered species.

The chronology of cloning would have been so much more romantic had Noah been the first creature to walk up the ramp of the ark of endangered species that scientists are currently attempting to clone. Plans are underway to clone the African bongo antelope, the Sumatran tiger, the reluctant-to-reproduce giant panda, and other endangered species. Cloning could also reincarnate some species that are already extinct – most immediately, perhaps, the bucardo mountain goat of Spain. The last bucardo – a female – died of a smashed skull when a tree fell on her in early 2000. Spanish scientists have preserved some of her cells.

---

[x] Hwa A. Lim, *Genetically Yours: Bioinforming, biopharming, and biofarming*, (World Scientific Publishing Co., New Jersey, 2002).

Though Noah the cloned gaur bull did not make it up the ramp of Noah's Ark of cloning of endangered and extinct species, we hope this book will carry on his legacy.

In a related, but much more controversial attempt closer to Noah the biblical character himself, several groups in various countries are in the race to clone the first human being. Within weeks after Italian scientist Severino Antinori declared two women in the Commonwealth of Independent States and another in an Islamic state were carrying human fetuses which he had helped clone, it had transpired that four other teams of experts were working to the same end. To date, none of these announcements have delivered on their promise.

Nonetheless, it is still likely that the biblical character Noah's own relatives will walk up the ramp of Noah's Ark of cloning before many endangered or extinct species will be cloned. In fact, while writing this book, the first "human clone," a 7-pound girl, was delivered by Caesarean birth on December 26, 2002 to an American couple.[δ] The first clone was followed by a second one, delivered by natural birth in Holland on January 3, 2003. A third was delivered to a Japanese couple on January 22, 2003. By February 4, 2003, there were five clones! These claims, made by the same group, have been received with skepticism and have yet to be verified. Most people believe these claims are hoaxes.[ε]

Thus we, humankind as a whole, have come along way. Since antiquity, we have been trying to affect the outcomes of natural births, even though the definition of perfection has never been properly understood. Then the method of affecting the outcomes of births became much more sophisticated. For more than three decades now, we are capable of manipulating our genetic stock; and for a decade now, we are able to clone animals from adult cells. Now we are encroaching on cloning our own self.[γ]

In spite of recent advances in heredity, reproduction, cloning, stem cell research and regenerative medicine, the average person knows little about these subjects, particularly cloning, stem cell research and regenerative medicine. The lack of understanding is not because the layperson is

---

[δ] Dennis Kelly, "Company claims world's first human clone", *USA Today*, December 27, 2002.

[ε] "Clones: And then there were three", *The Associated Press*, January 22, 2003.

[γ] One of the questions commonly asked is "how did Noah's Ark carry all those animals?" An instinctive reaction to the question is if cloning had been known, then only cell samples would have been needed. This is not correct. Even if cloning had been known before the Big Flood, the Ark would still have to carry at least half as many animals, if not more. This is why: the current cloning technology requires a cell nucleus, an egg and a surrogate mother to clone each of the animals. So we will need at least the female. But the rate of success is so low that we may need a number of surrogate mothers for each of the animals.

disinterested in these subjects. On the contrary, parents, students, friends and strangers continually approach me and ask: "Is intelligence inherited?" or "Is it not horrible to be able to clone humans?" Unlike me, the experts have detailed answers to many of these questions; yet, little of the information has been available to the average person. Hollywood movies are not helping either. It seems that much of the hard work has already been done, because so much of the story seems to be known, by moviegoers. Movies more often than not tend to convey the wrong information and thus mislead the public.

I hope to lay to rest many of the misleading issues, misunderstandings, misconceptions, and common wisdoms. As you leaf through the pages of this book, you will notice that the current commercialization of biotechnology, reproductive technology (reprotech), etc have their roots in the Enlightenment of the seventeenth century when the enclosure law was enacted. "Enclosing" means "surrounding a piece of land with hedges, ditches, moats or other barriers to the free passage of humans or animals." Privatization of land or real estate began a process of privatization across the world. Today, with the exception of Antarctica, every single square foot of land mass on the earth is either under private commercial ownership or government control. Antarctica has been partially preserved as a non-exploitable shared commons by international agreement.$^{\diamond}$

The enclosure of landmass has been followed in rapid succession by commercial enclosure of parts of the oceanic commons, of the atmospheric commons, and more recently of the electromagnetic commons by media stations and computer conglomerates. Having laid claims on "all" commercializable commons outside the human body, we now turn inwards toward the body commons: biotechnology companies are laying claims on the genetic commons and body part commons.

Lying in between, within and without the human body, is the reproduction commons. When it comes to reproduction, it is natural – we come equipped with all the hardware and software, called sex, for such a purpose. It is a natural instinct to propagate. Some people grieve over the inability to conceive; others use contraceptives so that they will not conceive. Historically, there are noteworthy interregna: in the 1960s, there were people who wanted to have "sex without procreation;" then *in vitro* fertilization came into being in the 1970s. Now in the new millennium, only decades later and extremely short in the evolutionary time scale, there are people who want to have "procreation without sex" by cloning.

$^{\diamond}$ Hwa A. Lim, *Change: In business, corporate governance, education, scandals, technology and warfare*, (EN Publishing, Santa Clara, 2003).

Cloning is, in a certain sense, biofacturing (biologically manufacturing). Still, there must be something about sex that people are willing to risk being caught in the Whitehouse, pink house, houses of other colors, outhouse, ill-repute house... If sex is so good, why do some people want to get rid of sex? If cloning is so good, will we not have sex anymore? Is cloning the right thing to do? If cloning is so god (no spelling error here), are we invading the sanctity of life? What does it mean as biology has transitioned from being the science of "what we are and how we came to be" to the science of "what we can become"?

This book is by no means an authoritative book on the subject matter. It is written for laypeople, students and young scientists who are considering embarking on the fields. It seeks to provide background information so that readers can form their own opinion. It covers all important areas pertinent to sex, sexual reproduction, cloning, stem cell research and regenerative medicine at a general readership and not too technical level: sex and sexual reproduction; the history of, the basics of the art of, the success to date of, applications of, the future of, and the bag of ethical implications of cloning and stem cell research; and the applications of stem cell research and cloning in regenerative medicine. I also provide "food for thoughts" for each chapter at the end of the book with the hope that these exercises will stimulate discussions among readers of reading groups, students in classrooms, and interested newcomers in workshops, or these exercises will beneficially fill the leisure time of interested readers and make good topics for water-cooler discussions.

Hwa A. Lim
hal_lim@yahoo.com
Spring of a prolific year

Contents

1 Prologue

"As
vital as sex may be, it is a
glorious, glittering puzzle. Why do peacocks
drag around such grand tails, but not peahens? Why is
it that when Australian redback spiders mate, the male hurls
himself onto the female's poisonous fang, becoming a meal for her
at the end of the act? Why do ant nests contain thousands of sterile
female workers, all serving a fertile queen? Why do males
always have small mobile sperm, while females
have giant, immobile eggs? Why are
there males and females
at all?"

- Carl Zimmer, *Evolution*, 2001

▶ About the Title of the Book

Friends and colleagues are not only instrumental in shaping the contents of this book, but also the title.

Originally, I had decided on a more innocent title, *"Mother Nature, Father Time and Children Author."* While doing research for the book, I came across numerous mildly disturbing facts: –

- Fact One: In performing human cloning, a researcher needs an egg, which has been hollowed out, and a donor cell. The nucleus of the donor cell is then transplanted into the egg.
- Fact Two: The resulting embryo is then implanted into the uterus of a surrogate mother. After months of gestation, of course, the clone would be born the same sex as the donor. Since cloning involves no fertilization, but requires an egg and a uterus, in principle, the male population is redundant, provided no male clones were desired.
- Fact Three: In aquaculture or fish farming, in order to maximize feed-to-body-mass conversion, certain cloned fish have only female population. Male fish are too aggressive and spend too much energy fighting each other![1]

---

[1] Hwa A. Lim, *Genetically Yours: Bioinforming, biopharming, and biofarming*, (World Scientific Publishing Co., New Jersey, 2002).

- Fact Four: Someone at a company has actually succeeded in human parthenogenesis, that is, the development of an egg into an embryo without sperm fertilizing the ovum. The trick is to emulate chemically the environment for the egg to start cell division.[2]
- Fact Five: The whiptail lizard population in the western United States has no males. The way they reproduce is having one female mount another female and mimic what a "male lizard" would do while mating. The eggs simply start dividing and growing into embryos without being fertilized by sperm. In this way, the lizards invariably give birth to only females, all of which are identical to their mothers. The offspring is in all respect a clone. This reproduction process is a form of parthenogenesis.[3]

There are other facts, but these facts suffice to show that the male population plays a rather redundant supporting role. These facts were bubbling in my head for a while: would the male human population eventually become "extinct" if cloning and parthenogenesis techniques were perfected? Would the male human population eventually "die out" if they do not cut out the time they spend fighting and spend more time being intimate with their "significant other halves?"

On a business trip abroad in September 2002, while crossing a bridge, an inspiration struck and I changed the title to "*If Sex Is So Good, Why Clone?*" to highlight what has been going through my head, and to bring out the antimetabole – "procreation without sex" and "sex without procreation." I do not know what the significance of that bridge is, but I can tell you I was then in a country where all forms of pornography are forbidden.

At a function in Silicon Valley, California in December 2002, I met a famed octogenarian from the San Francisco Bay Area. He suggested "Why 'If'?" "If" was subsequently dropped. Then in January 2003, I met another famed Broadway octogenarian in Los Angeles, who suggested why "So Good" when it should be "Sooo... Good" with quiver on "Sooo..." which I do not know if he intentionally pronounced it that way or that it was because of his age. Out of respect, I dare not ask him to clarify. I thought it would have been better to add the 'o's to "Good" rather than the "So." But this is only a matter of age difference. After all, the famed Broadway octogenarian is about twice as intelligent in age as I am. Since I am not able to put the quivering effect in the title, I decided not to take the suggestion. Quivering effect will have to wait until we have a VCD or DVD version of this book.

---

[2] Dan Vergano, "Technique might quell stem cell research concerns", *USA Today*, January 31, 2002.
[3] Carl Zimmer, *Evolution: The triumph of an idea*, (HarperCollins Publishers, New York, 2001).

During the months I had been tapping on the computer keyboard to complete this book, like other living things, I still needed to ingest but had inadvertently become more sedentary. One day, after a long day's work, I stood up to stretch. Just at the moment, an inquisitive seven-year old happened by. Innocently she asked, "Your stomach is so big. Are you pregnant?" I hope she was joking. As far as I know, male parthenogenesis is still not possible, even in the higher animal kingdom.[4]

When I was completing this book for submission to a publisher to publish into an imprint in 2003, a very intelligent eleven-year-old boy saw me busy tapping away on my computer keyboard. He approached and asked what I was doing. I told him I was putting finishing touches to a book. Out of curiosity, he took a peek at the cover, and then asked, "How much would this book cost?" I said, "This is for the publisher to decide." He then lamented, "My mom would never let me read this book."

He was right. Certain books are for adults, but this one is definitely NOT one of those books. Do not let the title deceive you; or should I say, "Do not let the title tempt you." If you are purchasing this book for colorful pictures, you will be very, very disappointed. But if you are purchasing this book to learn more about the many ramifications – actual and perceived – of sex, reproduction, cloning, stem cell research and regenerative medicine, I hope you will enjoy reading this book as much as I enjoyed tapping away on my computer keyboard writing this book.[5]

▶ Risks of Writing this Book

No book on cloning, and the related issue of reproduction, will be complete without a discourse into sex. A dictionary defines "sex" as a noun, meaning 1. either of the two main groups (male and female) into which living things are placed according to their reproductive function, the fact of becoming one of these. 2. sexual feelings or impulses, attraction between members of the two sexes. 3. sexual intercourse.[6]

From these dictionary definitions of what "sex" is and the explanation of how the title of this book comes about, it is clear why I am taking some risk writing this book. The following two scenarios would help explain.

---

[4] Fact Four mentioned above only led to the creation of an embryo of a few cells, not a complete organism.

[5] After HAL [HAL are the initials of the author] had submitted the book to the publisher to print into imprints, there was another suggestion for the title: why 'Sex is so good, why clone?' when it should be 'Sex is so great, why clone?' So the debate over the title continues.

[6] *Oxford American Dictionary*, Heald Colleges Edition, (Avon Books, New York, 1980).

> Scenario One: Whenever I go to a bookstore to purchase books and walk up to a cashier to pay for the books, the cashier would always say smilingly while scanning one of the books over the barcode reader, "This is a great book. I have read it myself." I suspect nine out of ten times the cashier has not read the book. It is just a courtesy.
>
> Scenario Two: Whenever I go to a bookstore to purchase books and one of the books happens to have the word "sex" in its title, the more conspicuous the word "sex" is the better. When I walk up to a cashier to pay for the books, the cashier would utter nothing, and even blush embarrassingly.

I am sure bookstore browsers can relate with such experiences. But what is the risk I am taking? Of course the risk has nothing to do with sex. Though purchasing books with the word "sex" in the title would create some awkward feeling, books with the word "sex" in the title are usually the most leafed books in any bookstore, a testimony to indicate that readers love books about sex (in its narrow third dictionary definition) – how to improve the technique, how to give and get pleasure, how to last longer, all that stuff.[7]

There is an anecdote in which a job applicant had to answer a list of questions. In one of the questions, the applicant was asked "sex:" the applicant's instinctive interpretation of the question was "human copulation" and wondered why the prospective employer would be interested in his or her lifestyle. So the applicant provided a rather reluctant, but honest answer "frequently" to indicate that he or she made love on regular basis. The question, as you may have guessed by now, asked whether the applicant was a male or a female, the first dictionary definition!

To avoid such uncomfortable situations, if I want to use the word sex to mean only "human copulation," I will say "coitus." It is precisely because most people interpret the word "sex" in such a narrow sense, that is, to mean coitus, that American sex researcher Shere Hite (1942-) writes, "I am suggesting we call sex something else, and it should include everything from kissing to sitting close together..." to indicate that sex encompasses more than copulation.

This anecdote and sexpert (or sex expert) Hite's redefinition of sex provide a hint of the risk I am taking: in this book I write about more than sex in its narrow sense. In the pages that follow you will find extraordinary wisdom on dozens of topics that go beyond copulation, the third definition in the dictionary. These topics merely dance along the periphery of sex. In fact

---

[7] Stephen C. George and Ken Winston Caine, *A Lifetime of Sex*, (Rodale, Inc., 1998).

sex in science encompasses a much wider scope than the dictionary definition of sex. "Sex" in science touches on coitus, sexual reproduction, asexual reproduction, altruism in sex, sexual attraction, the battle of the sperm in sexual reproduction, the role of sex in coevolution, disadvantages of sexual reproduction, evolutionary adaptations to sexual reproduction, maternal and paternal investments in sexual reproduction and many other topics related to sex and sexual reproduction.

In science, cloning is a form of assisted asexual reproduction; assisted with the help of technology. As such, when discussing the implications and ramifications of cloning, a thorough understanding of sex (in the scientific sense) will help. My discourse into sex thus does not qualify me to be a sexpert, but rather a necessity to make this book more complete.

For scientists to talk about sex and sexual reproduction in isolation is missing the point. Sex and sexual reproduction are intricately linked with all the other topics touched upon in this book. This takes us to one of the most important truisms: reproduction is more than just "the sperm meeting the egg" after some intimate moments. It is a simple concept, yet surprisingly few people have grasped it, even the experts. When you look at sex and reproduction books on the market, they treat sexual reproduction as sexual reproduction, or they deal with reproductive technology such as *in vitro* fertilization as reproductive technology, or they provide pros and cons of the very controversial cloning as ethical issues.

Then again these days, fundamentals in "the birds and the bees" are presented by a variety of instructors and learning institutions, including sitcom writers, school playground know-it-alls, Internet chat-rooms, magazines, and a seemingly endless faculty of sexperts (sex experts). Some of these are not necessarily bad places to get started, but they are not always credible.[8]

Hence the concept for a new book emerges: explain the basic concepts of all the related topics along the periphery of sex and reproduction, and interrelate them together in one volume. I take great pride in taking the risk of writing this book because over the years, I have given lectures on some of the topics and have on numerous occasions been accosted by people from different walks of life who ask questions on related topics while I was eating out, walking down a street or having a leisurely chat. This book provides many of the answers, and intends to help eliminate the misunderstandings.

Ultimately, this book is about sex in its broadest sense, reproduction, reproductive technologies, and new breakthroughs in stem cell research and regenerative medicine. If these are what you get out of the book, I will be very

---

[8] Felicia Zopol, *Let's Talk About Sex*, (Running Press, Philadelphia, 2002).

satisfied. But if I am truly successful in taking the risk, decades from now, people will still be reading this book or use it as a reference. This will make me feel very rewarded indeed.

▶ The Title Revisited

Just before the imprint came out, in late 2002, I had the opportunity to give a lecture to the public in California on stem cell research and related topics. I cited this forthcoming imprint. At the conclusion of the lecture, someone came up to me, after the Q&A session and told me in private, why "*Sex Is So Good, Why Clone?*" when it should be "*Sex Is So Great, Why Clone?*"

A few months later, in 2003, a flyer came out featuring the imprint. A fellow author (on tour books) saw the flyer. Jokingly, he said, "Dr. Lim just wrote a book on '*SISG*'..." In November 2005, I was at a university lecturing on a topic not related to cloning. During the introduction, my host introduced my background, and then went on to say, "...Dr. Lim had recently written a book. The title is '*If Sex Is That Good, Why Clone?*'..." There was laughter from the audience.

Now you see, when it comes to sex, people not only like to do it in different ways, but also like to say and express it in their own ways! Some awkward situations arose. In 2004, when I presented a copy of the imprint to an official of a Middle Eastern country, he decided to cover the word "Sex" when we posed for a photograph. When I presented a copy of the imprint to a chief minister of a Southeast Asian country, he looked at the title and said, "A very interesting book."

In order to avoid further confusions, and after consulting the marketing department of the publishing house, I decided to change the title to *Multiplicity Yours: Cloning, stem cell research, and regenerative medicine.* This not only brings us closer to a very successful book I have had with the publisher, *Genetically Yours: Bioinforming, biopharming, and biofarming* (published in 2002), but also begins a "Yours" series of books.

In a Webster dictionary, "multiplicity" is defined as: the quality of being multiple, or various; a state of being many. There are also technical definitions of "multiplicity" in mathematics and physics, a computer program of the same name, or a disease with the same name (but also variously known as multiple personality disorder or dissociative identity disorder). These are of less relevance to us. Of more relevance is a 1996 film starring Michael Keaton (as Doug Kinney) and Andie MacDowell (as Laura Kinney). In the movie, *Multiplicity*, Doug is a stressed-out family man. He meets up with a cloning enthusiast and makes a clone of himself in order to take over himself at work while he tries to spend quality time with his family. The clone turns out to suffer from residual quirks of the cloning process. More clones have to be made. Eventually Doug's cloning misadventures end up complicating rather than simplifying his life.

Now it is obvious why the project to clone a beloved mutt named Missy is called the Missyplicity Project. The project was backed by entrepreneur John Sperling and was based initially at Texas A&M University.

Of most relevance is this book is actually a collection of answers to questions raised at the end of my lectures, asked by friends and acquaintances in leisurely chats, over the phone, received in postal and electronic mail. I thus decide it is appropriate to title the book in a correspondence format, just like we usually pen off a letter with "Affectionately yours," or "Truly yours."

This book thus has a longer gestation period than the three years since I keyed in the first word on my laptop. During this period, my laptop has been replaced twice, and the contents of the book changed, and were revised and updated many times. The title has also evolved from *"Mother Nature, Father Time and Children Author"* to *"Sex Is So Good, Why Clone?"* to *"Multiplicity Yours: Cloning, stem cell research, and regenerative medicine."* This book was first published in 2004. This is an updated, revised and expanded new book.

What Has Gravity Got To Do with This Book?

This book may be a little heavy; gram for gram (we have to get metricated here, otherwise it should be "pound for pound") you are getting a great deal. But this is not I am getting at.

Albert Einstein (1879-1955) is by far the most famous and beloved scientist of all time. We revere him not only as a scientific genius but also as a moral and even a spiritual sage whose enduring aphorisms touch on matters from the sublime (such as "Science without religion is lame, religion without science is blind") to the playful (such as "Gravity cannot be blamed for people falling in love"). There is a guesstimate of 500 books about Einstein in print, of which at least a dozen were published in 2005.

Why 2005? The year 2005 was the "World Year of Physics." That year celebrated the centennial of the "miraculous year" when a young patent clerk in Bern, Switzerland, revolutionized physics with five papers on relativity, quantum mechanics and thermodynamics. With other scientific giants, Einstein contributed greatly to the science of the twentieth century.[9]

For the first half of the twentieth century, physics yielded not only deep insights into nature – which resonated with the disorienting work of creative visionaries like Pablo Picasso (1881-1973), James Joyce (1882-1941) and Sigmund Freud (1856-1939) – but also history-jolting technologies like the atomic bomb, nuclear power, radar, lasers, transistors and all the gadgets that make up the computer and communications industries. Physics also contributed greatly to the invention of tools that would prove essential for probing studies of life science today. Physics mattered.[10]

These days, biology has displaced physics as the scientific enterprise with the most intellectual, practical and economic clout. Biology has given us thrilling, chilling technologies like genetic engineering, cloning, stem cells, and regenerative medicine. Many of our most pressing problems are also biological: AIDS, SARS, avian flu and other epidemics, ailments, overpopulation, environmental remediation, species extinction, even warfare (particularly after September 11, 2001). We naturally look for answers to these problems not from physicists, but from scientists grounded in biology.

It is fair to say no modern biologists have come close to Einstein's extra-scientific reputation. Einstein took advantage of his fame to speak out on nuclear weapons, nuclear power, militarism and other vital issues through lectures, essays, interviews, petitions and letters to world leaders. When he spoke, people listened. After Israel's first president, the chemist Chaim Azriel Weizmann (1874-1952), died, the Israeli cabinet asked Einstein if he would consider becoming the country's president. Einstein politely declined – perhaps to the relief of the Israeli officials, given his avowed commitment to pacifism and a supranational government.[11]

It is hard to imagine any modern scientist, physicist or biologist, being lionized the way Einstein was. One reason may be that science as a whole has lost its moral sheen. We are more aware than ever of the downside of scientific advances, whether nuclear power or genetic recombination; moreover, as science has become increasingly institutionalized, it has come to

---

[9] John Horgan, "Einstein has left the building", *The New York Times*, January 1, 2006.

[10] Hwa A. Lim, "Bioinformatics and cheminformatics in the drug discovery cycle", in: *Lecture Notes in Computer Science, German Conference on Bioinformatics*, Leipzig, Germany, September/October 1996, Ralf Hofestädt, Thomas Lengauer, Markus Löffler, and Dietmer Schomburg (eds.), (Springer, Berlin, 1997), pp. 30-43.

[11] While awaiting Einstein's answer, David Ben-Gurion, then prime minister of Israel, reportedly asked an aide, "What are we going to do if he accepts?"

be perceived as just another guild pursuing its own selfish interests alongside truth and the common good.

Why this digression about Einstein? Gravity certainly has nothing to do with cloning, stem cell research, and regenerative medicine. As you will see, we shall be mentioning Einstein again in the chapter on cloning: will we get an Einstein if we clone Einstein? In discussing ethical and moral implications of cloning, stem cell research and regenerative medicine, we will find Einstein's aphorism "Science without religion is lame, religion without science is blind" pertinent.

Sir Edward Tylor introduced the concept of survival (NOT the same "survival" as in "survival of the fittest"). In this case, survivals are ancient customs or habits that had persisted long after their original purpose had been lost or forgotten. Many of our instinctive reactions and emotions are on occasions at odds with this modern age, but they continue to shape the way we live. As you will see, in cloning, stem cell research and regenerative medicine, science and religion will clash head on. In parts of the world (surprisingly, the U.S. tops the list) where modern mores often collide with ancient traditions and customs, what constitutes progress can be debatable, and in some cases, even divisive. The U.S. government has been debating on the issues for decades, with no definitive conclusions; political candidates and Supreme Court nominees have to answer tough questions on these issues; and these issues become platforms of political parties, stand of interest groups. These are also the issues debated in recent U.S. presidential campaigns. Once elected, the president usually takes a certain stand, at times counter to the campaign promises.

This is yet the greatest risk of writing this book, for when drafts of more controversial chapters of this book were posted on the Internet I received electronic email offering diverse opinions, suggestions and criticisms. I am not running for any public office. This makes writing this book a little easier. But I have to emphasize that this book is not a systematic inquiry of any kind. I possess no special wisdom, nor am I peddling a policy.

One-Handed Scientist

At the dawn of the twentieth century, science was recognized as the dominating force of the age. Objective scientific analysis promised to open everything to human control. This is when science and technology began to become key issues in public decision-making.

Harry Truman (1884-1972) is the 33rd President of the U.S. After listening to his scientific advisors propose a well-argued hypothesis, and then escape from it with an "on the other hand," he is reputed to have pleaded for a "one-handed scientist" and said that he could solve the world's technical problems if he could just find a "one-handed" scientist.

The 34th U.S. President, Dwight D. Eisenhower (1890-1969), not to be outdone, is known to have made the statement, "I'm a physicist and I just can't resist saying, 'on the one hand... but on the other hand'..."

Or was it U.S. Senator Edmund Muskie (1914-1996) who first pleaded for a "one-handed scientist?" It was a time when supersonic transport was first coming into the U.S. and the senate was trying to decide whether it would create giant holes in the ozone layer. Senator Muskie had a big hearing on Capitol Hill and he called expert scientists together. A blue ribbon panel presented their findings and said, "Our findings show this, that the preponderance of data is that it won't cause any danger. On the other hand, our data shows just the opposite, also, that we need to do more research." It was at this point that Senator Muskie got up and quipped, "Will somebody please find me a one-handed scientist?"

Researchers learn early in their careers about the fine line between cautious explanation and avoidance; this is especially so when it comes to the demand for simple answers to complex scientific questions. In the policy arena, for example, elected officials dream of the "one-handed" scientist. Researchers offering scientific expertise often state a likely option based on analysis of data, and quickly follow with "but on the other hand," proceeding to describe layers of factors that might rule out science's best guess at the moment. But people want answers, and become frustrated by responses that are a long list of options, potential factors and qualifiers.

Even Leon Kass, President George W. Bush's choice to head his council to monitor stem cell research, on August 10, 2001 National Public Radio's "All Things Considered" said:

> "I have opinions, but you cannot work in the field of ethics and be neutral. In fact, it's a disservice to have spent your time thinking about these things and say, 'on the one hand, on the other.' I mean, what you really want are people who have thought these things through and have a position."

Two days later, on CBS's "Face the Nation," he said:

> "I don't have a firm position on stem cell research myself. I think it's a terribly difficult and vexing question."

Now you see why Einstein is by far the most famous scientist. He was not only a genius, but also a "one-handed" speaker. When he spoke out on vital issues through lectures, essays, interviews, petitions and letters to world leaders, he spoke with "one-handed" authority. Recall that Truman was the U.S. president who decided to drop the two atomic bombs in Nagasaki and Hiroshima during World War II. Einstein had written Franklin D. Roosevelt (1882-1945), President Truman's predecessor, and warned him of the danger of atomic bombs. President Truman could not be referring to Einstein when he was pleading for a "one-handed scientist."

Uncertainty, probability, change and refinement are inherent aspects of science. The popular view of uncertainty, however real, carries images of bumbling, ineptitude, indecisiveness, weakness, and almost subversive, but all the word really means is that there is something we do not know. It may be an unknowable like the future, or it may be easy to find out like a phone number. Whether or not someone else knows it, if you yourself do not know it, you are harboring uncertainty. Find someone who claims to know everything, and you will have found someone to avoid. Someone once said that life can be understood backwards, but unfortunately it must be lived forwards. Probability goes both ways.[12]

Scientists are generally much more patient with the progress of knowledge – continually self-correcting, building on prior knowledge but never absolute – than nonscientists. Other activities, particularly politics and business, call for quick and definitive answers. In the legal profession, most laws – being an accumulation of unchanging law – are in fact irrelevant in a society that is constantly changing. Time-tested circumlocutions in the legal profession are also not uncommon. This tension between science, legal profession and non-scientific activities may be irreducible.

The journalist is often in the position of posing questions to which the research world can offer no authoritative answer, but can offer just a cautious account of the current state of informed belief, including the "on the one hand... on the other hand..." type of answers. Inevitably, some members of the public will ignore all the caveats (such as "this is just a preliminary study," "the sample isn't representative of the population," "we won't know for at least two more years," "it's only been tested on mice") and believe whatever they want to believe, or believe in one hand more than the other. Sensationalism will always build an audience in the short run. Responsible journalists recognize this tendency and resist the temptation to abet it.[13]

---

[12] H.W. Lewis, *Why Flip a Coin? The art and science of good decision making*, (Barnes & Noble Books, New York, 1997).
[13] Kenneth K. Goldstein, "Training for the medical information complex", *21ˢᵗC*, Special Section: Medicine and the Media, Issue 4.2, Columbia University, Fall, 1999.

Information Cornucopia

As most of us have noticed, with joy and some dread, we now live in an information-rich era, with abundant print and digital media clamoring to get our attention. Data, information, news and "facts" flash endlessly across pages, computer and TV screens, battering our minds with the flotsam flowing at us in the information river. Some of them credible, perhaps even true, but how is one to tell the difference? Often web logs, chat rooms and news groups snatch them up and soon they are all over the place, physical and cyber. Some of the sites and reportage are legitimate, but others may be specious and dangerous, or at least with insufficient verification.

The torrents of data and information include a special flow dealing with new scientific breakthroughs in biotechnology, healthcare and medicine. Such material is often given to the general public via the conduit we call the science journalist. The job of the science journalist is to evaluate that information mass (may as well be information mess) and translate it into user-friendly language appropriately balanced with what is and is not true, as far as we know.[14]

Who goes into this particular profession of science journalism? Quite a few science writers start as practicing scientists but find that scientific work holds less fascination than the chance to consider research from an outsider's vantage point. The traffic may also go the other way – some science journalists end up going back to get an advanced science degree. However, there is often an expertise gap between professional researchers and journalists without advanced scientific training. The journalist cannot always expect to grasp technical topics on the same level as the specialist. But being a layperson also carries a practical advantage: the journalist can be more fluent than the scientist in communicating to the general public.[15]

The information complex has grown powerful, and this power can be used either for beneficial or venal purposes. Infomercials may run with the news and stretch the claims, and the media communication may reduce all subjects to caricatures and sound bytes. In some cases, Hollywood will immaturely take breakthroughs or discoveries a step further to make them into movies, further sensationalizing them. As certain notorious instances have borne out, a distorted story can do a great deal of harm to the public, to the scientists

---

[14] Kenneth K. Goldstein, "Training for the medical information complex", *21ˢᵗC*, Special Section: Medicine and the Media, Issue 4.2, Columbia University, Fall, 1999.

[15] Science journalists are hybrid communicators – part journalist, part technologist or scientist – who may be self-trained or trained in a journalism school's science and technology writing curriculum.

involved, to the field itself, or to other parties unjustly accused of irresponsible conduct.

As we shall see later in the book, the Hwang Woo-suk clone-gate is one excellent example in which the media had played both the role of the spinmeister to take the breakthroughs and ran with them, and the role of a detective to unravel the scandal down to the last thread. But here as a sidebar, we will present the case of sex chromosome study as a case in point.

Will the Y Chromosome Decay and Take Men with It?

Down to the genomic level, the genomic sequences of the sex chromosomes can shed light on the behavioral and biological differences between the sexes (male and female).

Besides the 22 pairs of autosomes, a human male has an X and a Y chromosome, while a human female has two X chromosomes. Genetic mutations and diseases such as color blindness, autism and hemophilia that are linked to the X chromosome tend to affect males because they do not have another X to compensate for the faults.[16]

Containing 1,098 genes, or about 5% of the human genome, the X chromosome is linked to more than 300 human diseases. Its genetic code may help to explain why women are so different from men, along with information that may help to improve the diagnosis of illnesses ranging from hemophilia, blindness and autism to obesity and leukemia.[17]

The Y chromosome is an eroded version of the X chromosome, with only a few genes. The X chromosome is also bigger than the Y chromosome. Because females have two copies of X chromosomes, one of them is largely switched off or inactivated in each cell so that, like men, they operate with just one copy functioning. But scientists have long known that the inactivation is not complete, i.e., not all of the genes on the silenced chromosome are inactivated.[18]

A study has found that 15% of the genes on the inactivated copy continues to function, sending out chemical orders for the cell to manufacture specific proteins. A more surprising fact is that about another 10% of the genes, in which the activity level varies widely among woman, from zero in some to varying levels in others. This should be contrasted with the more consistent activity levels in X chromosomes from men, or in other chromosomes in either

---

[16] Patricia Reaney, "X chromosome shows why women differ from men", *Reuter*, March 16, 2005.
[17] Mark T. Ross, et al, "The DNA sequence of the human X chromosome", *Nature*, 434, March 17, 2005, pp. 325-337.
[18] "Study reveals new difference between the sexes", *The Associated Press*, March 17, 2005.

sex. The effects of these genes from the inactive X chromosome could explain some of the differences between men and women that are not attributable to sex hormones, and the variability in activity levels could explain the differences among women.[19]

Some 300 million years ago, the Y chromosome used to carry the same 1,100 or so genes as its partner, the X chromosome. Because the Y cannot exchange DNA with the X and update its genes, in humans, over time, it has lost all but 16 of its X-related genes through mutation or failure to stay relevant to their owner's survival. Over time the Y, however, has gained some genes from other chromosomes because it is a safe haven for genes that benefit only men, since the Y never enters a woman's body. These added genes, not surprisingly, all have functions involved in making sperm.

This evolutionary decline has led to predictions that the Y chromosome will be completely bereft of functional genes within ten million years. Although there is evidence of gene conversion within massive Y-linked palindromes which runs counter to this hypothesis, most unique Y-linked genes are not situated in palindromes and have no gene conversion partners. The "impending demise" hypothesis thus rests on understanding the degree of conservation of these genes.[20]

David Page of the Whitehead Institute in Cambridge, Massachusetts and an expert on the Y chromosome has been seeking to understand whether the Y will lose yet more genes and lapse into terminal decay, taking men with it. The idea of the Y's extinction is so tantalizing from the perspective of gender politics. But before you leap into conclusion and declare "death to the men," think DNA, and never underestimate the DNA.

In 2003, Page discovered a surprising mechanism that protects the sperm-making genes. These genes exist in pairs, arranged so that when the DNA of the chromosome is folded back on itself, the two copies of the gene are aligned. If one copy of the gene has been hit by a mutation, the cell can repair it by correcting the mismatch in DNA units. The 16 X-related genes are present in only single copies, and similar protection mechanism has not been found.

---

[19] Laura Carrel, and Huntington F. Willard, "X-inactivation profile reveals extensive variability in X-linked gene expression in females", *Nature*, 434, March 17, 2005, pp. 400-404.
[20] Jennifer F. Hughes, Helen Skaletsky, Tatyana Pyntikova, Patrick J. Minx, Tina Graves, Steve Rozen, Richard K. Wilson, and David C. Page, "Conservation of Y-linked genes during human evolution revealed by comparative sequencing in chimpanzee", *Nature*, 437, September 1, 2005, pp. 100-103.

A Goal of This Book

The X chromosome study can be put to venal ends as a gender politics; or it can be beneficially used to understand why women make higher doses of certain proteins than men, which could result in differences in both normal life and disease. This example – differences of the sexes – will serve to exemplify many of the touchy issues we will encounter in this book. The controversy surrounding the issue, in fact pales when compared with the controversies over cloning, stem cell research, and regenerative medicine.

To see human reproduction purely as a gift from God overlooks the many dangerous strings attached to that gift. Similarly, to see stem cell research, reproduction technology, and regenerative medicine as just a commercial evil would be to overlook the undeniable good that accompanies that evil. Failing to recognize both sides of these issues – the potential vast benefits and the ethical cost – misses the essential and complexities of these issues. This is an exemplary case that has all the earmarks of the always-Byzantine intersection of science, technology, business, religion and politics.

Particularly in this era in which sound bytes substitute for news, docudramas for history, and book reviews for books, one area where scientific and technological breakthroughs (for example, in biotech and medicine), journalism and scientific writing can act together to great social benefit is in reaching the curious and inquisitive populations – whether intelligent, or technologically and medically underserved populations – with accurate information. This is what I intend to achieve as the author of this book. I am not a journalist; this book offers my diligent notes as a scientist and a fascinated observer.

► Inside This Book

Do you know that:

- Humans have been tinkering with heredity for thousands of years?
- Infanticides of the past were ceremonially elevated to be a form of human sacrifice to dissimulate the main purpose?
- Shadows of our savannah past cast over our modern mores and ways of life?
- Sexual reproduction is evolutionarily inefficient?
- There is an invisible hand guiding the current commercialization of human body shop and human reproduction?
- Animals have been cloned with regularity?

- The cat has been cloned, and so has the dog?
- More than 100,000 assisted reproductive procedures are carried out each year, in the U.S. alone?
- Some thirty human "clones" are born each day, in the U.S. alone?
- Clones will never grow up to be exactly identical to their progenitors?
- Parts of our body can regenerate?
- Legally a clone can run for the U.S. presidency?
- Our first reaction to a human clone can be stranger than you might think?
- …

To find answers to these and many other fascinating and tantalizing questions, we note that on the one end of the human reproduction spectrum are sex and sexual reproduction; on the other end of the spectrum are cloning and stem cell research, and their applications in regenerative medicine. I will now take you for a leisurely "sightseeing" tour between these two extremities of the genetic world.

Unlike sexual reproduction, cloning reproduces identical copies using a form of reproductive technology rather than reproducing naturally. In between the two extremes there are a number of reproductive technologies, including *in vitro* fertilization. In sexual reproduction, we leave the outcome to genetic roulette; with the help of technology, for many people it means having a baby is much safer – for most people – more within their control. But whether technology always brings more good than harm is one big question; and whether the current state of reproductive technology is ready for cloning is yet another issue? Whether our current technological quests into the human body shop and human reproductive system "are imprisoned by the twin gods of technology and profit, a doleful tale of Faustian bargains wrapped in commercial paper, of promissory notes extolling the benefits of a coming technological utopia…" is an issue for the intelligent readers to decide.[21]

To be able to discourse on all these issues, I begin with a primer on genetics to bring all readers up to speed. I start by differentiating living and nonliving things and show that reproduction is very much a part of living things. Of note in this chapter is how our knowledge of genetics and heredity has been accelerating in the past century and a half since the seminal works of Charles Darwin (1809-1882) and Gregory Mendel (1822-1884). Particularly, in the past three decades technological breakthroughs in genetic manipulations

---

[21] Jeremy Rifkin, In: the Foreword of Andrew Kimbrell, *The Human Body Shop*, (HarperCollins Publishers, New York, 1993).

have been coming forth with breathtaking pace. As a sidebar, I present a case in history where genetics played a critical role to bring out its significance. Readers who are familiar with the fundamentals and timeline of genetics may skip this chapter without loss of continuity.

In Chapter Three I discuss the role of evolution in sexual reproduction by citing supporting evidences from nature and from computer simulations. I also discuss the advantages and disadvantages of sexual reproduction, including sex persists to purge of mutations or sex persists to fight diseases. Sex is also a means to pass on genetic materials. But if passing on genetic materials is the sole purpose of sex, then asexual reproduction will be more effective; cloning, another reproductive technique is also relatively more effective in passing on genetic material than sexual reproduction. In certain sense, our primal sexual instinct can be traced to our ancestral lives on the savannah.

Chapter Four delves into the mating dance, Cupid's chemicals, matrimony, and household arrangements.

As we move across the spectrum of reproduction, from natural sexual reproduction towards cloning, we come across the reasons why people seek the help of reproductive technologies and the various types of reproductive technology available. Thus in Chapter Five, I discuss the quest for perfection as reasons why some people seek the help of reproductive technology. I also go back in time to see that the search for perfection is not something of late, but rather something that had been practiced since antiquity. Nonetheless, "What is perfection?" "What is beauty?" are contestable issues. In this chapter, I also explain infanticides were practiced in ancient times to cull imperfect infants and the founders of Rome were actually survivors of exposure! Having shown that we are beginning to tinker with the genetics of reproduction, I warn of the potential rise of eugenics. In Chapter Six, I discuss *in vitro* fertilization (IVF) as a means to help infertile parents. I also go into detail of the human life cycle development to see how birth defects can be inflicted during the various critical stages of development in the womb.

Just as when I mention "sex," very likely the first thing that comes to mind is copulation; similarly, when I mention "clone," very likely the first thing that comes to mind is reproducing identical copies. In fact, there are at least two types of cloning: reproductive cloning and therapeutic cloning.

In Chapter Seven I provide examples of successful reproductive cloning efforts from the barnyard, from before the time of the celebrated Dolly the sheep to recent cases. Chapter Eight continues the successes of animal cloning to talk about reproductive cloning of something closer to our hearts, pet cloning. The reproductive physiology of each animal can be rather unique.

Thus, the cat has been cloned, and the dog only recently, but the cloning of some other animals is an ongoing quest.

This paves the way for Chapter Nine, which surveys the currently available cloning techniques that have been used for cloning these barnyard animals and pets. This chapter also compares and contrasts these techniques. Unbeknown to many people, twins are natural clones!

Chapter Ten is perhaps the main chapter of the book. It touches on the very controversial issue of human reproductive cloning. This chapter attempts to survey all pertinent aspects of human reproductive cloning, all the way from human development, how the male and the female are formed during development, the key players in human cloning, to health issues with human reproductive cloning. Not surprisingly, the rather successful cloning technologies used in the barnyard are being modified for human cloning. Current cloning technologies, as applied to humans, are still not perfect and human clones are believed to likely have health problems at birth or later in life.

Chapter Eleven returns to therapeutic cloning – a cloning effort much more acceptable than reproductive cloning. I discuss the significance of therapeutic cloning in curing human ailments, concentrating particularly on stem cell research and regenerative medicine. In Chapter Twelve I also take a cursory survey of the different social and political climates in different parts of the world to see why certain countries have laws that are more progressive than others when it comes to guiding therapeutic cloning. I will also dissect a cloning shenanigan of scandalous proportion.

Having discussed the current state of the art of cloning and the potential uses of cloning, I enumerate the legal and ethical issues of cloning in Chapter Thirteen. Issues arise not only within the family, but also in the wider circle of public acceptance. As far as cloning is concerned, our current laws are still wanting. But when the law and ethical concerns are very clear, scientists can always come up with innovative ways to overcome the legal and ethical impasse.

The book closes with Chapter Fourteen to trace the reason why technology has made such great inroads into our lives. It has not only thrown open the gate of natural resource commons, but also the gates of genetic commons and now bodily commons for potential commercial exploitation. The international effort to convert the genetic blueprints of millions of years of evolution to privately held intellectual property represents both the completion of almost a millennium of commercial history and the enclosing of the last remaining frontier of the natural world – the final frontier where no life scientists have gone before.[22,23] Along the envelope of this final frontier are bodily commons, and activities on the bodily commons such as genetic manipulations, reproductive technologies, cloning, stem cell research, regenerative medicine...[24]

▶ Best Use of This Book

Despite the fact that the chapters in this book have been organized in a certain logical flow, the chapters can be read or consulted independently. Interested readers will also find the index useful for looking up topics of interest. If you find that the topics of interest to you are spread among different chapters, congratulate yourself because this means that you are a complex, active, thinking, and living reader who has been able to interrelate the pertinent topics.

So without further fellatio and foreplay, I invite you to take this ecstatic odyssey to see why if sex is so good, people still attempt to clone. And why not? All sorts of people from all walks of life have their own perspective on the subjects: cloning, reproduction and sex, stem cell research and regenerative medicine, as well as an insatiable curiosity to read what others have to say about reproduction, the controversial cloning, and the world's oldest obsession – sex.

This is the first book that addresses all these issues all in one volume.

●●●●●　　●●●●●　　●●●●●

Sunt bona, sunt quaedam mediocria, sunt plura mala, quae legis hic:
aliter non fit, Avite, liber.[25]

If you are wondering what the sentences mean, they are just a Latin quip to say:

"Here are some good things, some so-so, and some bad.
There's no other way to make a book."

---

[22] Andrew Kimbrell, *The Human Body Shop: The engineering and marketing of life*, (Harper, San Francisco, 1993).
[23] Jeremy Rifkin, *The Biotech Century: Harnessing the gene and remaking the world*, (Tarcher/Putnam, New York, 1998).
[24] Hwa A. Lim, *Genetically Yours: Bioinforming, biopharming, and biofarming*, (World Scientific Publishing Co., New Jersey, 2002).
[25] Marcus Valerius Martialis (40–104 AD), *Epigrammata*, XV, 16.

Read on…

2 A Primer on Genetics

"We used to think our fate was in the stars. Now we know, in large measure, our fate is in our genes."

- James Watson, in *Time Magazine*, March 20, 1989

In 1953, Julian Huxley wrote that humanity – the product of evolutionary creativity – is now obligated to continue the creative process by becoming the architect for the future development of life. The *Homo sapiens'* destiny is to be the sole agent of further evolutionary advances on the planet.[26] This means all that Darwin asked of people in his "survival of the fittest" was that they compete for their own life. The new cosmology asks that people be the creator of life.

We take this as the starting point of the excursion of this book. All genetic manipulations are ways for shortening Father Time in the evolutionary process of the Children of Mother Nature. Since we are now creators of life, we are thus authors in this new cosmology of Mother Nature, Father Time and Children Author.

▶ Living Things

Surrounding us are living and nonliving things. Living things include all animals and plants, and animals include human beings. Biology is the study of living things.

---

[26] Julian Huxley, *Evolution in Action*, (New American Library, New York, 1953).

Prior to the 1600s many people believed that nonliving things, such as a bale of straw, could spontaneously transform into living things, such as a mouse. This, as we know it now, is not the case. It exists only in magic tricks.

Living and nonliving things can be deceptively difficult to differentiate. However, with proper definition, we can easily tell them apart. All living things exhibit seven characteristics. They ingest, excrete, respire, grow, respond to stimuli, move, and reproduce.

Living organisms eat or feed – they need food from their environment to convert into energy to perform all functions, to maintain growth or to stay healthy. Living organisms excrete – they remove waste from their bodies to avoid accumulation to a dangerous level. Living organisms breathe or respire – they exchange gases with their surroundings; animals take in oxygen and exhale carbon dioxide. Besides respiring (taking in oxygen and releasing carbon dioxide), plants also photosynthesize (taking in carbon dioxide and releasing oxygen) in sunlight.

Collectively, the biochemical process involving the intake and digestion of food, the combination of the food with oxygen to release energy for cellular functions, and the elimination of accumulated waste products is called metabolism. For certain organisms, metabolism can occur in the absence of oxygen. Some other organisms absorb nutrients in such a basic form that digestion is not required. Therefore, for these organisms the essential processes of metabolism are ingestion and excretion.

Living organisms grow – they become larger in size by converting the food they consume into body mass. Living organisms are organized into structures called cells. The cell is the structural and functional unit of all life as we know it. The smallest amount of life possible is the single cell. Unicellular organisms also grow; multi-cellular organisms have a more sophisticated growth process. Cells differentiate into different types of cells. The differentiation process also allows the development of and specialization into a variety of tissues and organs. The resulting multi-cellular organisms such as human beings can be rather complex. In increasing complexity, the smallest part of an organism is a cell, a number of which are organized into tissues, such as muscles, which have a common function. Tissues are organized into organs such as the heart. Organs are organized into organ systems such as the cardiovascular system. Organ systems functioning as a unit make up the living organism.

Living organisms respond to external stimuli – they must endure a constantly changing environment, that is to say, they are sensitive to changes around them, such as touch, light, heat, and sound. This ability to respond to environmental changes necessitates some sort of sensory mechanism to detect the changes and some process by which the organism can respond. Changes must be met with measured responses if the organism is to survive. These measured responses include movement or other observable behaviors. Similarly, homeostasis is the process by which organisms maintain their internal environments at optimal or ideal conditions to run the biochemical process they need for life. Injury, disease, cold, dehydration and other environmental stresses test homeostasis.

Living organisms move – they show internal movement, which means they have the ability to move substances from one part of the body to another; some organisms show external movement as well, which means they migrate from place to place in search of food, for example.

Living organisms reproduce – they produce young: humans make babies, cows produce calves, cats produce kittens, rabbits produce fawns, fish produce fries (for those of you who live to eat, this is not the type of fries you think it is), chickens lay eggs, and plants produce seeds. There are many forms of reproduction. Sexual reproduction requires the formation of eggs and sperms, usually in separately sexed – male and female – individuals. Asexual reproduction involving fission, budding or some other process requires only a single parent. Sexual reproduction is more advanced than asexual reproduction because it requires specialized structures. All living things must reproduce. Otherwise they face extinction. For all life as we know it, a set of instructions, called a genetic code contained in DNA, is transmitted from the parent to the offspring during reproduction.

Nonliving things may show none, or some of the seven characteristics. For example, sand shows none of the characteristics; ice crystals can grow bigger if the conditions are right, but they are nonliving things; machines can move and need fuel, but they are nonliving things as well.

Living things show all the seven characteristics. Some of these characteristics are less obvious in plants and microscopic organisms. Under certain circumstances these characteristics can be suspended for a time, as in seeds or eggs, and reappear later. Viruses are small because they lack most of life's usual parts and processes, such as metabolism and respiration. Barely alive, they behave more as robots than organisms. To thrive and reproduce, they invade a cell and take over its biochemical gear, often at the expense of the host. As is clear by now, a living thing does not survive by itself; it thrives in an environment with others like itself, and others unlike itself. Thus, a larger organization outside of an individual living thing is a population. A population is generally all of one species – a group of interbreeding individuals that produce fertile, viable offspring. Next comes a community, which consists of populations of different species interacting with each other in an area. A community includes all living things in the area.

An ecosystem not only includes living things in the area, but also the physical environment and the interrelated flow of energy. Desert ecosystems or forest ecosystems are examples. The biosphere is the zone surrounding the earth where life is located.

▶ Tinkering Using Technology

With knowledge accumulation and technological advances, we, humankind as a whole, have been tampering with ways to improve on each of these characteristics. For growth, we have made food or medication that will aid the process. We have invented modern transportation systems to help us move. Taking advantage of our response to stimuli, we have created artificial environments to thrill ourselves to enrich our lives, and have created virtual reality to make life more exciting. We take in oxygen, along with fragrance and perfumes. Cuisines are so tempting that instead of "we eat to live," for many, "we live to eat" until our weight issue becomes a health hazard. We have to excrete to avoid an accumulation of waste in our bodies. To make going to the stool a pleasure, we have invented smart toilets.

Not all the technological conveniences and improvements on our lives are completely unrelated, but in all these processes, we extract, excrete and produce so much waste that we have directly or indirectly exerted on Mother Nature. Breathing, for example, is a process we take so much for granted until we realize we have polluted the environment to the extent hazardous to our health. So, technology, which normally helps improve our life, is at times at odds with our survival.

Our lifestyle has also changed drastically, sometimes adversely. We mistakenly regard playing video games as a form of vigorous exercise; we watch TV so much that we become sedentary couch potatoes and forget how to move, only to burp once in a while; we indulge and imbibe so much that we, instead of growing lengthwise, grow sidewise in the mid-section, exchanging the six-packs for a layer of tire around the waist. Diamond Jim Brady, the millionaire railroad tycoon of the Gilded Age and the legendary New York gourmand, famously began his pre-theater dinner with three dozen oysters. Not everyone can afford such a luxuriously extravagant lifestyle, but for the many frugal middle-class, they ingest so many leftovers that instead of going to waste, the food goes to waist. After eating so much we have to excrete. We sometimes build outhouses as far away from our home as possible so that they become other people's discomfort.

Exploiting our response to stimulus, we thrill ourselves conquering Mount Everest; some of us even swim; and rather unfortunately, some of us violently beat up our spouses. We perspire so much that we constantly complain of foul odors; instead of being mutually attracted by natural pheromones, we are confused by all the perfumes and deodorants, thus contributing to a high divorce rate. We grow so tall that we are natural basketball players because the basket is within arm's reach; we grow so big that we are bulldozers in American football games, or are natural Sumo wrestlers, but at a costly price of lifelong subscription to steroids. We look up to the body builder Arnold Schwarzenegger – of Hollywood movies *Conan the Barbarian*, *Predator*, *Terminator* fame – and elect him governor (or should we say governator) of the great State of California.[27]

Yet in all these, we put all the blame on our genes when something goes wrong: there is a genetic cause for obesity, alcoholism, halitosis, violence… and best of all, there are also genetic causes for stupidity and for not reading this book. Our genes are to be blamed for everything, pleasant or unpleasant. There are people who just grow fat "thinking" about food because they believe they have obesity genes! There are people for whom the only exercise they have is moving their forks; and the world-famous overweight tenor Luciano Pavarotti is known to have said, "The best exercise I have ever done is to push the dinner table away."

Drug companies capitalize on this. Frustrated at the failure to come up with new drugs for curing diseases, the pharmaceutical industry has been looking elsewhere for profitable markets. And they have found one in human fear, ego and vanity.

"Lifestyle" drugs are drugs whose primary function is to restore social faculties or attributes that tend to diminish with age: Rogaine for treating baldness, Xenical for obesity, Prozac for depression, Viagra for impotence. Botox for migraines is now repackaged as a wrinkle reducer. Finasteride, scientifically designed to block the metabolism of testosterone to shrink the size of the prostate, is now repackaged as Propecia for hair loss. Some young students really suffer from attention deficiency disorder (ADD). However many parents, failing in educating their children, use ADD as an excuse and thus helping their children commit to life-long dependencies on ADD drugs.

Humans have become a drug chest, and in so doing, we abuse our body, knowingly, creating probably one of the largest preventable human-made diseases outside of nicotine addiction.

---

[27] Patrick May and Lori Aratani, "The state divided, Bay Area, L.A. gloomy, rest smiling over election results", *San Jose Mercury News*, October 9, 2003.

Other companies exploit human characteristics in other ways. For example, the telephone makes it possible for us to communicate with each other without first having to gather together in one place. Cellular phones take it one step further: we not only talk, we talk while driving and putting everyone else in danger. Sometimes crippling innocent victims, putting them on artificial limbs or support of other technologies. A convenience has become a nuisance: any person whose life is so empty that he or she spends a whole day talking nonsense into a black box is at best idiotic, at worst moronic.

Risks and benefits of each new technology have to be weighed against each other. Simple cost-benefit analyses can at times help decide which technology will stay and which will not. Abuses aside, to date, technologies have certainly make life much more safe, comfortable and enjoyable: we no longer live in caves and be exposed to the elements, but delve in the comfort of modern homes; we no longer have to cuddle in front of bonfires because we are better clad; we no longer have to hunt for every meal, but eat better and more nutritiously...

Life has been so comfortable with technologies that it is only of late that we do not blame every single problem on our genes. We are beginning to talk about diseases of the affluence – those diseases that arise because of changing lifestyles, changing for the better that is.

In other words, we age gracefully if we take good care of ourselves or we age prematurely if we neglect to maintain a healthy lifestyle. Some of us, including athletes, dose ourselves to premature death. Essentially, each person's life may be viewed as having a given maximum expectancy. What we are trying to do is trying to get the most out of the maximum, both in terms of quality and quantity. In quality, we want to have a good life, perhaps by using modern conveniences; in quantity, we want to have a long life. If we abuse our health, we survive for much shorter than the maximum allowable; if we take good care of ourselves, we live life to the fullest extent. If we unnecessarily take drugs we become living vegetables.

But now we are going beyond that. New technologies keep pushing the envelope. We are now beginning to tamper with the aging process by trying to renew life through stem cell research and regenerative medicine. We are now on the verge of extending that maximum.

Reproduction is slightly different in that sense. Reproduction is not so much for a living organism's bodily upkeep. Rather, it is how our species propagates. Thus it is not surprising that we would like to make the process safer and to have a handle on the process itself, instead of leaving it to chance. Indeed we as a species have been tampering with the process since antiquity.

Thus arises a hot contention: some view reproduction is sanctity of life not to be meddled with; others regard new technologies as more choices. For most of us, to reproduce means to first have sex, and hopefully the wonders of life will take over. But for many, sex is also a pleasure of life. While for still some others, sex is a commercial opportunity of a lifetime.

After all, prostitution is the oldest form of profession; and pornography is one of the best businesses with a multi-billion dollar market. Ask any Web surfers and they will testify that pornographic sites, spams and junk electronic mail from generic sex aids online suppliers are a great nuisance to the otherwise great Internet. Sex, perhaps is one of the few forms of entertainment the poor can afford. Rather unfortunately, they reproduce more than their better-to-do counterparts, and the global population is on the rise, most drastically in underdeveloped nations! The better-to-do counterparts have access to technology (contraception) and other luxuries; the less-to-do people have less access to technology, or seem to ignore technology because of their illiteracy. Infidelity is not uncommon, even in the White House, the White Hall, and houses of all colors.

In all these attempts – life-maintenance, life-prolongation, and life-propagation – to improve upon our survival as a species, when a new way is introduced, there are always initial resistance, though some more intense than others. Some of the resistance can even involve scare tactics. However, when all the mysteries, misunderstandings and misconceptions have been eradicated, the public soon accepts the new way when the technique for so doing has been improved to an acceptable level. This, of course, assumes that the new technology will bring more good than harm. Otherwise, the market force and other social forces will reject the technology.

In other words, as individuals, we need technology to help those less fortunate or those born challenged; some of us, however, have stretched the use of technology. As a species, we want to survive for eternity, and survival is guaranteed only for the fittest in the natural world. With technology, we need not be the fittest. In fact, if we as a species are willing to lead life constantly supported by machines, we can still survive as the unfittest. Or as some of the most brave life fighters (born challenged, or after a mishap or a long disease) have shown, we as a species can still lead a very useful and productive life as "revival of the fittest."

▶ Survival of the Fittest

The phrase "survival of the fittest" is commonly credited to Charles Darwin. In fact the phrase was not used in the first edition of *On the Origin of*

Species.[28] English philosopher Herbert Spencer first coined the phrase to refer to "those individuals best able to promote the progress and improvement of their species."[29] It was Alfred Russell Wallace, the co-developer of the evolutionary theory, who suggested to Darwin that he replaced the term "natural selection," which seemed difficult for people to understand, with "survival of the fittest." "Survival of the fittest" also provokes misunderstanding unless a universally accepted definition of "fitness" can be determined. Today, "fitness" is understood to imply reproductive success, not necessarily strength or dominance. Reproduction is one of the seven characteristics of living things. If a living thing can reproduce successfully, then the living thing is fit; if not, it is unfit.

The concept of "survival of the fittest" is still not complete without an understanding of heredity. Understanding of heredity and the use of genetic principles extend back more than 10,000 years.[30] Early humans noticed that many individual differences occurred among the wild animals and plants of nature. They discovered that if they selected those individual animals and plants with desirable traits and by interbreeding them, they produced offspring with more of the same traits.[31] Iterating the process for numerous generations, they gradually converted wild animals and plants into the domesticated varieties that made agriculture possible. This is generally regarded as the first instance of manipulating life to cater to human needs.

About 3,000 years ago, to ease backbreaking labor, early peasants bred female horses with male asses to produce mules. Mules are more desirable beasts of burden because of their endurance, surefootedness, intelligence, and longer life expectancy.

Ancient writings provide further evidence that early societies displayed a keen interest in heredity. The Chinese surnames originated about five thousand years ago, during a time of maternal society. This explains why the Chinese character for surname (姓) is made up of two characters: one meaning woman (女) and the other meaning to give birth (生). What interests us the most is that surnames originated from the name of the village in which one lived or the family to which one belonged. A man and a woman of the same surname could not marry. The Chinese had discovered that marriages of close relatives could be detrimental to future generations. Incestuous or consanguineous mating should be forbidden.[32]

---

[28] Charles Darwin, *The Origin of Species*, Gillian Beer (ed.), (Oxford University Press, 1998).
[29] After Darwin, directed by Martin Lavut, (Galafilm documentary).
[30] MSNBC Interactive, www.msnbc.com/news/49963.asp#BODY
[31] Benjamin A. Pierce, *The Family Genetic Sourcebook*, (John Wiley & Sons, Inc., New York, 1990).
[32] "The origin of Chinese surnames", www.chinavista.com

Hindu sacred books dating back 2,000 years attribute the characteristics of children primarily to the father, and that difference between the son and father were thought to result from the influence of the mother. The writings provide simple rules for choosing a spouse, suggesting women from a family with undesirable traits should be shunned.

Evidence of knowledge of human heredity permeates ancient Greek poetic and philosophical literature. The Spartan state determined whether children, both male and female, were strong when they were born; weakling infants were left in the hills to die of exposure. Exposing weak or sickly children was a common practice in the Greek world, but Sparta institutionalized it as a state activity rather than a domestic activity.

The Talmud, a book of Jewish civil and religious laws based on oral traditions dating back thousands of years, describes in detail the pattern of inheritance for hemophilia. It says that if a woman bears two sons who died of bleeding following circumcision, any additional sons should not be circumcised. Furthermore, the sons of her sisters must not be circumcised, but the sons of her brothers should. This is sound genetic advice based on the X-linked inheritance of hemophilia, which means that the gene causing the disease is located on a particular structure called the X chromosome.

How Genetics Changed the Course of History

Genes not only shape our personal lives, sometimes they influence the course of history as well.

Hemophilia or bleeder's disease, the disease mentioned in the Talmud, is a rare disorder in which the blood fails to clot. A bruise can lead to internal bleeding and death. The gene for hemophilia in the European royal families can be traced to the English royal family in 1819 when Victoria Alexandrina was born to the Duke of Kent and the Princess of Saxe-Coburg. Victoria, later Queen of England and ruler of the British Empire for over 60 years, was a carrier of hemophilia. In other words, she carried a defective gene for hemophilia but did not have the disease herself because she also carried a normal gene for clotting blood. Victoria had nine children. One, Leopold, was hemophilic and died at the age of 31. But before his death, he passed on the gene for hemophilia to his daughter. At least two of Victoria's daughters also inherited the gene and like their mother were unaffected carriers.

Through intermarriage of the royal families of Europe, the gene was eventually passed into the royal houses of Germany, Russia and Spain. In all, ten male descendants of Victoria suffered from hemophilia. Most of them bled to death early in their lives.

Perhaps the most famous of the royal hemophiliacs was Alexis, son of the Russian Tsar Nicholas II and Alexandra, granddaughter of Queen Victoria. Alexandra was a carrier of the hemophilia gene. Alexis was her first-born and heir to the Russian throne.

When Alexis was born in 1904, the royal family celebrated the arrival. But the rejoice was to be short-lived for soon Nicholas and Alexandra learned that their son possessed hemophilia. The disease is still incurable.

Alexis suffered much from the genetic disorder. Minor injuries frequently led to internal hemorrhaging and excruciating pain. The parents forbade him to participate in sports and other physical activities that might endanger his life, but scrapes and bruises were inevitable. The poor boy went through one crisis after another.

Nicholas and Alexandra worried constantly and were greatly distressed by Alexis' illness and suffering. Alexandra felt tremendous guilt. In desperation, the royal couple turned to a mystic and monk Gregory Efimovich Rasputin. Under Rasputin's care, Alexis recovered from several serious bleeding episodes and Rasputin consequently gained considerable influence over the royal family.

Historians argue that Rasputin's influence over the royal family and the Tsar's preoccupation with his son's hemophilia led to a neglected country. The Russian people revolted, eventually overthrowing the Tsar to usher in a Marxist state in Russia. The Bolsheviks executed Nicholas, Alexandra and the entire royal family, including Alexis, on July 17, 1918.

Had Alexis not been hemophilic, there could have been a more peaceful transfer of power, and Russia might still have a royal family.

▶ Timeline of Genetics

To pave the way for upcoming chapters and to fully understand humankind's quest to tinker with nature, we have to take a cursory journey along the historical path of heredity and genetics.

Though most textbooks trace genetics back to Gregory Mendel, available historical records show that genetics can be traced further back. Evidences show that Mendel was fully aware of his predecessors' work.

William Harvey (1578-1657), an English physician, is best known for publishing *Exercitatio Anatomica de Motu Cordis et Sanguinis in Animalibus*, his discovery of the circulation of the blood in 1628. In 1651 he published a second groundbreaking book, *Exercitationes de Generatione Animalium (On the Generation of Animals)*, in which he suggested that all living animals originate from eggs. The work is now regarded as the basis for modern embryology.

Joseph Gottlieb Kolreuter (1733-1806) conducted experiments that led him to conclude that parents contribute equally to the characteristics of their offspring in 1767. Jean-Baptiste de Monet, Chevalier de Lamarck (1744-1829) was the first to develop a comprehensive and systematic theory of evolution. In 1809, he suggested that species can change due to adaptive characteristics and that these characteristics can be inherited.

Rudolf Virchow (1821-1902) is the founder of cellular pathology and a leader in anthropology. He was an advocate of the cell theory. In 1858 he published an influential theory that cells arose from each other in a series of continuous regeneration (*Omnis cellula e cellula*). In other words, new cells result from the division of cells.

In 1859, Charles Robert Darwin (1809-1882) published *The Origin of Species*, a dense and difficult book to read.[33] Yet it makes such compelling argument for evolution as the means by which all life on Earth evolved to its present multitude of forms. The book has never been out of print since it was first published on November 24, 1859. The idea of evolution through greater survival and reproduction of those individuals with any slight advantage over their fellows, better known as "survival of the fittest," was not entirely new. Jean-Baptiste Lamarck and Darwin's grandfather, Erasmus Darwin, had espoused the idea. What Charles Darwin did differently, following decades of careful field observations coupled with scientific experimentation, was he was able to demonstrate with virtually unassailable logical reasoning that his theory was the most plausible means for accounting for the wide variety of species of life on the planet.

In 1866, a young Augustinian monk Gregory Mendel (1822-1884) published the results of his experiments with the common pea plant, *Versuche über Pflanzenhybriden* (Experiment on Plant Hybridization). Mendel outlined his conclusions that traits in pea plants were determined by two factors, one inherited from the female parent and one from the male parent. He explained how these two factors separated when the sex cells (eggs and pollens in plants) were formed, one factor going into each sex cell. When sex cells fused in the process of fertilization, the factor from the pollen united with the factor from the egg, and together they determined the trait of the offspring. Mendel also recognized that chance determined which one of the two factors an offspring inherited from its mother and which one of the two factors it inherited from its father. This role of chance produced distinctive ratios of traits in the offspring.

In 1871, Swiss chemist Johannes Friedrich Miescher (1844-1895) isolated the DNA from human white blood cells and called the substance "nuclein."

---

[33] Charles Darwin, *The Origin of Species*, Gillian Beer (ed.), (Oxford University Press, 1998).

Miescher thought DNA's function was for storing phosphorus and had no idea that it played a crucial role in heredity. Two years later, Friedrich Schneider first observed the stages of cell division or cell mitosis while studying transparent flatworms under a microscope. In 1883, August Weismann (1834-1914) determined the difference between body cells and sex cells.

There was considerable interest in Mendel's work of 1866, but surprisingly, none recognized the far-reaching implications of his conclusions. Mendel's discovery remained buried in the scientific literature for 35 years until around the turn of the century, three botanists – Hugo de Vries, a Dutch scientist; Erich von Tschermak working in Vienna; and Carl Correns in Tubingen – all began to conduct breeding experiments on plants. Working independently, in the process of analyzing and writing up their results, they discovered Mendel's work. All three scientists immediately interpreted their own results in terms of Mendel's rule of inheritance, and published their results in 1900, with reference to Mendel's work.

Rediscovery of Mendel's laws in 1900 clarified inheritance. But Mendel, de Vries, von Tschermak and Correns worked with traits of whole organisms. They did not investigate how characteristics were sorted and combined on a cellular level. German scientist Theodor Boveri (1862-1915) and American scientist Walter Sutton (1877-1916), working independently, hypothesized that chromosomes carried the cell's units of inheritance in 1902. They were the first to note the behavior of chromosomes during division of sex cells was the basis for Mendel's law of heredity.

In 1913, Alfred Henry Sturtevant (1891-1970) was the first scientist to develop a technique for mapping the locations of specific genes of the chromosomes in the fruit fly *Drosophila melanogaster*. George Beadle (1903-1989) and Edward Tatum (1909-1975) were the first to introduce fungus as a genetic model and show that each gene is a blueprint for a protein in 1941. They were co-founders of the one-gene-one-protein hypothesis. In 1944, Oswald Theodore Avery (1877-1955), Maclyn McCarty (1911-2005) and Colin MacLeod (1909-1972) were the first to show that the agent responsible for transferring genetic information was not a protein, as biochemists of the time believed, but DNA (deoxyribonucleic acid).

In 1950, Erwin Chargaff (1905-2002) discovered regularity in proportions of DNA bases for different species. In all organisms he studied, the amount of adenine (A) approximately equaled that of thymine (T), and the amount of guanine (G) approximately equaled that of cytosine (C). In 1953, James Dewey Watson (1928-) and Francis Crick (1916-2004) proposed the double helix structure of DNA.

T.T. Puck and colleagues succeeded in growing clones of human cells in the laboratory in 1956. Two years later F.C. Steward concluded that each cell of a multi-cellular organism must contain all the ingredients necessary to create a complete organism. The following year Jerome Lejeune (1926-1994) and colleagues determined the genetic basis of Down's syndrome. In 1966, Victor McKusick (1921-) published a catalog of 1,487 known genes, identified by the protein and genetic disorder associated with each gene.

A breakthrough came in 1972. In that year Paul Berg (1926-) successfully spliced together the DNAs of two different organisms. This discovery laid the foundation of genetic engineering and the modern biotechnology industry. In 1973, Herbert Boyer (1936-) and Stanley N. Cohen (1935-) developed recombinant DNA technology showing that genetically engineered DNA molecules might be cloned in foreign cells.

E.M. Southern developed the Southern blot method for analyzing DNA in 1975. In 1977, Walter Gilbert (1932-) and Frederick Sanger (1918-) devised techniques for sequencing DNA. Sanger also determined the complete nucleotide sequences of a bacteriophage (a virus that infects bacteria), ϕ-χ 174 (5,375 nucleotides), human mitochondrial DNA (16,338 nucleotides) and bacteriophage lambda (48,500 nucleotides).

On a topic pertinent to our discussion, on July 25, 1978, the world's first test-tube baby, a.k.a. *in vitro* fertilization (IVF) baby, Louise Brown was born at Oldham and District General Hospital in England. The birth was the culmination of many years of work by Patrick Steptoe (1913-1988) and Bob Edwards (1925-). Refinements in the technology have increased pregnancy rates and it is estimated that in 2004 about 1.5 million children have been born by IVF. Their breakthrough laid the groundwork for further innovations such as intracytoplasmatic sperm injection (ICSI), embryo biopsy (PGD), and stem cell research.

On a different front of legal significance, in 1980, the U.S. Supreme Court ruled that a genetically modified microorganism could be patented.

Kary Mullis conceived and developed polymerase chain reaction (PCR), a technology for rapidly multiplying fragments of DNA, in 1983.

From 1986-1990, the effort to sequence the human genome was launched. The Human Genome Project started in earnest in the early 1990s. On June 26, 2000, Bill Clinton of U.S. and Tony Blair of U.K. announced the completion of the first draft of the human genome, one of the greatest achievements of human scientific endeavors. In July 2000, an orange bug genome was published.[34] In December 2000, a first plant genome was sequenced.[35] In February 2001, a draft of the human genome sequence was published.[36,37] In January 2002 a deadly plant bug genome was published.[38] In April 2002, a draft of the rice genome was published.[39,40] In December 2002, a draft of the mouse genome was published.[41] In 2004, the complete genome of *Methyloccocus capsulatus* was published.[42] In this same year, the chimpanzee DNA sequence was also published.[43,44] The dog genome was sequenced in 2005.[45]

The sequencing quest continues...

On December 27, 2002, Clonaid claimed the first human clone – a 7-pound girl – had been born a day earlier to an American couple at 11:55am at an undisclosed location.[46] Later the same group announced four other clones. The claims have yet to be verified... On February 14, 2003, Dolly the cloned sheep was euthanized...

In May 2003, Idaho Gem, a mule, was cloned and successfully delivered, followed immediately by Prometea, a horse.

Bearing a striking resemblance to the 2003 announcement by Clonaid, Panos Zavos announced on January 18, 2004 that he had transplanted a cloned embryo into a 35-year-old-woman less than two weeks earlier.[47] To date, there has been no announcement of any results from the transplant.

---

[34] A. Simpson, et al, "The genome sequence of the plant pathogen *Xylella fastidiosa*", *Nature*, 406, July 2000, pp. 151-157.

[35] Arabidopsis Genome Initiative, "Analysis of the genome sequence of the flowering plant *Arabidopsis thaliana*", *Nature*, 408, December 2000, pp. 796-815.

[36] J.C. Venter, et al, "The sequence of the human genome", *Science*, 291, February 16, 2001, pp. 1304-1351.

[37] E.S. Lander, et al, "The Genome International Sequencing Consortium, Initial sequencing and analysis of the human genome", *Nature*, 409, February 15, 2001, pp. 860-921.

[38] M. Salanoubat, "Genome sequence of the plant pathogen *Ralstonia solanacearum*", *Nature*, 415, January 2002, pp. 497-502.

[39] J. Yu, et al, "A draft sequence of the rice genome (*Oryza sativa L. ssp indica*)", *Science*, 296, April 2002, pp. 79-92.

[40] S.A. Goff, et al, "A draft sequence of the rice genome (*Oryza sativa L. ssp japonica*)", *Science*, 296, April 2002, pp. 92-100.

[41] R.H. Waterston, et al, "Initial sequencing and comparative analysis of the mouse genome", *Nature*, 420, December 5, 2002, pp. 520-562.

[42] Naomi Ward, et al, "Genomic insights into methanotrophy: The complete genome sequence of *Methylococcus capsulatus* (Bath)", *PLoS Bio*, 2(10), October 2004, e303.

[43] The International Chimpanzee Chromosome 22 Consortium, "DNA sequence and comparative analysis of chimpanzee chromosome 22", *Nature*, 429, May 2004, pp. 382-388.

[44] The Chimpanzee Sequencing and Analysis Consortium, "Initial sequence of the chimpanzee genome and comparison with the human genome", *Nature*, 437, September 2005, pp. 69-87.

[45] Kerstin Lindblad-Toh1, et al, "Genome sequence, comparative analysis and haplotype structure of the domestic dog", *Nature*, 438, December 2005, pp. 803-819.

[46] Dennis Kelly, "Company claims world's first human clone", *USA Today*, December 27, 2002.

[47] "Experts demand human cloning proof", *CNN News*, January 18, 2004.

A cloned dog by the name Snuppy was announced in August 2005.[48]
The cloning quest continues…

•••••   •••••   •••••

In this chapter we also provided a cursory survey of the milestones in genetics and cloning. The list is by no means exhaustive. There is one noteworthy point: the rate at which technological breakthroughs have been coming forth is accelerating. Armed with this backdrop, we are now ready for the rest of this book.

---

[48] Editorial, "A dog's life", *Nature*, 436, August 2005, pp. 604-604.

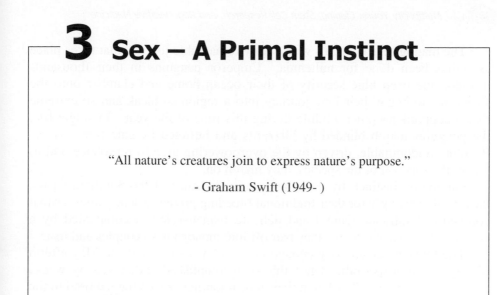

3 Sex – A Primal Instinct

"All nature's creatures join to express nature's purpose."

- Graham Swift (1949-)

▶ The Primal Instinct for Sex

I used to have a professor who when asked "Why?" would immediately recast the question into "Why not?" So why do we have sex? The answer is why not?

Sex lives vary: some animals mate for life while others spend only seconds with their sex partners. The household arrangement also varies: in some cases, the male and female share the work; in others, the male takes care of the offspring, or the female does all the work; in yet others, neither does and the offspring is left to fend for themselves right from birth. A few examples will suffice.

Emperor Penguins

Emperor penguins are the largest of the 17 species of penguins, and one of the most biologically interesting. Concentrated in the Weddell Sea and Dronning Maud Land, Enderby, Princess Elizabeth Lands, and the Ross Sea, they spend their entire lives on the cold Antarctic ice and in its waters. They survive – breeding, raising young, and eating – by relying on a number of clever adaptations.

Each year in the winter (March in the Antarctica), nearly all creatures leave except for the emperors, the only animals that spend the winter on Antarctica's open ice. These flightless birds breed in the winter, unlike most birds, which breed in the springtime.

The mating season begins with a truly remarkable journey that takes place as it has been done for millennia. Emperor penguins in their thousands abandon the deep blue security of their ocean home and clamber onto the frozen ice to begin their long journey into a region so bleak and so extreme that it supports no other wildlife during this time of the year. In single file, the penguins march blinded by blizzards and buffeted by gale force winds. Resolute, indomitable, driven by the overpowering urge to reproduce and to assure the survival of the species, they march on.

Guided by instinct, by the otherworldly radiance of the Southern Cross, they head unerringly for their traditional breeding ground where – after a ritual courtship of intricate dances and delicate maneuvering, accompanied by a cacophony of ecstatic song – they pair off into monogamous couples and mate.

The females remain long enough only to lay a single egg (in May after a 63-day gestation period). Once this is accomplished, exhausted by weeks without nourishment, they begin their return journey across the ice field to the fish-filled seas. Unlike most penguins, which feed on surface krill, emperor penguins live on fish, squid, and crustaceans caught on long, deep pursuit dives. The 60-mile journey to the sea is hazardous, and rapacious leopard seals a predatory threat.

The male emperors are left behind to guard and incubate the precious eggs, keeping them warm on their feet, enveloped by the stomach, in a "brood pouch" – a very warm layer of feathered skin designed to keep the egg cozy – for 65-75 days during the coldest part of the Antarctic year. Subjected to subzero temperatures that drop to as low as −80°F and winds that reach velocities of up to 112 miles per hour, they too face great dangers.

To keep warm, the emperors clump together in huge, huddled masses. They take turns moving to the inside of the group, where they're protected from the icy cold temperatures and wind. Once they've had a chance to warm up, they take their turns back on the circle's edges, giving fellow penguins time in the warmer center.

After two long months during which the males eat nothing except incubating, the eggs begin to hatch. Once they have emerged into their ghostly white new world, the chicks cannot survive for long on their fathers' limited food reserves. They feed the chicks with "milk" produced by a gland in their esophagus; during this time, the males lose one-third to one-half their body weight. The young chicks will remain in the brood pouches for a short time until they are able to regulate their own body temperatures. If their mothers are late returning from the ocean with food, the newly hatched young will die.

When the females return from the sea, bringing food they regurgitate to feed the now hatched chicks. The males, exhausted and starved, eagerly leave for the long journey for their own fishing session at sea. The mothers take over care of the chicks. The youngsters stay sheltered in their mothers' brood pouches for two months. If a young chick falls out of that warm spot, it can freeze to death in as little as two minutes.

As the young penguins grow, adults leave them in groups of chicks called crèches while they leave to fish. While the adults fish, the chicks face the ever-present threat of attack by prowling giant petrels, which have now returned.

There is a reason for the timing of emperor penguins' hatching. By December or January, when the Antarctic weather has warmed somewhat, the ice floes the penguins occupy begin to break up, bringing open waters closer to the nesting sites. Now the chicks are at the age of independence – old enough to take their first faltering dive into the deep blue waters of the Antarctica at a time when food is most plentiful and predators are few.[49]

Salmons

Prior to returning to their spawning grounds from the ocean, adult salmons convert much of their body mass to eggs and sperm production. The spawning process involves mature salmon (male and female) swimming upstream against hard river currents, waterfalls and other obstacles to their home creek where they will spawn. Salmon will usually spawn in the creek they were born in and can find this creek through a remarkable sense of smell.

The effort to migrate up the river does not allow for adequate feeding periods for they only have a limited time. They use their body mass as their source of nutrients and energy. In the spawning process the female releases her eggs and the male fertilizes them. Salmon spawning areas usually have high quality sand and gravel because during the spawning process the females will bury the fertilized eggs in the gravel. This focused effort depletes their body resources and after egg laying and sperm milting, adult salmons die.

Pacific Grunions

Similarly, each spring millions of Pacific grunions fling themselves up the Pacific beaches to deposit their eggs in the wet sand. They time their spawning season with the spring tide, one of the highest tides of the year. This choice ensures that their eggs will be high enough on the beach to be out of reach of scavenger fish. The high tides will also ensure that they will have enough waves to get back to the ocean. Mistiming means both the little fish and their eggs will perish.

---

[49] *March of the Penguin*, Documentary narrated by Morgan Freeman, 2005.

Other Aspects of Sex

The moon's pull creates high tides. But the moon is not the only timepiece for sexual activities. Many mammals become amorous during the lengthening days of spring. Long days signal warm weather is coming and adequate food supply for offspring. For other animals, particularly those in arid plains, such as the gazelle, rains mean the food supply will be more abundant to feed the young. Not surprising, predators such as jackals time their births to coincide with their prey.

In the U.S., the Census Bureau statistics indicate the number of births typically peaks during the summer, that is, there are more people born under the astrological sign of Cancer (June 22-July 22) than any other signs. If this hearsay is reliable, then a plausible explanation is probably during winter months couples cuddle more, or they time their babies to arrive in warmer summer months.

Animals themselves produce other signals as well. Many insects and mammals produce chemicals called pheromones, subtle perfumes to attract the opposite sex. Other animals change color, shape, develop special body parts, such as sharp curled horns, to signal they are ready to mate. The colorful plumage of peacock, or fish that turns as bright as a neon sign are examples.

We human beings, as a sexual animal, take sex for granted. The question of why we have sex is not one that occurs to most of us. We have sex because we want to have children, or because it gives us pleasure, or both. Sex plays an important role in reproduction. The conventional reproductive strategy is to have two individuals – the father and the mother – combine their genes in an offspring, with each parent typically contributing half of the genetic material. Thus offspring produced by sex carry unique sets of genes or variety. Unique sets of genes offer the offspring a survival advantage in a changing environment. Variety not only brings spice to life, but it also is a key to survival. Imagine a neighborhood of identical clones?

But sex is not the only way to have children. Many organisms can reproduce without sex. Bacteria and many protozoa, for example, can simply divide themselves into two identical copies without the help of a partner.

Though rare, asexual animals – those without sex or sexual organs – do exist. For example, whiptail lizards in the western United States have no males. The way they reproduce is having one female mount another female, bite her neck and wrap around her like a doughnut, and mimic what a "male lizard" would do while mating. Herpetologists believe the mounting female makes the other female ovulate. The eggs simply start dividing and growing into embryos without being fertilized by sperm. The "impregnated mothers" return the favor to their pseudo-mates by playing the part of the male. In this way, the lizards invariably give birth to only females, all of which are identical to their mothers.[50] The offspring is in all respect a clone. The reproduction process is a form of parthenogenesis.[51] Reproducing without the benefit of a partner produces genetically identical clone offspring which can be much more vulnerable to disease or succumb to changing environments.

An analogy will perhaps bring out the difference between sexual and asexual reproductions: The genetic makeup of an organism is like a soccer team.[52] Everybody has a full team, with different players playing for each position. There are good players and bad players just like there are good genes and bad genes. Clones merely pass their own rosters on to their offspring, while sexual species create new rosters.

Sex Reigns – Birds and Bees Do It, But Why?

Thus sex is not only unnecessary, but it also is a potential recipe for evolutionary disaster. For one thing, it is an inefficient way to reproduce. In a population of asexual whiptail lizards, every lizard can bear baby lizards of her own; in contrast, in a population that reproduces sexually, only half of them can. In a community in which only sexual and asexual species are living side by side, assuming no mortalities from accidents and diseases, the asexual population would quickly swamp the sexual population with their explosive birthrate.

By all rights, any group of animals that evolves by sexual reproduction should be promptly out-competed by nonsexual ones. And yet sex reigns: peacocks drag around grand tails – an announcement to their predators – to attract the opposite sex. Still, peacocks show no sign of evolving away their colorful plumage.

Sex carries other cost as well. When males compete for females by locking horns, they expend a huge amount of energy; when males compete for females by singing, they put themselves at risk of being attacked by predators. Humans ingest unnecessary hormones and steroids to develop the body physique to attract the opposite sex that they suffer from long-term health problems. And many consume sex potions or pills, not to mention Viagra. Humans also commit infidelity and adultery. To impress on their mates, some

---

[50] Carl Zimmer, *Evolution: The triumph of an idea*, (HarperCollins Publishers, New York, 2001).

[51] "Partheno" is a Greek word meaning virgin.

[52] This analogy works if we ignore the fact that the number of players in a soccer team is only 11, while the number of genes in different organisms is different, and can be as high as tens of thousands.

macho ones plunge to their untimely death; yet others commit murder out of jealousy.

Spiders provide another striking example of the cost of sexual reproduction. The Australian male redback spiders do not produce a web, they are found on the fringe of a female's web. During the summer mating season, the male has to make overtures to the female to discover whether she is ready to mate, which can prove fatal if she mistakes him for prey. In order to occupy the female's attention during mating, the male spider offers her his abdomen by standing on his head and somersaulting his abdomen towards her mouthparts. The female squirts digestive juices onto the male's abdomen. Most males do not survive this process. Yet new generations of redback spider males are hurling themselves onto the female's poisonous fang, becoming a meal for her at the end of the sexual act, just as their fathers did.

If sexual reproduction is so costly, then why only a fraction of 1% of all vertebrates reproduces asexually like the virgin births of whiptail lizards? Why is sex so successful despite all its disadvantages?

▶ Models of Co-Evolution

Let us compare asexual reproduction and sexual reproduction at the genetic level. In asexual reproduction, only one parent is required and all of the parent's genes are passed on to its progeny, that is, the parent's genetic endowment is passed in a chunk. In sexual reproduction, two parents are required and only half of each parent's genes are passed on to the next generation, that is, the parents' genetic material is passed on after having been shuffled. At first blush, it would seem that sex just gets in the way of passing on genetic materials and it makes some sense to have self-reproducing females and just dispense with males altogether. But before we hastily jump to the conclusion that asexual reproduction is a better evolutionary strategy, we should perhaps ask, "since sex persists, then why sexual reproduction exists at all?"

For many years, the general proposition that sex is good for evolution because it creates genetic diversity to cope with changing and challenging environments was widely accepted. Is there any other evolutionary edge sex may provide to the sexual organism? A number of theories have been put forward to explain why sexual reproduction may be more evolutionarily advantageous than asexual reproduction. After more than half a century of debate and some 20 published theories, scientists are now trying to pin down what the payoff is.

By the 1980s, two dominant hypotheses remained: the deleterious mutation hypothesis and the Red Queen hypothesis.

The Deleterious Mutation Hypothesis

The champion of this theory is Alexey Kondrashov. The idea is that sex exists to purge a species of damaging genetic mutations. The argument is that in an asexual population, every time a creature dies because of a mutation, the mutation dies with the creature. In a sexual population, some of the creatures born have lots of mutations and some have few. If the ones with lots of mutations die, then sex purges the species of the mutations. Since most mutations are harmful, sex provides an advantage.[53]

Then the question arises why eliminate mutations this way, rather than correcting most of them by better proofreading? This is a matter of genetic economy: it is cheaper to allow some mistakes through and remove them later. Mathematically, the cost of perfecting proofreading mechanism escalates as the mechanism becomes more and more perfect.[54]

If such a hypothesis is to make any sense at all, the rate of deleterious mutations must be high enough to warrant sex purging mutations a necessity. An estimate is that the rate of deleterious mutations must exceed one per individual per generation if sex is to earn its keep eliminating mutations. In most creatures, the rate of deleterious mutations is right at the threshold, providing evidence to support the theory.

Even if the rate is higher than the threshold for sex to earn its keep to purge mutations, all that proves is that sex perhaps plays a role in purging mutations. It does not explain why sex persists. In a community of only sexual and asexual individuals, the sexual population will be driven extinct by the much greater productivity of asexual clones. Unless the clones' genetic drawbacks will appear in time, this will be the expected outcome. Deleterious mutation hypothesis works too slowly in this sense.[55]

William Rice of University of California, Santa Barbara, has been pursuing the question of sex in evolutionary biology by endowing *Drosophila* fruit flies bogus chromosomes, then checked the eye color of thousands of descendants. Rice reported that his experiments bolster the idea that sex helps good genes spread through a population. The work shows that a favored gene spread more quickly if it appears on a chromosome that participates in gene-

---

[53] Malcolm Ritter, "Birds do it… but why?" *ABC News*, February 14, 2002.
[54] Nikolay A. Kolchanov, and Hwa A. Lim, *Computer Analysis of Genetic Macromolecules: Structure, function and evolution*, (World Scientific Publishing Co., New Jersey, 1994).
[55] Matt Ridley. "Is sex good for anything?" *New Scientist*, 140 (1902), 1993.

shuffling (recall the new roster of the soccer team analogy). The experiment also shows that harmful mutations accumulate faster without gene-shuffling. In fact, it appears that humans naturally produce harmful mutations so often that the species would go extinct if sex does not purge the harmful mutations. However, it is not clear how often that situation occurs in other animals and therefore how widely the deleterious mutation theory can be applied.

The Red Queen Hypothesis

In the late 1980s, the Red Queen hypothesis emerged and has been garnering support steadily. The hypothesis is that sex fights parasites. Parasites take a tremendous toll on their hosts, and any adaptation that helps the hosts escape them may become hugely successful.

Imagine a pond of fish that reproduce by cloning so the offspring of each fish is an identical copy of its mother. The fish are not all carbon copies of one another for a mutation may arise in one fish and be passed down her descendants. They will form a strain that can be distinguished from other strains by unique mutations.

Suppose a parasite invades the pond. The parasite mutates as it spreads, forming new strains of its own. Some strains or variants of the parasite carry mutations that make them good at attacking certain strains of fish. The strain that can attack the most common fish (call them Fish A) has the most hosts to attack and that variant soon becomes the most common of all parasite strains (call them Parasites A). The other parasite strains, limited to fewer hosts, dwindle to low levels.

But Parasites A undermine their own success. They thrive so successfully in their strain of fish that they kill Fish A faster than the hosts can reproduce. The population of Fish A crashes, and as it disappears, its parasites have a harder time finding new hosts to infect. The number of Parasites A crashes as well.

This attack on Fish A gives the rarer strains of fishes an evolutionary edge. Unburdened by parasites, their numbers swell. Eventually another fish strain, call it Fish B, becomes most common. As Fish B grows more successful, it becomes fertile ground for the rarer parasites (Parasites B) that are best adapted to them. Parasites B begin to multiply and catch up with the explosion of their host. Like Parasites A, Parasites B undermine their own success. Now the population of Fish B crashes, to be replaced by Fish C, and so on.[56]

---

[56] Carl Zimmer, *Evolution: The triumph of an idea*, (HarperCollins Publishers, New York, 2001).

Biologists call this model of co-evolution the Red Queen hypothesis. First coined by Leigh van Valen, it refers to the character in Lewis Carroll's *Through the Looking Glass* who takes Alice on a long run that brings them nowhere. Hosts and parasites experience a huge amount of evolution, but it does not produce any long-term changes in either of them. It is as if they are evolving in place.

So the Red Queen hypothesis is very simple: sex is needed to fight parasites. Parasites, which include fungi, bacteria and viruses, specialize in breaking into cells as if they have the keys to unlock the cells' locks. The parasites then either devour the cells (as fungi and bacteria do), or subvert the cell's machinery for the purpose of making their own clones (as viruses). The molecular mechanism to do so is molecular recognition, like in one protein recognizing another, or a key recognizing a lock: parasites use protein molecules that bind to other protein molecules on cell surfaces of the host. The race between the parasites and the hosts is then parasites invent new keys, hosts change the locks! If one lock is common in one generation, parasites with the key that fits it will flourish.

In summary, according to the Red Queen hypothesis, sexual reproduction persists because it enables host species to evolve new genetic defenses against parasites that attempt to live off them. Sexual species can call on a library of locks unavailable to asexual species.

A sexual animal is not a clone of its mother, nor is it a simple blend of its parents' genes. Instead it carries a combination of genes from both its mother and father. When cells divide into eggs or sperms, each pair of chromosomes warp around each other and swap genes. Thanks to this sexual dance, the genes of a father and a mother can be shuffled into multitudes of different combinations to confer the offspring a better survival chance.

The key-lock analogy is evident in heterozygosity. In sexual reproduction, an organism inherits a particular gene from each parent. When an organism carries two different forms of the gene, it is said to be heterozygotic; if it carries only one form of the gene, it is said to be monozygotic. An example is sickle cell anemia. Sickle cell anemia is caused by an error in the gene that produces hemoglobin, the oxygen-carrying component of the red blood cells. The defective hemoglobin is a sickle shaped, long, sticky polymer instead of the normal round shape. The abnormally shaped cells clog up blood passageways, leading ultimately to oxygen starvation of vital organs.

The sickle gene, however, helps to defeat malaria. So where malaria is common, the heterozygotes – those with one normal gene and one sickle gene, are better off than homozygotes – those with a pair of normal genes or a pair of sickle genes. Homozygotes may be infected by malaria if they have only normal genes, or suffer from sickle cell anemia if they have only sickle genes.

▶ Computer Simulations

William Hamilton (1936-2000), an Oxford biologist, proposed in the early 1980s that sex might bring an advantage to animals struggling through the Red Queen race because it makes it harder for the parasites to adapt to them.[57]

Interestingly, Hamilton died of malaria contracted on an expedition to Congo. In January 2000, he with two brave companions was in the depth of the Congo jungle seeking evidence to bolster a radical hypothesis for which he was a strong proponent. He believed that the AIDS epidemic could be traced to contaminated oral polio vaccines tested in Africa in the 1950s. Way into the expedition he was rushed back to London, apparently with severe malaria, seemed to recover, but then collapsed into complications and coma from which he never recovered. He passed away on March 7, 2000.[58]

During his industrious career, one of the things Hamilton sought to understand is the existence of sex, and why most species propagate sexually rather than asexually. He suggested that sexual reproduction, by continually shaking up the genome, helps to keep organisms a step ahead of their parasites, which might be capable of annihilating a population of asexually spawned clones. The parasite-avoidance hypothesis of sex remains debatable, but many studies of how animals choose their mates at least partly support it.

In the late 1970s, with the help of two colleagues, Hamilton performed computer simulation of sex in artificial life. The simulation began with an imaginary population of 200 creatures, some sexual and some asexual. Death was randomly assigned. As expected, the sexual population died out. The inference is that other things being equal, in a game of sexual and asexual, asexual always wins. Asexual reproduction is more efficient and the genes are guaranteed to pass on to the next generation.[59]

Hamilton and colleagues next introduced several species of parasites, 200 of each. Virulence genes defined the virulence of the parasites; similarly, resistance genes in the hosts defined the resistance to match parasite virulence. The least resistant host and least virulent parasites were eliminated in each

---

[57] Natalie Angier, "William Hamilton, 70, dies, an evolutionary biologist", *The New York Times*, March 20, 2000.
[58] Richard Dawkins, "On W.D. Hamilton", *The Independent*, March 10, 2000.
[59] Nikolay A. Kolchanov, and Hwa A. Lim, *Computer Analysis of Genetic Macromolecules: Structure, function and evolution*, (World Scientific Publishing Co., New Jersey, 1994).

generation. Sexual population won the game most often when there were lots of genes that determined resistance and virulence in the creature.

In the simulation, as resistance genes that worked became more common, so would the virulence genes. Then these resistance genes would grow rare, followed by the corresponding virulence genes. Hamilton concluded, "The essence of sex in our theory is that it stores genes that are currently bad but have promise for reuse. It continually tries them in combination, waiting for the time when the focus of disadvantage has moved elsewhere."

Using fish as an example again, sexual fish do not evolve into distinct strains; their genes scatter throughout the pond's population, mingling with genes of other fish. Fish genes that have lost their protective and combative powers against the parasite can be stored away in the DNA of the fish population, among other effective genes. These obsolete genes may later provide more protection against new parasite strains, at which point they will spread once more through the pond's populations. Parasites can still attack fish that reproduce sexually, but they cannot force them into boom-and-bust cycles as dramatic as the ones suffered by their asexual or clonal cousins.

Supporting Evidence

Empirical support for the Red Queen hypothesis has been growing. First, asexual is more common in species that are little troubled by diseases. These include boom-and-bust microscopic creatures, arctic and high altitude plants and insects. But the best support of the Red Queen hypothesis perhaps comes from a study on a little fish called topminnow by Curtis Lively and Robert Vrijenhoek.

Topminnow is common in Mexico. The fish is under constant attack by a parasite that causes black cysts in their flesh, thus the name black-spot disease. The fish sometimes crossbreeds with other similar fish to produce an asexual hybrid.

Lively and Vrijenhoek studied topminnows in several ponds and streams in Mexico. In one pond, the researchers found that asexually reproducing topminnows harbored many more black-spot worms than did their sexually reproducing counterparts. In another pond where two strains of clones lived, the more common strain was subject to more infections. This is in agreement with the Red Queen hypothesis which predicts that sexual topminnows could devise new defenses faster by recombination than asexual topminnow could.

In a third pond, the topminnows seemed to contradict the Red Queen hypothesis because the sexual fish were more vulnerable than the asexual ones. Closer examination revealed that the pond had dried up in a drought a few years earlier. When the pond returned, only a few fish recolonized it. Consequently, the sexual fish were highly inbred and were thus deprived of the genetic diversity. The researchers added more sexual topminnows to the pond to increase the genetic diversity. Within two years the sexual fish were immune to the parasites. This provided even stronger confirmation of the Red Queen hypothesis.

Computer simulations and real-world evidence aside, it could well be that the deleterious hypothesis and the Red Queen hypothesis are both true, that is, sex serves to help purge a species of mutations, and sex is needed to fight diseases. Or that it is matter of time scale: deleterious mutation hypothesis may be true for long-lived creatures like mammals and trees, but not for short-lived creatures like insects.

Scandal without Sex

There are always exceptions to the rule.

Bdelloid rotifers are the commonest of rotifer varieties. They are soft-bodied and are found in fresh ponds or on mosses. The name describes their habit of moving about: extending the body and attaching the head, then catching up with the foot and again extending the body, like a leech.[60]

They are all females for no males have ever been observed among them. In reproducing, they produce eggs requiring no fertilization. For some *bdelloids*, the young develop to maturity within the mother's body cavity and are born live. For others, eggs are produced and deposited in the vegetation of the pond's edge. In either case, the invariably female offspring clones in turn produce eggs requiring no fertilization and so on. This is another example of parthenogenesis or virgin birth.

Since there is no sexual stage to introduce change and variety to the genome, it is something of a mystery as to how *bdelloid rotifers* have survived for so long as a thriving and stable animal group. Indeed, they have been reproducing through asexual cloning for tens of millions of years. This puzzles scientists for they believe no creatures should be able to keep doing so for so long. *Bdelloid rotifers* have been nicknamed an evolutionary scandal, a scandal without sex!

---

[60] www.micrographia.com/index.htm

▶ Selfish Genes

So far we have concentrated our discussions mainly on sex persists to purge of mutations and sex persists to fight diseases in the deleterious mutation hypothesis and the Red Queen hypothesis, respectively. We have only alluded to sex as a means to pass on genetic materials. If passing on genetic materials is the sole purpose of sex then asexual reproduction will be more effective; cloning, another reproductive technique that is the subject matter of a following chapter is also relatively more effective in passing on genetic material than sexual reproduction. Let us examine passing on genetic legacy in further detail.

Hamilton burst into the field of evolutionary biology when he was still a graduate student at Cambridge University. In 1963 and 1964, he published two papers based on his doctoral thesis that have proved so seminal to evolutionary biology. He coined the term inclusive fitness, also known as kin selection.[61,62]

Through the model of inclusive fitness, Hamilton proposed an elegant and mathematically sophisticated way of understanding altruistic behavior, a problem that had baffled naturalists from Darwin onward. If organisms are inherently selfish, and supposedly devoted to personal survival and reproduction, why do so many species display seemingly self sacrificial behavior? Why do worker bees forsake the opportunity to breed in favor of caring for the queen's young? And why will those infertile auntie bees commit suicide in defense of the hive?

Hamilton realized that the unusual genetic structure of the bees resulted in the workers being so closely related to one another that, in slaving for the hive, they were essentially slaving for the persistence of their own gene pool. In other words, although they appeared altruistic, they were, from a gene's point of view, behaving with characteristic selfishness.[63]

This explanation recasts the concept of fitness – an individual's success in reproducing, as defined in "survival of the fittest" – to incorporate the survival and reproductive success of the creature's close relatives. This is the origin of the term inclusive fitness. Also in so doing, Hamilton merged Darwin's focus on individual animals competing for the privilege of siring the next generation with Mendel's studies of how distinct genetic traits are transmitted over time.

---

[61] W.D. Hamilton, "The general evolution of social behaviour", Part I, *J. Theor, Biol.* 7, (1964), pp. 1-16.

[62] W.D. Hamilton, "The general evolution of social behaviour", Part II, *J. Theor, Biol.* 7, (1964), pp. 17-52.

[63] Richard Dawkins, *The Selfish Gene*, (Oxford University Press, 1989).

Selfishly Altruistic Kin Selection

The Origin of Species is universally regarded as one of the greatest landmarks in evolutionary biology. It, however, leaves a number of problems unsolved and these problems continued to vex Darwin until his death.

To understand the problems vexing Darwin, let us look at genetic inheritance. Inheritance is the basis of evolutionary change for constant randomization of information carriers obviously does not lead to meaningful information. Without a safe mechanism for transmitting genetic information from one generation onto the next, there would be random arrangements of genetic building blocks. Thus the cornerstone of evolution is genetics. Only by selectively conserving well-tried genes can there be competitions between new-yet-untried ones introduced by mechanisms such as mutations. Darwin was able to formulate a theory of gradual evolutionary change caused by adaptive mutation, selected out of multitudes of other random variants. In this relentless struggle for existence, everyone fights everyone else: if the trait that leads to successful reproduction is safely transmitted to the offspring, the trait will spread and eventually be represented as a feature in the species. Further, if the trait leads to procreation at the expense of the competitor, the animal not only increases its fitness, but also reduces the fitness of the competitor.[64]

So far Darwin's natural selection is fine until the concept is applied to social behavior. A lifelong observer of animals, Darwin wrestled mightily with sociality and the intrinsic apparent contradiction between intensive competitive struggle for survival of individual organisms and the widespread observation that many animals exhibit richly cooperative social behavior: shoaling fish, cooperatively hunting wolves or lion packs, the symbiosis between the fungus and the plant, the subterranean colonies of the naked mole rat, coalition forming in primates and the social insects are, but some excellent examples of cooperative social behavior. Hymenopteran colonies, including honeybees, wasps, bumble bees, ants and termites, usually consist of one reproducing queen and a multitude of sterile workers and soldiers. They practice a form of sociality known as eusociality. In the wonderful world of eusociality, in addition to sterile individuals cooperatively helping the fertile animals raising the offspring, at least two generations overlap in their life stages capable of contributing to colony labor so that the offspring can assist their parents during part of their life cycle.

Before Hamilton introduced the concept of kin selection, evolutionary theory emphasizes on the maximization of individual fitness, that is, the

---

[64] N.A. Kolchanov and Hwa A. Lim, *Computer Analysis of Genetic Macromolecule: Structure, function and evolution*, (World Scientific Publishing Co., New Jersey, 1994).

number of surviving offspring of the individual. The central component of Hamilton's theory is inclusive fitness, that is, one's offspring plus the number of extra offspring relatives bear to help propagate the genetic legacy. This concept seems to be able to reconcile Darwin's "survival of the fittest" with eusociality.

There is another supporting evidence that makes kin selection such a beautiful theory – the haploid-diploid sex determination of the eusocial hymenopterans. Most animal genera have a hetero- and a homogametic sex, that is, a different set of sex chromosomes for the different sexes: males are the heterogametic sex in mammals while females are the heterogametic sex in birds; females are the homogametic sex in mammals while males are the homogametic sex in birds, butterfly and moths. For example, in humans, males have an X- and a Y-chromosome while a female has two X-chromosomes.

Hymenopterans universally produce males from unfertilized haploid (one set of chromosomes) eggs and females from fertilized diploid (two sets of chromosomes) eggs! Such a system skews relatedness in an almost perfect way for eusociality to evolve. As an example, consider a female worker bee. Half of its genome (its genome is its complete set of genetic material) comes from its haploid father and half from the diploid mother. What this means is that she carries all of her father's genes and half of that of her mother's genes. Her sister carries a similar proportion of genetic makeup. The two sisters thus share the entire genome of the common father, and on average, a quarter of their mother's genome, yielding a coefficient of relatedness of 75%. Thus, altruistically helping their queen mother and her offspring can help spread their traits through the population.[65]

Though in humans there is no such skewing of relatedness since the sperm of the father and the egg of the mother are each haploid, the idea can be roughly understood by one biologist's remark in a bar, "Gladly die for two brothers, four cousins or eight second cousins," because each of them carries the requisite percentage of the individual's genes to compensate for the mortal deed. Of course, human altruism is more complicated than the understanding of our drunken friend, as we presently show.

Kin selection may explain many observed social behaviors, but it is widely known that animals (particularly primates) direct cooperative social behavior toward individuals who are not biological kin. Robert L. Trivers (1943-) provided a theoretical explanation of this form of altruism, known as

---

[65] Bjorn Brembs, "Hamilton's Theory" in: *Encyclopedia of Genetics*, Sydney Brenner and Jeffrey H. Miller, (eds.), (Academic Press, San Diego, 2001).

reciprocal altruism. This behavior is predicted when the cost of an altruistic act to the giver is smaller than the benefit to the recipient in a long-term interaction between the giver and the recipient so that the opportunities of giving and receiving even out, and the individuals keep score of their giving and receiving.[66]

••••• ••••• •••••

In this chapter, we discussed why humans are capable of behaving with profound self-sacrifice for kin and for nonrelatives. Research has shown that the general principles of inclusive fitness and reciprocal altruism apply throughout the natural world and in many human transactions as well. As for our drunkard friend, he will die for other drinking buddies as well even when he is sober, provided he and his friend will have a net benefit in his mortal deed.

In the next chapter, we will talk about mating and sexual reproduction.

---

[66] Robert L. Trivers, "The evolution of reciprocal altruism", *Quart. Revs. Biol.*, 46, 1971, pp. 35-57.

4 Mating Dance

> "The strangest things determine our choice of mate – subtle body odors, a symmetrical face, the precise proportions of waist to hip."
>
> - Robert Winston, In *Human Instinct* (Bantam, 2002)

▶ Basic Instinct

Sir Edward Tylor (1832-1917) is credited with sparking interest in anthropological science in England as a result of his extensive research. In 1871, Tylor drew on his travel experience and his research in archaeology, linguistics, history, geography, paleontology, and folklore to develop a grand theory of social development.[67] He concluded that all modern societies originated as primitive, savage groups and went through the same stages along the way to modern social forms.[68] During the stages of progress, these societies went through the same stages of development of knowledge and tool use: from shaped pebble to fire to bows and arrows. They also asked similar fundamental question of things about them and constructed myths to answer them. Tylor came up with the concept of survival – ancient customs and habits that had survived eons, long after their original purpose had been lost or forgotten.

We, humankind, take pride in sitting on the apex of the pyramid of intelligence, but we must not forget that we have gone through the same stages of development. Up to 10 million years after the appearance of our

---

[67] Edward Tylor, *Religion in Primitive Culture*, (Harper Torchbook, New York, 1958).
[68] Hwa A. Lim, *Change: In business, corporate governance, education, scandals, technology and warfare*, (EN Publishing, Santa Clara, California, 2003).

earliest ancestors, *Homo sapiens* not only look, move and breathe like an ape, they also think like one. Many of our instinctive reactions and emotions are unnecessary in modern age, but they continue to shape the way we live. Consequently, what we do, what we find appealing, how we react and many of our instincts can be traced back to the days on the savannah.

Of all the human instincts, sex stands out, partly because we are obsessed with sex. Even when we are not expressively sexual, we spend a lot of time in activities that are connected to sex and reproduction: money, career, appearance, friendship, and competition. These activities may provide an answer to why professionals spend a large fraction of their income on appearance. For example, professional American women spend up to one-third of their income on appearance.

Sex begins with searching for a mate. Because both men and women face a tremendous burden of internal fertilization, a nine-month gestation, and lactation, they would have benefited greatly by selecting mates who possess resources. Most *Homo sapiens* have a complex series of tests to see if a candidate is a potential mate. Men from different cultures are attracted to different types of female body, but there seems to be one universal attraction: the preferred waist-to-hip ratio is 0.7. This ratio holds true for ancient Venus figurines (Venus is the goddess of love and beauty), and stone female figurines found in Europe and Asia, all of which have huge breast and rotund buttocks, and can be enormously fat, but they all adhere to the golden section of 0.7.

External appearances aside, there are other chemistry at work.

Scents of Love

Pheromones are already well understood in other mammals, especially rodents. Rodents have vomeronasal organs (VNO) in their noses. They use VNO to detect pheromones in the urine of other rats and use the sense to find a mate. When rats choose a mate, they must avoid a mate with an immune system too similar to their own so that their offspring can fight off a wider range of diseases.

In 1970s, Kunio Yamazaki at the Monell Chemical Senses Center in Philadelphia began to study the genetics of mating of mice. He and colleagues concentrated on a group of genes call MHC genes, which are present in nearly all the cells of mammals and play a major role in the immune system. Remarkably, they found that mice are more likely to look for mates with dissimilar MHCs. A plausible explanation is survival of offspring: we all carry genetic defects in our genetic systems which could be fatal to our children, but if we mate with someone who does not have the same genetic defect, our children will nearly always be free from such defects and will have a better chance of surviving.

In 1985, Bruce Jafek, an otolaryngologist at the University of Colorado, Denver and David Moran, who is now at the University of Pennsylvania's Smell and Taste Center in Philadelphia found evidence that VNO exists in most adult humans. This was not the first discovery of VNO in humans. In the early 1800s, L. Jacobson, a Danish physician, detected likely structures in a patient's nose, but he assumed they were non-sensory organs. Others thought that although VNO exist in human embryos, they disappear during development or remain vestigial – imperfectly developed.

As well as lurking in urine, pheromones are also found in sweat. In 1995, Claus Wedekind at University of Bern, Switzerland tested the MHC gene hypothesis with T-shirt sniffing. The researchers arranged a number of female students according to their MHC gene types Male students, whose MHC genes had been typed, were asked to wear cotton T-shirt so that their body odor permeated the fabric. The T-shirts were then taken to a laboratory to be sniffed by the female students. They consistently picked out T-shirts worn by male students with dissimilar MHC gene types.

Martha McClintock of the University of Chicago in a separate T-shirt experiment showed that women want most is a man who smells similar to her father. This makes sense because her offspring will get genes similar enough to the tried and tested immune system of her dad while at the same time, different enough to ensure a wide range of genes for immunity. There seems to be a drive to strike a balance between reckless out-breeding and dangerous inbreeding!

Cupid's Chemicals

When two people do finally fall passionately in love, a period that normally lasts for 18 months to three years, the state of mind has a lot to do with the so-called love-drug phenylethylamine (PEA). PEA is produced in the brain in large quantities and the effects are somewhat similar to amphetamines or speed. Note that PEA is also produced in thrills, like during bungee jumps or during free fall of parachute jumps. The feeling of thrills is the kind of feeling that someone falling crazily in love can relate to.

Helen Fisher of Rutgers University in New Jersey proposes that we fall in love in three stages, each involving a different set of chemicals.

The first stage is lust. Lust is driven by the sex hormones: testosterone and estrogen. Testosterone is not only found in men, it also plays a major role in the sex drive of women.

The second stage is attraction. When people fall in love, they can think of nothing else. In this stage, a group of neuro-transmitters called monoamines play a major role: dopamine, also activated by cocaine and nicotine; norepinephrine, also known as adrenalin gets the heart racing; and serotonin, one that makes us temporarily insane. Love-struck people might even lose their appetite and need for sleep while daydreaming about their new flame. They have to thank Cupid for all the strange feelings they are experiencing.

As far as attraction goes, it can take between 90 seconds to 4 minutes to decide if we fancy someone. A survey shows that attraction is communicated:

- 55% through body language
- 38% through the tone and speed of our voice
- 7% through what we say.

So we have only 7% success rate in sweet-talking someone into our arms and we have to be brief, in less than five minutes.

The third stage is attachment. People cannot stay in the attraction stage forever. Otherwise they will be constantly daydreaming and will never get any work done. Cupid's chemicals have to wear off at some point. Attachment is a longer lasting commitment and it is the bond that keeps couples together to go on to have children. Two hormones released by the nervous system are important in this stage. The first hormone is vasopressin, which controls the kidney. A suppressed level of vasopressin in animals, for example prairie voles that indulge in more sex than is strictly necessary for the purpose of reproduction, shows that the bond between the animal couple deteriorates immediately. The second hormone is oxytocin, released by the hypothalamus gland during childbirth. One of its functions is to make the mammary gland lactate. It helps cement the strong bond between the mother and the child. It is also released by both sexes during orgasm and is believed to promote bonding when adults are intimate. A theory is that the more sex a couple has, the stronger the bond. And as the experiment on prairie voles shows, the bonding will be sustainable as long as the level of vasopressin remains high.

► Mating Dance

Like humans, our close genetic relative chimpanzees live in family groups; like humans, they can spend years nurturing a newborn till maturity. But when it comes to mating dances, there are key differences. Most importantly, female chimpanzees mate every few years. When they are fertile, they advertise with bright red genitalia. Attracted and aroused by the sight, male chimpanzees will crowd around the female, competing to copulate as often as possible.

In contrast, it is not obvious to men when women are fertile. This difference may have helped forge closer bonds between women and their male partners. Since men do not know when their women are ready to conceive, they hang around in a bid to improve their odds of becoming fathers. To maintain a male's interest, human females may have evolved attractions such as curvaceous breasts. Human females, in fact are the only primates to have permanent swollen breasts; males interpret swollen breast with fertility. Any wonder there is a continued fascination with cleavage among both men and women?

Of Man and Ape

The genomic sequences of the human and the chimpanzee (*Pan troglodytes*) hold a trove of information about human evolution, and shed light on the behavioral and biological differences between a human and a chimpanzee.[69]

In the course of evolution, the human and the chimpanzee split from a common ancestor some six million years ago. A comparison of the human and chimp genomes reveals a vast number of differences – some 40 million – in the sequence of DNA units in the genomes. Most of the differences are caused by a random process known as genetic drift and have little effect. For now, their large numbers make it difficult for scientists to find which of the changes were caused by natural selection.

But a different aspect of the comparison has yielded insights into another question – the evolution of the human Y chromosome. A new finding indicates that, in comparison with the flamboyant promiscuity of chimpanzees, humans have led sexually virtuous lives for the last six million years.

In the human, the Y chromosome has lost all but 16 of its X-related genes. However, the Y chromosome has gained other genes that benefit only the male. Just like the human Y, the chimp Y chromosome also has 16 X-related genes, except that it has lost the use of 5 of the 16 X-related genes. The genes are there, but have been inactivated by mutation. The explanation lies in the chimpanzee's high-spirited sexual behavior: female chimps mate with all males around, so as to make each refrain from killing a child that might be his.

In this mating system, the alpha male nonetheless scores most of the paternities, according to DNA tests. This can be explained in terms sperm competition, primatologists believe – the alpha male produces more and better sperm, which out-compete those of rival males. This polyandrous mating system puts such intense pressure on the sperm-making genes that any

---

[69] The Chimpanzee Sequencing and Analysis Consortium, "Initial sequence of the chimpanzee genome and comparison with the human genome", *Nature*, 437, September 1, 2005, pp. 69-87.

improved version will be favored by natural selection. All the other genes will be dragged along with it, even if an X-related gene has been inactivated.[70]

If chimps have lost five of their 16 X-related genes on their Y chromosome in the last six million years because of sperm competition, and humans have lost none, humans presumably had a much less promiscuous mating system. A different study by experts who study fossil human remains believes that the human mating system of long-term bonds between a man and woman evolved only some 1.7 million years ago. Males in the human lineage also became much smaller at this time, a sign likely of reduced competition.

Another study implies that even before the time of long-term bond, but during the first four million years after the chimp-human split, the human mating system did not rely on sperm competition. It is a reasonable inference that the human-chimpanzee common ancestor might have been gorilla-like rather than chimpanzee-like. The gorilla mating system has no sperm competition because the silverback maintains exclusive access to his harem.

The scientists who have compared the whole genomes of the two species – the human and the chimpanzee – say they have found 35 million sites on the aligned genomes where there are different DNA units, and another five million where units have been added or deleted. Each genome is about three billion units in length. This would lead to the conclusion that the human and the chimpanzee are 98.8% similar, genetically. At the level of the whole animal, primatologists have uncovered copious similarities between the social behavior of chimpanzees, bonobos and humans, some of which may eventually be linked to genes.[71]

But this rich vein of discovery may be choked off if the great apes can no longer be studied in the wild. Great apes are under harsh pressures in their native habitat. Their populations are dwindling fast as forests are cut down and people shoot them for meat. They may soon disappear from the wild altogether, primatologists fear, except in the few sanctuaries that have been established.

The difference between the human and the ape is, ultimately, both cultural and genetic.

Sexual Matrimony and Polyamory

Men have also evolved to have assets to attract the opposite sex: large muscles may signal men's prowess as a hunter and defender. But strength

---

[70] Nicholas Wade, "In chimpanzee DNA, signs of Y chromosome's evolution", *The New York Times*, September 1, 2005.

[71] Ze Cheng, et al, "A genome-wide comparison of recent chimpanzee and human segmental duplications", *Nature*, 437, September 1, 2005, pp. 437, 88-93.

alone probably is not enough to attract and keep a mate. To be successful, men also have to show that they are creative and dependable providers, clever enough to find food and shelter for families in hostile environments.

These are nature's extravagance. With advances in science and technology, humans take the whole thing one step further. Humans use perfumes, consume steroids to grow bigger, use life-style drugs whose primary functions are to restore social faculties that diminish with age such as impotence and baldness. Ever wonder the divorce rate is so high nowadays? Perhaps people use perfumes too much and give the wrong "pheromones" until they get married and get very close to realize that the body odors are different; similar arguments may apply for other "attributes," such as finding out the partner is actually bald, and not the "nice-looking" person he or she was.

In the modern world, physical attributes such as muscles and breasts may have little meaning since most people work in offices and restrict their foraging to the local supermarkets, restaurants, or ready-to-serve TV dinners. Brawn power, physique and external appearance would seem to have been substituted by intelligence, yet these attributes still have sexual appeal, a fact fully exploited by advertising agents who use exposed bodies of supermodels and muscular athletes to sell everything from luxury items like cars to household goods like soaps.

Sex still sells. Period.

With so many distractions, some natural (like pheromones), some unnatural (like perfumes) and other superficial (like implanted hair), sexual partnerships are strained. In addition, the modern world requires incomes from both partners to maintain a family, and with more than forty years slated for productivity on the job, it is only a matter of time before someone pairs up with someone else at work.

Indeed, monogamy – two partners staying sexually faithful to one another – is one of the most rare forms of sexual partnerships. Less than 3% of mammals and bird species practice monogamy. In many of these species, cheating or alternative arrangements are common. Infidelity and adultery aside, men in some cultures have many wives (polygamy or polygyny), yet in others, many men share a single wife (polyandry). Polyamory is a nonpossessive, honest, responsible and ethical philosophy and practice of loving multiple people simultaneously.

In the animal kingdom, the jacana is a bird in which a female will have four or five male partners, each care for eggs laid by the female. African mole rats and termites live in underground colonies with a single queen that produces all the offspring in the colony, resulting in almost all the colony's inhabitants are brothers and sisters. Male elephant seals, in contrast, play king and rule over a harem that includes every female on a long stretch of beach.[72]

Whatever the sexual matrimony arrangements, they have the same goal: to increase the number of offspring carrying the parents' genes, or in the case of termites, the common genes of the entire colony. And in each case, the partners work to ensure the survival of a new generation.

► And Less Love

From the point of view of Cupid's chemicals, love is literally a drug and is highly addictive. Like all addictions, there is a law of diminishing returns. After the initial period of PEA, the brain starts to pump out endorphins, a brain opiate that is like morphine, serving to calm the mind, kill pain and reduce anxiety. Evolutionary psychologist Helen Fisher explains why we appear to be preprogrammed to eventually lose interest in a sexual partner.[73] Fisher suggests that humans pursue a similar strategy to animals such as foxes. Foxes are serially monogamous: they pair up for one breeding season and stay long enough to help raise their young before splitting up. Humans too are serially monogamous and stay for the time it takes to raise a single child through infancy, which is about 4 years. Fisher backed up her claims by conducting research in nearly 60 countries and by showing that divorce rates peak at around four years into marriages. Consequently, every marriage is a divorce waiting to happen. The following table shows that the U.S. leads in divorce rates.

Table 1. Divorce rates of countries, compared to that of the U.S. (extracted from *Miami Herald*, Thursday, January 22, 1998).

Country	Divorce Rate (Per 1000 population)	Rate as % of US
Sri Lanka	0.15	3.03
Brazil	0.26	5.25
Italy	0.27	5.45
Mexico	0.33	6.67
Turkey	0.37	7.47
Mongolia	0.37	7.47
Chile	0.38	7.68
Jamaica	0.38	7.68

[72] The Nature of Sex, Public Broadcasting Station. www.pbs.org/wnet/nature/sex.
[73] Helen Fisher, *Anatomy of Love: A natural history of mating, marriage and why we stray*, (Fawcett Columbine, New York, 1995).

Table 1. (*Continued*)

Cypress	0.39	7.88
El Salvador	0.41	8.28
Ecuador	0.42	8.48
Mauritius	0.47	9.49
Thailand	0.58	11.72
Syria	0.65	13.13
Panama	0.68	13.74
Brunei	0.72	14.55
Greece	0.76	15.35
China	0.79	15.96
Singapore	0.80	16.16
Tunisia	0.82	16.57
Albania	0.83	16.77
Portugal	0.88	17.78
Korea	0.88	17.78
Trinidad	0.97	19.60
Qatar	0.97	19.60
Guadeloupe	1.18	23.84
Barbados	1.21	24.44
Finland	1.85	37.37
Canada	2.46	49.70
Australia	2.52	50.91
New Zealand	2.63	53.13
Denmark	2.81	56.77
United Kingdom	3.08	62.22
Russia	3.36	67.88
Puerto Rico	4.47	90.30
US	4.95	100.00

There are several possible household arrangements as well: married couples, unmarried couples of the opposite sex, or unmarried couples of the same sex living together. Since 2002, in certain states in the U.S., same sex marriages are legal. This leads to another possible household arrangement: married couples of the same sex living together.

Table 2. Cohabiting in perspective: The 2000 census in the U.S. counts 4.9 million households in which the head of the household has an unmarried partner of the opposite sex. That is an increase from 3 million in 1990. Note the statistics was taken in 2000 when same sex marriages were still not allowed.

Description	Millions
Household in the U.S.	105.5
Married couples	54.5
Unmarried couples with opposite-sex partners	4.9
Unmarried couples with same-sex partners	0.6

To understand the divorce rates, we can seek explanation in contemporary hunter-gatherer societies, the closest modern-day comparisons to ancient human life on the savannah. In traditional societies such as the Australian aborigines and the Eskimos, there is a much longer period of breast-feeding than in the West, until the child is about three to four years old. The whole point of monogamy is, from a utilitarian perspective, that a partnership provides protection and resources for its own biological children.

Homo sapiens are an unusual species in that the female's monthly menstrual cycle is hidden, unlike species such as baboons or bonobos. Sexual interactions occur more often among bonobos than among other primates. They engage in sex in virtually every partner combination, although such contact among close family members may be suppressed. They are not shy about broadcasting their fertile condition to the entire troop: the entire genital area swells and turns a bright shade of pink. The human female, on the other hand, often tend not to be aware when they are ovulating. Concealed ovulation is a clever ruse: it goes hand in hand with internal fertilization; concealed ovulation and internal fertilization evolved as mechanisms to ensure that a female's mate will be attentive all month long! This could be the very beginning of the trend towards monogamy in human culture?

This fact nonetheless, some estimates suggest that around 50% of married men and women, presidents and first ladies are no exceptions, are having extra-marital affairs. It is possible that infidelity and adultery are an evolutionary adaptation that has grown up alongside monogamy and long-term commitment.

If reproducing young is not the goal, may be men and women take on lovers to replenish the levels of Cupid's chemicals: love-drug phenylethylamine (PEA) during the lusting stage, monoamines during the attraction stage, and vasopressin during the commitment stage. May be men and women take on lovers to improve the chances of passing on their genetic legacy.

If a man wants to have affairs because he wants to pass on his genes and the man has a chance to impregnate another female, especially a female who is already married and need not be provided for, then he may well have stumbled across the best of evolutionary bargain – all the benefits of continuing his genetic legacy without having to worry about bringing up a child.

Then what is the evolutionary advantage of a woman taking on a lover? Indeed genetic studies in rural parts of some countries show that up to 15% of children are not the offspring of their legal fathers. A likely explanation for the female infidelity is that it is a woman's way of hedging her bets. There is an added sense of security in having more than one provider for herself and the children. In addition, a lover provides an insurance policy, the compensation being that if the husband dies prematurely, there is someone else to help take care of the children. For a woman, adultery is also about dipping into the genetic pool. The current husband may be infertile, or may simply carry poor genes. Taking on a lover is a way of introducing different DNA into the family without disrupting the stability of the family structure.

Men of all culture find younger women more attractive. A woman who is young and healthy has a better chance of bearing a number of children. For some men, the combination of the loss of sexual interest often found in a long-term monogamous relationship and the aging of the partner prompts the so-called "mid-life crisis." A small minority of men, who are sufficiently attractive or who have high status, end up marrying younger women, and in certain cases, a succession of younger women. These men who marry these younger women in their fertile prime, on an unconscious, biological level, may be striving to maximize their genetic legacy. In fact, this is a form of serial monogamy.

Because of infidelity and extra-marital affairs, there is a stumbling block for men: how does he ever know for certain that the children are their own?

When a child is born, parents have all the excitement and want to know what the child looks like. They may be cautious initially before giving away their hearts. The child's look is most important in the first days of life. Most, if not all, parents believe that their baby is better looking than anyone else's. Parents and family members also peer curiously into the baby's face to figure out whom it resembles. Mothers are likely to claim that the baby resembles the father.

Psychologists Margo Wilson and Martin Daly of McMaster University, Canada sent questionnaires to hundreds of new mothers and fathers and their relatives. From the responses, they found that claims of paternal resemblance were significantly more common than claims of maternal resemblance. Indeed, in many families everyone in the family commented on the baby's resemblance to the father.[74]

A plausible interpretation is: a mother has no doubt that the baby is hers, but the father always runs some risk of being duped. The father tends to look for confirmation that the child is really his. He has two sources of information to turn to: his knowledge of the mother's fidelity, and any physical evidence from the baby. Since facial features are highly heritable, emphasizing the baby's resemblance to the father can only help to erase any doubts and stoke his affection and encourage his investment in the newborn. Seeing some reflection of his own features in the baby's face is a powerful trigger of paternal feelings.[75]

---

[74] Martin Daly, and Margo Wilson, *The Truth About Cinderella: A Darwinian view of parental love*, (Yale University Press, 1999).
[75] Nancy Etcoff, *Survival of the Prettiest*, (Doubleday, 1999).

▶ Tug of War, *In Utero*

After the mating dance is over and the subsequent sex that follows, spermatozoa are out swimming for their lives to try to get to the egg first.

Robin Baker and Mark Bellis of the University of Manchester have an interesting theory that human spermatozoa come in different shapes and sizes. According to them, the most common sperms are the egg-getters, with conical heads and long tails, designed to swim for their lives. A different type of sperm is also ejaculated. These sperms have coiled tails, designed to wrap around foreign egg-getters to hamper their progress. These researchers believe that sperm competition is the main force to shape the genetic program that drives human sexuality.

When a spermatozoon meets the egg, the egg has been fertilized. *In utero*, mothers and fathers can use evolutionary tactics to increase their success. In mammals such as humans, a fertilized egg lodges itself in the mother's uterus and begins to develop into an embryo. A few days later it starts growing a placenta. The placenta pushes blood vessels into the mother's uterus to draw in blood and nutrients.

A growing embryo needs a huge amount of energy, which can drain on the mother's resources, sometimes to a dangerous level. If a mother lets an embryo grow too fast, she may cause herself grievous harm, threatening her future fertility or even her life. Evolution should therefore favor mothers who can keep embryonic and fetal development in check. But fathers have a different evolutionary agenda. A fast-growing, healthy embryo or fetus is an unalloyed good for the father. After all, the rate at which the embryo or fetus grows cannot threaten his own health or his ability to have more children.

A Harvard University biologist, David Haig, interested in conflicts and conflict resolution within the genome, has proposed that the genes the father and the mother give to the baby resolve the conflicting interests. Maternally inherited genes and paternally inherited ones do different things. For example, a gene called insulin-like growth factor 2 (IGF2) produces a protein, which stimulates the embryo to draw in more nutrients from its mother. In pregnant mice, the mother's copy of IGF2 gene is silent, while the father's copy is active. Mice also carry another gene that makes proteins whose job is to destroy IGF2 proteins. And the mother's copy of this IGF2 destroyer is active, while the father's is shut down.

It is thus plausible to argue that the father's genes are trying to speed up the growth of the mouse embryos while the mother's genes are trying to slow it down. This argument is corroborated by experiments: if the father's copy of

IGF2 is shut down, mice are born only 60% of their normal size; if the maternal copy of the gene that destroys IGF2 is shut down, they are born 20% heavier.

If we can push the argument for the roles of IGF2 gene and IGF2 destroyer gene further and apply it in cloning, a subject of interest in later chapters, we can argue that in the enucleation or the transplantation processes, the IGF2 and IGF2 destroyer genes may not properly be taken care of. This may explain why clones are born larger, sort of like acromegaly. Acromegaly originates in the Greek words for "extremities" and "enlargement" and reflects one of its most common symptoms – the abnormal growth of the hands and feet. An early feature is soft tissue swelling of the hands and feet. Gradually, bony changes alter the patient's facial features: the brow and lower jaw protrude, the nasal bone enlarges, and spacing of the teeth increases.[76]

▶ Back into the Past and Future

Marriage, childbearing, monogamy, polygamy, polygyny, polyandry, infidelity, adultery and divorces, all these relationships are inextricably intertwined in our genetic heritage, and shadows of the savannah will always be present, casting over modern mores and ways of life. In parts of the world where modern mores often collide with ancient traditions, what constitutes progress can be very divisive.

It seems that the desire to reproduce is an atavistic need, and sociobiologists will stress that everything humans do is to fulfill this need. Under the most rudimentary scrutiny, the argument is not self-consistent: humans have, throughout history, engage in activities that reduce their evolutionary fitness. Generations of men have viewed soldiering as the *sine qua non* of manhood, all the way from the Spartans to the Americans. How can these people give as many of their genes to the next generation as possible when they perish at their tender age on battlefields?

In many aborigine communities, far less emphasis is placed on biological links to children; in polyandrous societies, women may not even know who the father of a child is. Even today, many Pacific Islanders readily give their children away to siblings or friends who have none, or to strangers who might take the child to advanced countries such as the U.S. They consider farming out a child beneficial not only to the child but also to the entire family or clan.

---

[76] NIH Publication No. 02-3924, June 2002.

Ironically, these aborigine children are being farmed out to countries in which couples practice a very different way of preserving genetic legacy: Western couples tend to be more self-centered and focus on the making of persons, instead of a larger network of people with whom they have interactions and to whom they have obligations. In these advanced countries, what is apparently at stake is what reprotech (reproductive technology) produces is parental choice – the children will embody the desire of their parents, and not what the children should have to be better off.[77] Once children have been commodified in this manner, the next desire is to incorporate additional aspects of parental desires: they want not just any biological babies, but the best baby technology can provide. A 1992 March of Dime Birth Defects Foundation survey poll of 1,000 Americans found that 43% approved of prenatal genetic manipulations that would improve a baby's looks, while 42% like the idea of manipulations that would improve intelligence. Though the definition of "perfect" was not given, over a third fancied a perfect baby! Concepts of "what perfect is," "what "normal is" and "what bad is" are often merely a reflection of cultural stereotypes and prejudices of ourselves and the society at large. The whole notion of perfect, normal and abnormal can change and the issue becomes not the ability of the child to be happy but rather our ability to be happy with the child.[78]

•••••   •••••   •••••

Mating, marriages and household arrangements are only part of sexual reproduction. The quest for perfection, the subject matter of the next chapter, is also a major part of sex and sexual reproduction.

---

[77] Judith Lorber, and Lakshmi Bandlamudi, "The dynamics of marital bargaining in male infertility", *Gender and Society*, 7, 1993, pp. 33.
[78] Geoffrey Cowley, "Made to order babies", *Newsweek*, Winter/Spring, 1990, pp. 98.

5 Quest for Perfection

"The 1960s brought sex without procreation; the '80s brought procreation without sex."

- Lori B. Andrews, American Bar Foundation

► Beauty Lies in the Eyes of the Beholder

We all recognize beauty when we see it, and we even get so envious and jealous. Beyond envy and jealousy, what makes a beautiful face is something few can agree upon. Is there a mathematical equation of beauty or pulchritude? Is there some combination of angles, ratios and proportions for beauty? Or is beauty, like a work of art, a matter of opinion, taste and culture? Does the notion of beauty change with time?

As innocent as it sounds, these questions were familiar to the Greek philosophers 2,500 years ago. The face of Helen launched a thousand ships, just as Lady Diana (1961-1997) launched a thousand paparazzi. Interestingly, the Greeks had the same ideal of beauty for men and women. But in modern society, male beauty is considered less important since male attractiveness is measured in terms of power and social standing rather than by facial features. However, when women hold economic power, like the Wodaabes in northern Nigeria, then it is the males who become preoccupied with beauty and who dress up and compete in beauty pageants. In today's household arrangement, there are an increasing number of househusbands. It may not be long that "handsome" pageants become as popular as beauty pageants.

Ancient Definition of Perfection

In historical hindsight, Pythagoras (569-475 BC) drew up a precise formula for the perfect female face: for a female to be beautiful, the width of the mouth to the width of the nose should be 1.618. The same ratio holds for the width of the cheekbone to the width of the mouth.

When Greek writings of the Renaissance era (1260-1648) were rediscovered, tracts defining beauty in mathematics appear: an author defined absolute beauty for a face is one that is one-eighth the length of the body; another that the corners of the lips must be decline, forming an obtuse angle; and others.

In the 18th century, Dutch naturalist Pieter Camper (1722-1789) gave geometry a racist twist with "facial angles" – the angle between the tip of the chin and the nose. A profile of 100° represents divinity; the Europeans rated a high 90°, while the 70° scored by Africans was considered to be less ideal. Camper's facial angle was a combination of comparative anatomy and racial anthropology. It prescribed that the more vertical a line drawn from chin to forehead, the closer to the ideal head. The classical head was asserted to bear a perfect right angle and to be the epitome of aesthetic and anatomical perfection.[79,80]

Sir Francis Galton (1822-1911) was arguably the most respected and influential British scientific advocate of schemes of hereditary improvement in the half-century after 1860 and an advocate of eugenics. In 1878, he discovered that if photographs of a number of faces were superimposed, most people considered the resulting composite to be more beautiful than the individuals.

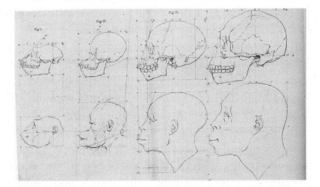

Figure 1. In the 19th century, geometry of facial angles was given a racial twist. The angle φ is the angle subtended by the tip of the chin and the nose. Racial types in the prognathous or simian jawline: 1). tailed monkey, φ=42°; 2). orangutan, φ=58°; 3). Negro, φ=70°; 4). Kalmuck, φ=70°.

---

[79] P. Camper, *Über den natürlichen Unterschied der Gesichtszüge*, Berlin, 1792.

[80] John van Wyhe, "The history of phrenology on the web", pages.britishlibrary.net/phrenology, 2002.

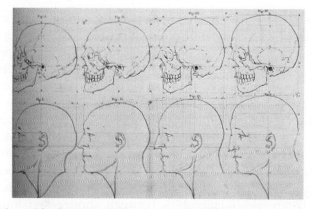

Figure 2. Racial types in the prognathous or simian jawline: 5). European, φ=80°; 6). Grecian bust, φ=90°; 7). Roman bust, φ=95°; 8). a case of hydrocephalus, φ=100°.

Search for the Perfect Face

David Perrett of the University of St. Andrews, Scotland has developed a computerized morphing system that can endlessly adjust faces to suit his research needs. He compiled composite photographs of Europeans and Japanese faces and his finding is challenging Galton's finding. Perrett found that individual faces were preferred to the composites. But when he used computers to enhance the composites away from the average, the resulting "face" is also preferred, lending credibility to Francis Bacon (1561-1626) who said, "There is no excellent beauty which hath not some strangeness in the proportion." Perrett found that attractive faces have higher cheekbones, a thinner jaw, and larger eyes relative to the size of the face. Beauty also transcends culture: the Japanese found the same European faces appealing as the European did, and vice versa.

In another experiment, Perrett took images of students' own faces and morphed them into the opposite sex faces. Of all the faces on offer, student subjects seem to pick their own morphed faces. Students could not recognize these morphed faces as their own, but they just like them. Perett explained that we find our own faces attractive because they remind us of the faces we looked at constantly in our early childhood years, our mom and dad.

Alfred Linney of University College Hospital, London has been using lasers to make precise measurements of the faces of beautiful fashion models. Linney found that the features of models are just as varied as those of everyone else. Interestingly enough, some models have protruding teeth, some models have long faces, and some other models have jutting chins. Equally interesting, many common cosmetic surgeries these days are performed to correct these features. For example, a common cosmetic surgery is orthodontics in which one type of problem people seek help for is teeth that stick out. As a general rule, an orthodontist would consider operating when teeth project more than 5 millimeters, but some of the models have teeth sticking out 8 millimeters and still look gorgeous. This casts doubts on the effectiveness of orthodontics as a form of cosmetic surgery for beautification.

Let us return to the very influential idea of golden section, a pleasing ratio of 1.6, which was used by Pythagoras and ancient Greeks to define beauty. Mark Lowey of University College Hospital, London, has made detailed measurements of fashion models' faces. His finding is that beautiful faces come closer to the golden section proportions than the rest of the population: on beautiful hands, the relationships between shorter and longer joints is 1.618, just as the distance from the hairline to the tip of the nose on a beautiful face is 1.618 times the distance between the nose and the chin.

According to Judith Langlois of University of Texas, even three-month-old babies prefer beautiful faces to plainer ones. As if we were born with a beauty detector, three-month-old babies stare longer at faces adults rated as beautiful than they do at faces adults deem unattractive. All of these suggest that there must be some evolutionary advantage to be beautiful. Could it be that beauty is an indication of a woman's fertility? Until recently, evolutionary theories of beauty and fertility opt for the law of average. Anthropologists propose that beauty represents no more than the average face in a human population. Evolutionary pressure operates against extreme features: people with average physical build stand the best chance of surviving to pass their genes on to future generations. In other words, average features serve to explain why the possessor is likely to be fertile. A fertile mate is one who has glossy hair and good skin since they show that they are healthy and free of parasites.

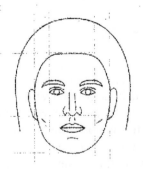

Figure 3. The mathematics of beauty: critical measurements of what a beautiful face is. (Figure adapted from *Beauty: Making sense of sex appeal*).

Yet another theory cites that women with baby faces who have big eyes, a small, full mouth and a small nose are attractive because they trigger the warm protective feelings we have towards small children.

As mentioned earlier, computer simulations by Perrett have yielded different attractive faces. Yet another computer simulation shows that the ideal face is that of a fertile young woman. Recently, Victor Johnston of the University of New Mexico linked perceptions of beauty with the effects of estrogen on the bodies of adolescent girls. Johnston ran computer programs to simulate the process of evolution: faces were randomly selected and rated according to attractiveness by volunteers. The most attractive faces were combined to breed a second generation of faces, iterating the process on the third and fourth generation, and so on. Gradually, a shorter, fuller-lipped face resulted.

The two key measurements on the resulting face are: the distance from the eyes to the chin, which is shorter, the length normally found in a girl aged eleven and a half; and the size of the lips, which are fatter, the size normally found on a girl fourteen years old. Combining the attractive features yields the ideal face as that of a woman of 24.8 years.

The proportions point to fertility. Up to puberty the faces of boys and girls are similar. After puberty, the rise of estrogen in girls gives them the fuller lips, while testosterone in boys gives them a fuller jaw. The age of 24.8 years old coincides with the age when most women are at their most fertile and the levels of estrogen are highest.

In male the features that are considered attractive are the mature ones: small eyes, large nose, thin lip and prominent chin, in contrast to the big eyes and small mouth and jaw of the attractive female young face. In addition, men have more flexibility in the proportion for beauty.

Any wonder why there are lipsticks out there to make one's lips look fatter, or why there are cosmetic injections to help make one's lips fuller?

Search for Hour-Glass Figure

It is also subconsciously thought that asymmetrical features are a sign of underlying genetic problems. Numerous studies in humans have shown that men in particular go for women with symmetrical faces. The preferences in women for symmetry is not quite as pronounced since women are also looking for men with ability to offer food and protection. These attributes are not in the genes, but in the hierarchical ranks and status. Because hierarchies are universal features among human groups and resources tend to accumulate to those who rise in the hierarchy, women solve their adaptive problem of

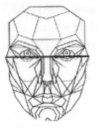

Figure 4. Men prefer women with symmetrical faces. Asymmetrical features are associated with genetic defects. (Figure adapted from *The Science of Love*, BBC).

acquiring resources in part by preferring men who are high in social status.[81] Any wonder why wealthy people attract more women? Or people in power have extramarital affairs?

The face, of course, is only a part of the whole human. Is there also a mathematical formula for a beautiful body?

Devendra Singh of University of Texas points out that while testosterone can lead to weight around the stomach, estrogen lays weight around the hips and buttocks. Full buttocks and narrow waist send out the message of fertility. Singh found that male students rated as attractive women with waist to hip ratio of between 0.67 and 0.8 (note the reciprocal of 1.618, Pythagoras' perfect ratio, is 0.618). Interestingly, women with this ratio are viewed as humorous, healthy and intelligent; those with thicker waist are viewed as faithful and kind; and women who are thin are viewed as aggressive and ambitious.

The famous Mona Lisa, sexy Marilyn Monroe, and the more *au courant* Cindy Crawford all seem to have the same proportion, though over the years, the slimmer look has taken over the chubbier look. In a survey of primitive people in South America, the people were shown photos of some fifty women. The beautiful people they picked out fall into the same proportion. The average ratio of Miss America winners over the years is also 0.7.[82]

So we are still far from a generally accepted definition of what beauty, pulchritude and perfection are, though the perfect ratio of 0.7 has been a leitmotif. We must also not forget that in surveys, photographs were used and that a lot of people are not photogenic at all. But there is some indication that evolution may have something to do with what face and shape we find beautiful.

---

[81] David Buss, *The Evolution of Desire: Strategies of human mating*, (Basic Books, New York, 1994).
[82] *The Science of Love*, BBC, November 2004.

▶ The Myth of Beauty

Much of the science of evolutionary psychology is controversial, yet it provides thought-provoking answers to the modern mores and ways of life.

If we ask a person what he or she wants in a mate, the range of answers is bewildering and can be frustrating for anyone interested in pinpointing that "what:" "Someone who is happy-go-lucky," "Kind hearted," "Great provider," "Great legs," "Voluptuous breasts," "Someone with great physical build," "Someone who loves to dance..." In these days of materialistic world, I have seen ads looking for mates in newspapers and magazines in which the seekers would only answer to people with citizenships of certain countries, or to Ph.D.... May be they are looking for someone who is intelligent, may be they think that all Ph.D. are rich since the odds that a Ph.D. is poor, <u>h</u>ungry and <u>d</u>esperate are low!

Evolutionary psychology proposes that we all show instinctive preferences and what we humans find alluring in a mate has roots in our evolutionary past. Anthropologists claim that culture plays an important role in what each group defines as beautiful. Tastes also change over time. What was beautiful, fashionable and attractive a quarter of a century ago may look rather silly today.[83] For example, a fair skin used to be associated with wealth since only the wealthy did not have to labor in the sun; now a tanned skin is associated with wealth because the rich can afford to go on vacations in sunny places.

There is no evolutionary precedent for the slim ideal – an ideal promoted by Hollywood, and to the extreme by Barbie dolls. As a matter of fact, selection should work against such a preference. It has been known for some time that women with eating disorders such as *anorexia nervosa* suffer disruptions in fertility and reproduction. Extreme slimness is a fashion trend set by the highest social classes, and Barbie figure is known to be not attainable.

Though fatness is not particularly attractive in our modern society, it is a sign of wealth and prosperity in the past, and in other cultures. In Niger, as in many other parts of Africa, being fat is regarded as beautiful for women. At the Hangandi festival, women of the Djerma ethnic group compete to become the heaviest, to be corpulent. To be competitive, they gorge on food, especially millet, and drink lots of water. The heaviest woman is the winner and is given a prize – and more food!

In Maradi, a sleepy town just north of the border with Nigeria, women take steroids to gain bulk, or pills to increase their appetites. To gain weight,

---

[83] Meredith Small, *What's Love Got To Do with It?* (Anchor Books, 1995).

some women even ingest animal feeds or vitamins. Besides animal feeds, the most popular product is dexamethasone, a kind of steroid easily bought on the streets. It is usually married women who buy these fattening products. If the women are too thin, they worry that their families and friends will think that their husbands are not taking care of them or that the husbands have abandoned them. If the beauty concept here is the reverse of the West's, then its motivations appear the same: seeking men's approval.

Dr. Ousmane Batoure, who is used to women coming to his general clinic desperately wanting to gain weight, comments, "The world is a funny place. In America, [if] you are rich, you have everything, and the women there want to become so thin as if they had nothing. Here in Africa... the women who buy these products have nothing, but they want to become fat as if they had everything."[84]

The idea that beauty is unimportant or a cultural construct is the real beauty myth itself. If we do not understand beauty, we will be enslaved by it. Whatever it is, beauty is the universal part of human experience: it provokes pleasures, rivets attention, and impels actions that help ensure the survival of our genes.[85]

The Cute Factor

Cuteness is not synonymous to beauty. In fact, it is distinct from beauty. Cuteness emphasizes rounded over sculptured, soft over refined, clumsy over quick. While beauty attracts admiration and demands a pedestal, cuteness attracts affection and demands a lap. Beauty is rare and brutal, despoiled by a single pimple; cuteness is commonplace and generous, content on occasion to co-segregate with homeliness.[86]

The science of evolutionary psychology also studies the evolution of visual signaling. It has identified a wide and still expanding assortment of features and behaviors that make something look cute: bright forward-facing eyes set low on a big round face, a pair of big round ears, floppy limbs and a side-to-side, teeter-totter gait, among many others.

These cute cues are those that indicate extreme youth or baby-like, vulnerability, harmlessness and need, and attending to them closely makes good Darwinian sense – survival of the cutest. As a species whose youngest members are so pathetically helpless they cannot lift their heads to suckle

---

[84] Norimitsu Onishi, "In Africa, fat is fashionable", *The New York Times*, February 13, 2001.

[85] Nancy Etcoff, *Survival of the Prettiest*, (Doubleday, 1999).

[86] Natalie Angier, "The cute factor", *The New York Times*, January 3, 2006.

without adult supervision and aid, human beings must be wired to respond quickly and gamely to any and all signs of infantile desire. But babies themselves, evolutionary scientists believe, did not really evolve to be cute. Instead, most of their salient qualities stem from the demands of human anatomy, and became appealing to a potential caretaker's eye only because infants would not survive otherwise.

Human babies have unusually large heads because humans have unusually large brains. The heads are large and round to accommodate the brain growth throughout the first months of life; and the plates of the skull stay flexible, unformed and unfused to allow for this development. Baby eyes and ears are situated comparatively far down the face and skull. It is only later when the bones in the cheek and jaw areas have fully developed that the eyes and ears appear to have migrated upward.

The cartilage tissue in an infant's nose is comparatively soft and undeveloped, giving most babies button noses. According to Paul H. Morris, an evolutionary scientist at the University of Portsmouth in England, baby skin sits relatively loose on the body, rather than being taut. This serves better the purpose of stretching during growth spurts. The lax packaging accentuates the overall roundness of form.

Baby movements are notably clumsy, an amusing combination of jerky and delayed steps. Learning to coordinate the body's many bilateral sets of large and fine muscle groups requires years of practice. When starting to walk, toddlers struggle continuously to balance themselves between left foot and right, and so the toddler gait consists as much of lateral movement as of any forward (and backward) momentum. Baby eyes are also notably forward-facing – a sort of binocular visions likely a legacy of our tree-dwelling ancestry.

The human cuteness criterion is set at such a low bar that it sweeps in and deems cute practically anything remotely resembling a human baby or a part thereof. This includes the young of virtually every mammalian species, penguins, panda bears, fuzzy-headed birds like Japanese cranes, woolly bear caterpillars, a bobbing balloon, a big round rock stacked on a smaller rock, a colon, a hyphen, a close parenthesis typed in succession or symbols such as :-) or ☺.

Scientists are just beginning to map cuteness' subtleties and source. New studies suggest that cute images stimulate the same pleasure centers of the brain aroused by sex, a good meal or psychoactive drugs like cocaine. Expectedly, an animal or object that possesses the greatest number of cute cues or most exaggerated cues provokes the loudest squeals.

Like beauty, primal and widespread though the taste for cuteness may be, researchers say it varies in strength and significance across cultures and eras. Something that was cute in the 1960s may not be cute today. In certain sense, the cute response can be compared to the love of sugar: everybody has sweetness receptors on the tongue, but some people and some countries eat a lot more candy than others.

The cuteness craze is particularly acute in Japan, where it goes by the name *"kawaii."* Truck drivers display Hello Kitty-style figurines on their dashboards; the police enliven safety billboards and wanted posters with two perky mouse-like mascots, *Pipo kun* and *Pipo chan*. Brian J. McVeigh, a scholar of East Asian studies at the University of Arizona, believes that behind the *kawaii* phenomenon is the strongly hierarchical nature of Japanese culture. Cuteness is used to soften up the vertical society, to soften power relations and present authority without being threatening.

In the U.S., the use of cute imagery is geared less toward blurring the line of command than toward celebrating America's favorite demographic – the young. Dr. Miles Orvell, professor of American studies at Temple University in Philadelphia, traces contemporary cute chic to the 1960s, with its celebration of a perennial childhood, a refusal to dress in adult clothes, an inversion of adult values, a love of bright colors and bloopy, cartoony patterns, the Lava Lamp.

Studies on animals beloved by the public reveal a human impulse to nurture anything even remotely baby-like, and a rather surprising tendency to identify with their preferred species. Some people are so wild for the penguin that they think it is a mammal and not a bird. They love the penguin's upright posture, its funny little tuxedo, the way it waddles as it walks. How like a child playing dress-up! Dr. Michel Gauthier-Clerc, a penguin researcher in Arles, France explained that the apparent awkwardness of the penguin's march had nothing to do with clumsiness or uncertain balance. Instead, the penguin waddles to save energy because from kinesiology consideration, a side-to-side walk burns fewer calories than a straightforward stride. For a bird that fasts for months and lives in a frigid climate, every calorie counts.

As for the penguin's maestro garb, the white front and black jacket is well adapted to its aquatic way of life, just like a fish. While submerged in water, the penguin's dark backside provides a camouflage from potential predators of air or land. The white chest, by contrast, obscures it from below, protecting it against predators while allowing it to sneak up on prey.

The giant panda is another case in accidental cuteness. Although it is a member of the highly carnivorous bear family, the giant panda specializes in eating bamboo. As it happens, many of the adaptations allowing the panda to get by on this vegetarian diet contribute to the panda's cute form. Inside the bear's large, rounded head are the highly developed jaw muscles and the set of broad, grinding molars for chewing some 40 pounds of fibrous bamboo plant a day. When it sits up against a tree and starts picking apart a bamboo stalk with its distinguishing pseudo-thumb, the panda looks just like a baby shucking a corn. Yet this humanesque posture and paws again are adaptations to its particular diet – the bear must have its "hands" free to be able to shred the bamboo leaves from their stalks.

The panda's distinctive markings further add to its appeal: the black patches around the eyes make them seem winsomely low on its face, while the black ears pop out cutely against the white fur of its temples. As with the penguin's tuxedo, the panda's two-toned coat very likely serves a twofold purpose. First, it helps a feeding bear blend peacefully into the dappled backdrop of bamboo; second, the sharp contrast between light and dark may serve as a social signal, helping solitary bears to locate each other when the time has come to find the perfect, too-cute mate.[87]

Cuteness may be adaptations in the natural world. In the business world, entertainment industry in particular, cuteness is exploited for maximal tweaking of our inherent baby radar. All our favorite cartoon characters such as the sloth, the wooly mammoth and saber-tooth tigers in *Ice Age* (2002), and Disney characters all sport forward-facing eyes, including the ducks and mice, species that in reality have eyes on the sides of their heads. Recent computer animated cartoon characters, in for example, *Lilo & Stitch* (2002), *The Incredibles* (2004) and *Madagascar* (2005), all have these cute features and walk in teeter-totter gait.

Advertisers also adopt and adapt their strategies for maximal marketeasing (amalgamated word derived from "market" and "tease") of our inherent weakness by using baby-like cute characters.

▶ No Choice for Baby

The quest for perfection may begin as early as in childbearing. Not only the quest for external appearances of choice is an important deciding factor in childbirth, but also the quest for good physique and the choice of sex.

In 1993, a Nigerian man with a Ph.D. flew to California with his wife to visit an assisted-reproduction clinic. The couple already had five daughters,

---

[87] Natalie Angier, "The cute factor", *The New York Times*, January 3, 2006.

but no sons. The man blamed the wife and wanted the clinic to perform *in vitro* fertilization using his sperm and an egg taken from another woman. The physician explained to him sex of offspring was determined by the sperm. Eggs carry only female-determining X chromosomes whereas sperms are ambidextrous, each carrying an X and a Y chromosome. The man listened but insisted on his request. The physician ultimately refused to acquiesce to his wishes. This story only goes to show that even in modern times, highly educated people understand reproduction poorly.

Acts to affect birth outcomes or the fate of offspring exist not only among humans, but also in the animal kingdom. Neither are they something new, for there are historical records and evidences to show that such acts have been practiced since antiquity.

Virtually all animals, from spiders that have evolved for more than 100 million years to monkeys to humans, are known to kill their offspring. Adult fish regularly gobble up their own spawn; birds of many species routinely ignore the youngest hatchling – usually the smallest – at feeding time or will not interfere when older siblings push weaker nest mates out of the nest to certain death. This may be a way of culling to make sure that abundant food at times of food shortages will be available for those most likely to survive.

It is also not uncommon that prides of lions are raided by outside lions, which, if triumphant, promptly slaughter cubs sired by the predecessors. Young silverback gorillas – probably the strongest of all primates – when intruding older males' troops and coming away victorious will often go after departing silverbacks' offspring. Within groups of canines and primates, dominant females who have just whelped are known to destroy a lesser female's newborn litter. This may be a primal instinct to ensure that one's genetic legacy will be passed on with minimal challenge from other competitors.

Humans also have a history of exterminating their young. By about seven thousand years ago, within the earliest cities, infanticide appeared to have been practiced by all extant groups of people. In certain instances, the act was ceremonially elevated to be a form of human sacrifice to dissimulate the main purpose.

Anthropologists who have studied aborigines as a window into the past argue that infanticide constituted the oldest form of birth control, and served to limit the growth of the hominid population for hundreds of thousands of years. Joseph Birdsell, who died in 1994, argued that archaic humans were cognizant of the consequences of overtaxing the environment in which they lived. They practiced population control as a guarantee that their population number would not overtax the resources. They were sophisticated enough to have realized the role of the female in population control.

For the sake of argument, we shall use a year as the duration from conception to birth of a child instead of the well-known nine months. If there are x number of fertile females, they can simultaneously produce x children in a year with the work of one fertile male; however, y fertile males cannot father y children off one fertile female in a year; they will need at least y years to do so. Thus a society desiring to limit its number might well settle upon the strategy of routinely killing off a certain proportion of its females at birth.[88]

Interestingly, computer models indicate that a population practicing selective female infanticide is destined to go out of existence in a time of about a hundred years. A later Monte Carlo simulation shows that as long as several groups come together to form a larger community exceeding 175, the effective breeding unit, for the purpose of mate selection, survival will be sustained.[89]

These evidences suffice to show since antiquity, humans have endeavored, in however haphazard and illogical fashions, to affect the sex of offspring or to modify the pool of offspring through control of mating, sterilization, and selective infanticide. Thus the aims of reducing birth defects through genetic manipulations and of creating a better baby are not new; but the means to achieve these aims have improved: assisted reproduction, genetic engineering, selective breeding, and cloning.[90]

Selective Paternal Reproduction

We thus see that for as long as humans have been human, we have attempted to influence birth outcome. From historical records, humans have done so on the verge of religions: to the end of ensuring we would get the babies we wanted, we humans developed a panoply of rituals, some of which are already out of practical purposes, but are still practiced among the aborigines. We wore masks and protective amulets, performed elaborate ceremonies, uttered incantation, built bonfires or lit candles, chanted and prayed. Some couples did some of the strangest things: they matched and bled each other believing that bloodiness would endow their offspring superior characteristics. In the midst of campaigns, Spartan men would leave the battlefield at opportune times to hurry back to make love with their wives, convinced that the heat

---

[88] Joseph B. Birdsell, "Some population problems involving Pleistocene Man", *Cold Spring Harbor Symposia of Quantitative Biology*, 22 (1957), pp. 68.
[89] Clive Gamble, *The Paleolithic Settlement of Europe*, (Cambridge University Press, Cambridge, 1986).
[90] Gina Maranto, *Quest for Perfection: The drive to breed better human beings*, (Lisa Drew Book, New York, 1996).

stirred in their blood by the fight would be imparted to their offspring, who would grow up lean, strong and battle-ready.

In modern days, some of the practices of the past are still being practiced: couples make love in prescribed positions or consume certain seafood to guarantee a child's gender; expecting mothers avoid gazing on certain objects, eating certain food or running into certain beasts believed harmful to the fetus. In trying to educate the fetus, expecting mothers listen to classical music, staring for long duration at paintings to stimulate the fetuses.

Through these and other procedures, both regarded as superstitious and scientific, we seek to make sure that our offspring would be brave, talented, attractive, and of utmost desire, healthy. When such efforts fail to give birth to desirable babies, we are capable of ruthless actions. For the most part of history, the choice has been less of whether to conceive than to keep them. Upon giving birth to babies of the undesired sex, mothers would get rid of them by smothering them, burying them or drowning them alive. When the babies are deformed, parents desert their babies out into the wild. It is almost an innate drive and desire among both individuals and groups to shape the results of pregnancies and to control the nature of the children.

The Spartans were not alone in condemning infants to death; they were merely more rigorous about it. Throughout ancient Greece, infanticide was seen as amply defensible: a father has an absolute prerogative to decide whether his children would live or die, expressed formally in a ceremony of *amphidromia*. On the fifth, seventh or tenth day after birth, a nurse would carry the infant around the hearth in the presence of the father, who either gave it a name or consigned it to death. Parents took unwanted (may be not cute enough) babies out into the countryside and abandoned them to the elements. This is the practice of "exposure." Exposure carries the psychological advantage of uncertainty so that fantasies of an infant's survival might be nursed as a compensation for its loss.

In *Politics*, written by Aristotle (382-322 BC) and in *The Republic*, written by Plato (428-347 BC), weak and deformed infants were to be weeded out ruthlessly. Plato, a firm admirer of Spartan customs in regard to reproduction, elaborated upon his conviction that the state would regulate which men and which women could have children, preventing those regarded as mentally feeble or physically inadequate from procreating.

It is a characteristic of the historical perspective that actions taken in the immediacy of the moment may, with the accretion of time, assume the appearance of trends. Thus, shaping pregnancies, controlling the nature of the baby and other efforts to affect birth outcomes may be regarded as a part of an overarching quest by humans to perfect themselves to conform to the trends. Indeed it was only within the last century when contraception was more effective and abortions medically safer that helped reduce unwanted births and thereby lessen infanticide.

The Founding of Rome

As a romantic historical footnote, in the lingo of pop psychology, the founders of Rome were exposure survivors. The founding of Rome is rather tangled and embroiled in myth. A legend has their parents Mars, the war god, and his consort Rhea Silvia; another legend holds that their father was unknown, their mother a handmaiden to the daughter of Tarchetius, king of Alba Longa.

The most romantic legend is the story attributing maternity to a daughter of Aumulius, a descendant of the brave Aeneas, who was to have been a hero fighting the Greeks in the Trojan wars in 1220 BC. According to Plutarch and other accounts, traces found by archaeologists of early settlements of the Palatine Hill date back to 750 BC. This date ties in very closely to the legend that Rome was founded on April 21, 753 BC. This date is traditionally celebrated in Rome as the festival of Parilia.

According to the legend, Numitor, king of Alba Longa, was overthrown by his younger brother Aumulius. To do away with any possible claims to his usurped throne, Aumulius murdered Numitor's sons and forced Numitor's daughter, Rhea Silvia, to become a vestal virgin. Vestal virgins were priestesses to the goddess Vesta and were expected to guard their virginity in the goddess' honor.

However Mars, the god of war, came to her in the temple, became enchanted by her beauty and had his way with her while she was asleep. As a result, she gave birth to two twin boys "of more than human size or beauty" in 753 BC. The enraged Aumulius bade his servant Faustulus to expose the boys. Faustulus tucked the boys into a small reed basket and abandoned them in the Tiber River outside of town. The river flooded and the basket landed downstream by a wild fig tree. A she-wolf with a highly sense of trans-species altruism happened by. The she-wolf (wolves are sacred to Mars) nursed the infants, and a woodpecker also fed and watched over them until a shepherd found them. Together with his wife, the shepherd raised the boys to adulthood.

Once grown, the boys took the names Acca and Larentia (or Romulus and Remus). They were also told of their true origins. The twins gathered a band of hardy, adventurous companions and killed Aumulius to return the grandfather Numitor to the throne. The twins also decided to found a new empire close to where they had been washed ashore. Reading omens from the flight of birds, they decided on a location on the southern flank of the Palatine Hill and that Romulus should be king. Romulus took to marking the city boundary with a plough drawn by a white bull and a white cow. Remus leapt over the furrow, either in jest or derision. The brothers thus had a falling out and Remus was killed. Romulus then named the city after himself.[91,92]

As a Roman law, Romulus took steps to spare at least some infants from the trauma he had experienced. A modern reconstruction of his proclamations reads:

> Romulus compelled the citizens to rear every male child and the first-born of the females, and he forbade them to put to death any child under three years of age, unless it was a cripple or a monster from birth. He did not prevent the parents from exposing such children, provided that they had displayed them first to the five nearest neighbors and had secured their approval.

However, the Romulan dictates appear to have been ignored. Romans continue to commit infanticide through the Roman Empire (27 BC–476 AD) with regularity.

Selective Maternal Reproduction

Selective paternal reproductions are attempts to preserve desirable traits through the fathers. There is another obvious way to preserve desirable traits – through the mothers.

Greeks, Romans, and other Westerners through the Middle Age drew the parallel between husbandry of domesticated animals and humans. In the early days of embryology, researchers or medical specialists held the same view. As far as they were concerned, species were interchangeable so that what was applicable to cattle is also applicable in humans. In other words, if they could boost the fertility of cows or improve birth outcomes among pigs, they could do the same for their own kind. With the advent of scientific breeding in the eighteenth century, the parallel was fully exploited and by the nineteenth century, techniques that had been developed for manipulating animal sperm had been adopted for humans. Over the years, each ensuing advance on the farm found its way into the clinic and the hospital. F.H.A. Marshall (1878-1949) is known to have remarked, "Generative physiology forms the basis of

---

[91] Plutarch, "Life of Lycurgus" in *The Lives of the Noble Grecians and Romans*, translated by John Dryden, Great Books of the Western World Series, ed. Robert Maynard Hutchins, Vol. 14 (Encyclopedia Britannica, Inc., Chicago, 1932).

[92] *Illustrated History of the Rome Empire*, www.roman-empire.net.

gynecological science, and must ever bear a close relation to the study of animal breeding."

The technique to concoct life in the laboratory was developed through years of experimentation with domesticated species. Like in most cases of scientific development, initially, the work was undertaken for the sake of acquiring pure knowledge, out of the desire to elucidate different aspects of biological wonder of sexual reproduction. As the knowledge matured, the interest became pragmatic and sometime utilitarian: reproductive animal scientists want to amplify the reproductive capability of "superior" females, just as they had done in promoting "superior" males.

To this end, there are two obvious schemes. A first scheme was to inseminate selected females, remove resulting embryos and implant them in another animal for gestation. In this way, the exceptional female would not have her uterus tied up for months and could be inseminated as many times as she came into heat. Consequently more offspring carrying her desirable traits would be produced than she could have gestated herself *seriatim*. A second scheme was to retrieve eggs from the ovaries from first-rate female for fertilization. Then bring the resulting embryos to term in lesser females. It is also possible to achieve the same end by taking an egg from a lower-rate stock and replace some or all of the genetic materials in its nucleus with higher-rate materials.

The first scheme involves embryo transfer and the tricky business of reinserting the embryo in another receptacle and to ensure the second receptacle would take the foreign embryo. The second scheme, which came to be known as *in vitro* fertilization, required that eggs be obtained when they are mature enough to undergo fertilization, but at the point just prior to their release from the ovaries. The eggs would have to be kept alive outside the body for some period of time, a few hours to a few days. The eggs would then have to be fertilized while cultured in a medium. The resulting embryo would then have to be implanted in another receptive uterus.

▶ Eugenics

Plato (428-347 BC) was an ardent advocate of the state regulating which men and which women could have children to prevent undesirable procreations. Pieter Camper (1722-1789) gave geometry a racist twist with "facial angles." Whatever the vicissitudes of history, engineering humans and human parts has always been intimately intertwined with politics. From the seventeenth century on, issues having to do with childbearing have spun around larger issues regarding population, class and caste, race and ethnic groups, and evolution. The European nation-states rose in maritime power and via force colonized the globe. The subsequent fear of decline and fall obsessed the imagination of educated elites, particularly those in England, France and Germany. Dreams of world hegemony turned to a paranoia fantasizing about the destruction of the whites, Nordic, and superior race through miscegenation, through interbreeding with the dark, southern, and inferior race. Ultimately emerging from this deranged fixation on bloodlines and supremacy came eugenics.[93]

Eugenics, as described by one American leader of the movement, is "the science of the improvement of the human race by better breeding." [94] Eugenics is a word adapted from the Greek word "eugene," meaning "noble, good in birth, or wellborn." The British polymath Francis Galton (1822-1911), who is generally regarded as the father of eugenics movement, introduced it into the English language in 1883. A cousin of Charles Darwin (1809-1882) and born in the same year as Gregory Mendel (1822-1884), Galton proposed a scheme of selective breeding for humans who he felt would halt or even reverse the downward drift. Julian Huxley (1887-1975), grandson of Thomas H. Huxley (1824-1895), publicly endorsed eugenics.

The "science" of eugenics has two tools: negative and positive eugenics. The purpose of negative eugenics is to prevent the unfit, the cacogenic ("kakos" in Greek means "bad") – historically those viewed as insane, feebleminded, criminal or in anyway inferior, from propagating; the purpose of positive eugenics is to encourage the propagation of the aristogenic – those viewed to have superior traits.

First Stage

The eugenics movement can be roughly divided into four stages. The first, from 1870-1905, was a time of preparation as the ideas of breeding humans in a more efficient manner became more widely accepted. Eugenics proponents supported restrictions on marriage and sexual behavior of the unfit. In 1896, Connecticut, U.S. became the first state to regulate marriage for breeding purposes. The law provided that "no man and woman either of whom is epileptic, or imbecile, or feebleminded shall marry or have extramarital relations when the woman is under forty-five years of age. The statute sets a

---

[93] Gina Maranto, *Quest for Perfection: The drive to breed better human beings*, (Lisa Drew Book, New York, 1996).
[94] Mark Haller, *Eugenics: Hereditarian attitudes in American thought*, (Rutgers University Press, New Brunswick, New Jersey, 1963).

minimum penalty of three years imprisonment for violations. Within a decade, five other states followed Connecticut's lead and passed eugenics marriage laws.[95]

Second Stage

The second stage of eugenics movement lasted from 1905 to about 1940. Over this four-decade period, eugenics became official state policy for countries around the world, and the U.S. became unquestionably the leader of the movement.

The eugenics movement of this stage also contained class politics, for most ardent advocates were also members of the American ruling families. The bluebloods – a European term having roots in the fact that as the people were so fair that their veins, carrying deoxygenated blood back to the heart and thus appear blue, showed – such as the Harrimans and the Dodges, were concerned over their potential loss of political control to various immigrants groups. After the immigration waves at the end of the nineteenth century, the WASP (White Anglo-Saxon Protestant) hegemony was being challenged by the Irish, the Jews, the Italians and other immigrant groups who were beginning to elect their own leaders to power. Eugenics was effectively used throughout the period to enact laws restricting immigrants.

After 1900, the Harrimans, the family that gave the Prescott Bush[96] (1863-1948) family its start, along with the Rockefellers granted more than $11 million to create a eugenics research laboratory at Cold Spring Harbor, New York, as well as eugenics studies at Harvard University, Columbia University, and Cornell University. The First International Congress of Eugenics was convened in London in 1912, with Winston Churchill (1874-1965) as a director. Obviously, the concept of "bloodlines" was highly significant to these people.[97]

The main method for the implementation of negative eugenics in the U.S. was the sterilization of the "unfit," sterilization of those incarcerated in mental institutions and in prisons. The first law for incarceration was passed in Indiana in 1907, about a decade after the first law to regulate marriages for breeding purposes passed in Connecticut. Fifteen other states followed suit in the following ten years and by 1930, twenty-eight states had passed

---

[95] Andrew Kimbrell, *The Human Body Shop: The engineering and marketing of life*, (HarperCollins Publishers, New York, 1993).

[96] Prescott Bush is the grandfather of George W. Bush, 43rd president of the United States.

[97] Jim Marrs, *Rule by Secrecy: The Hidden History That Connects the Trilateral Commission, the Freemasons, and the Great Pyramids*, (HarperCollins Publishers, 2000).

sterilization laws, 15,000 men and women had been sterilized and by 1958, 61,000 "unfit" Americans had been involuntarily sterilized in thirty states. The statutes were based on the Model Eugenical Sterilization Law drafted by Harry H. Laughlin (1880-1943), superintendent of the Eugenics Record Office (ERO) which was underwritten by Mary (Mrs. E. H.) Harriman. The law called for sterilization, regardless of etiology or prognosis, of criminals, mental patients, the feebleminded, inebriants, the blind, the diseased, the deaf, the deformed, and the dependent (that is, the homeless, including orphans, tramps, and paupers).[98]

Sterilization and negative eugenics did not go unopposed. Powerful religious institutions strenuously resisted state interference with the rights of the "unfit" to marry and to bear children. The state-sponsored sterilizations reached a fevered pitch when the U.S. Supreme Court resolved the doubt about sterilization of the "unfit." The case involved the proposed sterilization of a Virginian woman Carrie Buck, a daughter of a supposedly feebleminded mother Emma. Carrie was adopted at the age of four by a family in Charlottesville, Virginia and attended school until the sixth grade, but became mentally impaired and was unable to continue schooling. She was pregnant when she became unmanageable and like her mother before her, was institutionalized at the Virginia State Colony for Epileptics and Feebleminded, Lynchburg, Virginia at the age of seventeen, where she gave birth to Vivian, judged to be feebleminded at seven months old. Carrie and Emma were both judged to be feebleminded and promiscuous, primarily because they both had borne children out of wedlock.

Months later, Carrie became the first victim of the newly enacted Virginia sterilization law (1924). The law was challenged and made its way through the courts: experts testified that Carrie had a mental age of a nine-year-old, her daughter was defective, and her feeblemindedness and immorality were inherited and passed on; while the state maintained if she was sterilized, she could be deinstitutionalized to save the state significant expenses. Both the circuit court and the Virginia Supreme Court upheld the order to sterilize Carrie. The infamous case, *Buck v. Bell*,[99] reached the U.S. Supreme Court in 1927. The U.S. Supreme Court upheld the law and the sterilization of Carrie on May 2, 1927. The oft-quoted opinion, written by Justice Oliver Wendell Holmes (1841-1935), an avid supporter of eugenics, reads "...It is better for all the world, if instead of waiting for their imbecility, society can prevent those

---

[98] George J. Marlin, and Richard P. Rabatin, "G.K. Chesterton and Eugenics", *Fidelity*, June 1990, pp. 30.
[99] *BUCK v. BELL*, 274 U.S. 200 (1927).

who are manifestly unfit from continuing their kind... Three generations of imbeciles are enough..."[100]

The American model of sterilization became very influential: within five years after *Buck v. Bell*, Denmark, Finland, Sweden, Norway and provinces in Canada initiated forced sterilization. In 1933, Adolph Hitler (1889-1945) put into effect his notorious sterilization program, as is evident in Rudolph Hess's (1894-1987) infamous statement, "Nazism is applied biology"[101] and credited the U.S. eugenics program as the basis for their own as in a citation, "...[the U.S. as a country] where racial policy and thinking have become much more popular than in other countries... German racial hygienists looked to the United States for inspiration; in this sense, German racial hygiene followed the American lead..."[102]

Interestingly, England, the birthplace of the eugenics movement, was never able to pass sterilization laws. While British eugenicists continued to focus on positive eugenics to encourage people to improve their family's genetic endowment, American eugenicists fostered negative eugenics legislation to prevent the contamination of the American gene pool with supposedly unfit traits. And by the 1930s, eugenics critics became more vocal. In 1934, a special blue ribbon committee of the British government, headed by Laurence G. Brock (1879-1949), issued a report undercutting the scientific merits of eugenics, "...the more closely individual records are examined the more difficult it becomes...to say with certainty that the genetic endowment of any individual is such that it must produce a given result..." A committee of the American Neurological Association, headed by psychiatrist Abraham Myerson (1881-1948) corroborated the Brock report.[103] The Myerson report, appearing in 1936 and in just under a decade after *Buck v. Bell*, flatly contradicted the U.S. Supreme Court's decision, "...There is at present no scientific basis for sterilization on account of immorality or character defect. Human conduct and character are matters too complex a nature, too interwoven with social conditions...to permit any definite conclusions to be drawn concerning the part which heredity plays in their genesis..."[104]

Although it was a *bona fide* medical diagnosis in its day, "feeblemindedness" was clearly a catchall term that had virtually no clinical

---

[100] Mark Haller, *Eugenics: Hereditarian attitudes in American thought*, (Rutgers University Press, New Brunswick, New Jersey, 1963).

[101] George J. Marlin, and Richard P. Rabatin, "G.K. Chesterton and Eugenics", *Fidelity*, June 1990, pp. 30.

[102] Robert N. Proctor, *Racial Hygiene, Medicine under the Nazis*, (Harvard University Press, Cambridge, MA, 1988).

[103] Abraham Myerson, "A critique of proposed 'ideal' sterilization legislation", *Arch Neurol Psychiatry*, 33, 1935, pp. 453-466.

[104] Daniel J. Kevles, *In the Name of Eugenics*, (Alfred A. Knopf, New York, 1985).

meaning. It is no longer a medical terminology in use, having been reclassified as mildly retarded, learning disabled, or simply underachievers. Although the eugenicists saw the Buck family as a pedigree of degeneracy, today we would say that the Buck family had problems that a bit of financial assistance, education, and opportunity would have helped solve. Their only sin was to have been born poor women in the impoverished South of the U.S.

It is impossible to judge whether or not Carrie was "feebleminded" by the standards of her time, but she was patently not promiscuous. Vivian was a conception as the result of a rape by the nephew of Carrie's foster parents. Carrie, probably like many unwed mothers of that time, was institutionalized to prevent further shame to the family. Vivian was no imbecile either. Her first grade report card from the Venable School in Charlottesville showed that this daughter of a supposed social degenerate got straight As in deportment (conduct) and even made the honor role in April 1931. Unfortunately, Vivian died a year later of an intestinal disorder.[105]

Growing awareness of the heinous ends the Nazis pursued in the name of racial purity led to a popular abandonment of eugenics in the U.S., and the ERO closed on December 31, 1939, marking the end of the most influential stage of the eugenics movement.

Third Stage

The third stage of eugenics movement lasted from the 1940s to the 1960s.

Lower courts in the U.S. continued to uphold sterilization statutes under the precedence of *Buck v. Bell* until a significant blow to the law came in 1942. In *Skinner v. State of Oklahoma*,[106] convicted felon Jack T. Skinner had been convicted for stealing chickens at age nineteen and twice for armed robbery in later years. Under a 1935 law of Oklahoma, he was to be sterilized. The law prescribed involuntary sexual sterilization of anyone convicted of a felony three times, but allowed exemptions for white-collar felonies such as embezzlement, liquor law violations, or political offenses. The U.S. Supreme Court unanimously struck down the Oklahoma law as unconstitutional in the opinion written by Justice William O. Douglas (1898-1980). The opinion also resonated with the horror of the Americans who were just beginning to become aware of the eugenics nightmare ongoing in Adolph Hitler's Germany: "…When the law lays unequal hand on those who have committed intrinsically the same quality offense and sterilizes one and not the other, it has

---

[105] David Micklos, "None without hope: *Buck v. Bell* at 75", Dolan DNA Center, Cold Spring Harbor, 2002.
[106] Skinner v. State of Oklahoma, Ex. Rel. Williamson, 316 U.S. 535 (1942).

made as invidious a discrimination as if it had selected a particular race or nationality for oppressive treatments..."[107]

After World War II, eugenics movement lost most of its allure when horror stories of the German extermination programs of the Jews, gypsies and other undesirables leaked out; the "science of eugenics" – with its simplistic association of facial features with inferiority, unsophisticated view of genetics and inheritance, and the bizarre use of pseudosciences like phrenology, also lost its credibility. Citizens and scientists alike were disgusted by the application of eugenics to exclude immigrants, to justify sterilization of criminals and the mentally ill, to prohibit marriages between different racial groups, and to exterminate those viewed as genetically inferior.

With this backdrop, have you ever wondered why Angel Island, off the coast of California, was a retention ground for Asian immigrants while its counterpart, Ellis Island, off the coast of New York, was a pass-through for immigrants from Europe, until the 1960s?

Fourth Stage

In the 1960s, eugenics resurrected as prominent scientists reiterated eugenics as the main hope for humankind's survival and prosperity. This stage of eugenics was based on the revitalization of the science of genetics, particularly knowledge of genetic abnormalities and their relationship to hereditary diseases, such as identification of genes responsible for sickle-cell anemia, thalassemia, and other single-gene hereditary disorders. As late as 1968, Linus Pauling (1901-1994), the Nobel laureate for chemistry (1954) and peace (1962), proposed that anyone found to carry the genes for sickle-cell anemia or phenylketomuria disease should be refrained from falling in love with another diagnosed with the same disease.[108]

So a new eugenics has begun, but this time it is not based on political or racial prejudices, but on genetic predispositions to physical or mental disorders. The tools for this new eugenics are not sterilization or extermination, but rather the advanced biotechnological technique of genetic screening, embryo manipulations, genetic surgery, and cloning.

There is currently no cure for the vast majority of diseases and disorders that have been identified. The new eugenics can thus be negative or positive eugenics: negative when selective abortion is used; positive when genetic

---

[107] Skinner v. Oklahoma, 62, U.S. 110, 114 (1942).
[108] Troy Duster, *Backdoor to Eugenics*, (Routledge, New York, 1990).

surgery is used to ensure that undesirable genes are not passed on to future generation. In this sense, reproductive cloning also offers positive eugenics.

As we begin to know the genes that are involved in how we think and behave, we must not forget most human behaviors are the combined products of many genes acting together, further modified by early childhood experiences. Each parent brings to his or her children a unique legacy of tens of thousands of gene variations accumulated since the dawn of humans. Two entirely different gene worlds collide when a sperm meets an egg and the gene combinations are almost infinite: working in one complex combination, a specific gene variant might predispose a person to autism; in another combination, it might predispose the person to become a genius. After all, many a Nobel laureate and Pulitzer Prize winner are the offspring of modest parents. This is the beauty of sexual reproduction, in wedlock or not. One can never predict where genius will arise. No human lineage is without hope.

Sadly, despite the 1942 Skinner case, sterilization of people in institutions for the mentally ill and mentally retarded continued through the mid-1970s, and the 1927 *Buck v. Bell* precedent allowing sterilization of the so-called "feebleminded" has never been overruled.

•••••   •••••   •••••

Whenever there is a need, there seem to always be a supplier willing to cater to the need. The body shop business is no exception to this rule. In the next chapter we will see how an invisible hand is guiding the market of the body shop.

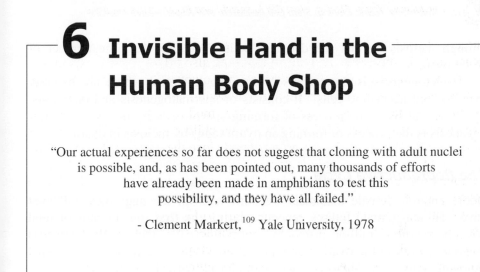

6 Invisible Hand in the Human Body Shop

"Our actual experiences so far does not suggest that cloning with adult nuclei
is possible, and, as has been pointed out, many thousands of efforts
have already been made in amphibians to test this
possibility, and they have all failed."

- Clement Markert,[109] Yale University, 1978

▶ The Human Life Cycle

For better understanding of this chapter and following chapters on cloning, we shall introduce a few terminologies commonly used in the discussion of human reproduction:

Ploidy is a term referring to the number of sets of chromosomes: a haploid organism has only one set of chromosomes; a diploid organism has two sets of chromosomes; and a polyploid organism has more than two sets of chromosomes. Human beings are diploid, that is, with the exception of our germ line cells (sperm and eggs) – which are haploid having 23 chromosomes – somatic body cells are diploid having 46 chromosomes.

Both somatic and germ line cells divide. There are two forms of cellular divisions – mitosis and meiosis. Mitosis maintains the cell's original ploidy number. Meiosis, on the other hand, reduces the ploidy number by half. Most cells in the human body are produced by mitosis. These are the somatic (or vegetative) cells. In a human somatic cellular division, the two daughter cells have 46 chromosomes, just like the parent cell. Germ line cells may go

---

[109] The author, HAL, has worked and co-edited books with this scientific giant. Markert (1917-1999) is commonly known as the co-founder of the concept of isozyme. In fact, in the International Congress on Isozyme, San Antonio, Texas, April 1997 to mark the eightieth birthday of this great man, six Nobel laureates came and the issue of Dolly the sheep's birth was one of the themes.

through meiosis so that the daughter cells have only half the number of chromosomes.

Gametogenesis is the process of forming gametes, which are haploid, from diploid germ line cells. It consists of spermatogenesis and oogenesis. Spermatogenesis is the process of forming sperm cells by meiosis in testes; oogenesis is the process of forming an ovum (egg) by meiosis in ovaries.

The Reproductive Cycle

Oddly enough, female mammals are born with all the eggs they will ever have. Human female fetuses generate, during the first four months of fetal life, about 6 to 8 million primordial germ line cells, technically known as oogonia, stored in the ovaries, and are diploid. They begin the chromosomal shuffling known as meiosis but spontaneously arrest themselves partway through the process and enter a dormant state in which they just sit until just prior to ovulation.

These germ cells start an attrition process or artresia in which eggs start their development only to degenerate and be lost even before birth so that at birth, the ovaries of female girls contain only about 1 or 2 million oocytes, which continue to decay and die, due probably to fatal errors during manufacture. By puberty and at the time of first menstruation, about four hundred thousand oocytes remain. As a woman enters her menstrual cycle each month, a number of oocytes regain momentum, completing a few more steps in meiosis and undergoing unequal division: the larger of the two resulting parts is the secondary oocyte; the smaller the first polar body. At the same time, the oocyte readying for ovulation requires a carapace of cells and forms what is known as primary follicle. As the cells surrounding the oocyte multiply, the follicle enlarges and fills with fluid. Gradually, the speck of egg migrates toward the surface of the ovary. Around day fourteen, the follicular bubble bursts, ejecting egg clad in the corona of cells. Hairlike cilia lining the fallopian tube propel the egg toward the uterus as the final chromosomal regroupings of meiosis take place. The oocyte divides once more in a second meiosis, producing a second polar body. These oocytes are now haploid. Oogenesis places most of the cytoplasm into the large egg; the smaller polar bodies do not develop. Thus a female normally produces one egg each menstrual cycle. An egg is released each month from puberty until menopause, a total of about 400-500 eggs.[110]

---

[110] Gina Maranto, *Quest for Perfection: The drive to breed better human beings*, (Lisa Drew Book, New York, 1996).

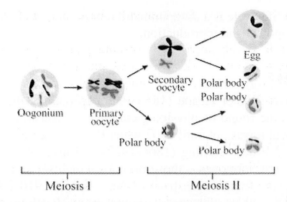

Oogonium   Primary oocyte

Secondary oocyte   Polar body

Egg

Polar body

Polar body

Polar body

Meiosis I   Meiosis II

Figure 1. Oogenesis is a form of meiosis. Oogenesis is an egg cell division in which the chromosomes replicate, followed by two nuclear divisions. The other form of meiosis is spermatogenesis in which a sperm cell divides. The difference between oogenesis and spermatogenesis is that during the second meiosis, the primary spermatocyte divides into four equal sperm cells; in oogenesis, during the secondary meiosis, both the secondary oocyte and first polar body divide, producing only one viable egg. (Figure adapted from various sources including lecture notes of Biology 103 at University of Georgia).

For a male, sperm production begins at puberty and continues throughout life, with 200,000,000 sperm produced each day! During sperm production (spermatogenesis), the germ line cells of a male undergo the same process as oogenesis in a female's germ line cells. The only difference is that in spermatogenesis, during the second meiosis, all the four meiotic products develop into haploid gametes (spermatocytes).

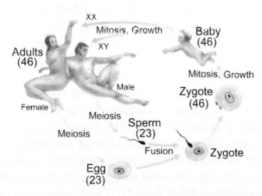

Figure 2. The life cycle of a human being: note that there are 23 chromosomes (haploid) in germ cells (egg and sperm), 46 chromosomes (diploid) in other cells; a man has an X and a Y chromosome, a woman has two X chromosomes. (Figure adapted from various sources including the lecture note of Biology 103 at University of Georgia).

The human life cycle is a diagrammatic representation of the events in the human's development and reproduction.

Sex Hormones

In the late 1800s, Walter Heape (1855-1929) speculated that a "generative ferment" drives the round of physiological changes constituting the monthly female cycle. But it was not until 1902 when British scientist William Bayliss (1860-1924) and Ernest Starling (1866-1927) at University College London discovered the first hormones – those proteins which serve as messengers or provocateurs in the female menstrual cycle. And in 1910 F.H.A. Marshall collected evidence that substances issuing from developing follicles are responsible for estrus (sexual heat).

Reproduction as we currently know it, involves a whole class of sex hormones, or gonadotropic hormones or simply gonadotropins, instead of just one circulating "generative ferment," as Heape had thought. By 1929, researchers in Germany and the U.S. had identified a hormone manufactured and released by the ovaries, estrone. In that same year Adolf Butenandt (1903-1995) explored the sex hormones and isolated the first pure form of this sex hormone from the urine of pregnant women. He also isolated 15 mg of a pure substance from a large quantity of policemen's urine. Within a few years, two steroids – estriol and estradiol – had been identified and synthesized. Collectively, estrone, estriol, and estradiol are all estrogens that are natural to the human body and belong to the steroid hormones. Only trace amounts of these steroids circulate in the blood and they perform important functions: in fetuses, estrogens trigger the growth of the uterus and vagina; in mature females they monthly prime the uterus and mammary glands for possible pregnancy. Estrogens also signal the hypothalamus in the brain and pituitary gland to produce two key gonadotropins: follicle-stimulating hormone (FSH) and luteinizing hormone (LH).

At the beginning of a menstrual cycle in a normal ovulating woman, the hypothalamus sends out a chemical signal to the pituitary gland in the form of gonadotropin-releasing hormone (GnRH), and in response, the pituitary disperses FSH. At this point, the ovaries have up to thirty follicles poised to mature, the largest of which measures about 5-8 millimeters. As the menstrual cycle proceeds, the amount of estradiol entering the blood during a twenty-four-hour period ten-folds from 0.03 milligram to about 0.4 milligram, and follicles increase in size to 12 millimeter or larger. After fourteen days, estrogen levels reach their peak. Cued by the high levels of estrogen, the pituitary gland releases LH in a surge, the pressure within a follicle becomes too great so that it bursts, spewing out its entire contents. The remaining follicles regress, having fulfilled their lifetime mission. Occasionally, more than one follicle will reach this stage simultaneously, leading to possible multiple conceptions. If no pregnancy results, a whole new cohort of follicles will be involved in the next menstrual cycle.

After ovulation, the menstrual cycle enters a luteal phase and progesterone is released, which incites changes in the uterine lining (endometrium), preparing the uterus for the possible arrival of a fertilized egg. Progesterone also inhibits the secretion of LH.

In older women, with the onset of menopause caused by a depletion of eggs, FSH levels skyrocket. As the egg supply dwindles, the number of FSH receptors falls and the FSH wander in the blood stream without ports to dock to, in contrast to pre-menopause younger female who have molecular docking sites for FSH.

Inside the Womb

During intercourse, some 300 million sperm may enter the vagina, but only one sperm (haploid or having 23 chromosomes) will penetrate the egg (haploid or having 23 chromosomes) yielding a nucleus with a full set of chromosomal endowment (diploid or having 46 chromosomes), resulting in a zygote.

The full set of chromosomal endowment contains the instruction to grow. Within hours after fertilization, the zygote undergoes cleavage (or cell division) or more technically, mitosis in which the ploidy is maintained (all resulting cell having 46 chromosomes). If cell division continued in this fashion, the hapless mother would give birth to a tumorous ball of cells of literally astronomical proportions in nine months.

Instead of endlessly dividing, the zygote's cells progressively take form: on day four or five, a first striking change occurs when the 32-cell clump (a morula, meaning mulberry in Latin) gives rise to two distinct layers wrapped around a fluid filled core. This spherical structure is called a blastocyst whose outer cells will turn into the placenta and the amniotic sac; its inner cells will become the embryo. By day six or seven after conception, the ball of cells has reached the uterus, where it may or may not attach itself to the uterine lining. If it fails to, no pregnancy occurs; if it attaches itself to the uterine wall, the ball of cells continues to divide. The inner layer of cells of the blastocyst balloons into two layers: the first layer, called the endoderm, will become the cells that line the gastrointestinal tract; the second layer, called the ectoderm, will become the neurons that make up the brain and spinal cord along with the epithelial cells that make up the skin.

By week two, the ectoderm produces a thin layer of cells known as the primary streak, forming a new layer called the mesoderm. The mesoderm will develop into the heart, the lungs and all the other internal organs. As the mesoderm forms, it interacts with the cells in the ectoderm to trigger another transformation. These cells roll up to become the neural tube, a rudimentary precursor of the spinal cord and brain. The embryo already has distinct clusters of cells at each end: one will develop into the mouth and the other the anus. Though still no more larger in size than a grain of rice, the head-to-tail axis has thus been formed, and other body parts will be arrayed along this axis.[111]

An interesting question is how this tiny cluster of cells knows what to do from here on? More than two decades ago, most developmental biologists thought different organisms developed according to different set of rules. Understanding how a fly, a worm, a chicken, a fish or a pig developed would not illuminate the development process in a human. Then in the 1980s, researchers found remarkable similarities in the development of all creatures, large and small.

Evolutionarily, fruit flies and humans separated 600 million years ago from a common ancestor. They both use the same genes to divide their bodies between front and back. Humans and zebra fish, both vertebrates and separated 420 million years ago, share even more genetic similarities. Having backbones, the zebra fish models human skeletal structure more closely than the invertebrate fruit fly.

Scientists studying model organisms have found that a special set of genes is in charge of forming the axis during fetal development. This set of genes is the *homeobox* or *HOX* genes. *HOX* genes were first discovered in the fruit fly in the early 1980s when scientists noticed that their absence caused striking mutations: head for example, grew feet instead of antennae, and thoraxes grew an extra pair of wings. *HOX* genes have been found in all animals, though the number of *HOX* genes may vary from one animal to another. The fruit fly has nine *HOX* genes; the human has thirty-nine.

---

[111] Alexander Tsiaras, and Barry Werth, *From Conception to Birth: A life unfolds*, (Doubleday & Co., 2002).

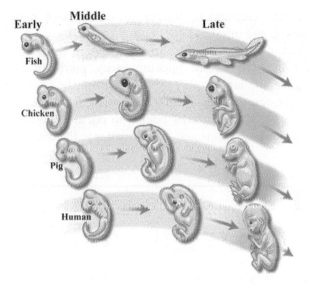

Figure 3. Scientists have long known that most creatures bear an uncanny resemblance in their embryonic development. (Figure adapted from *Genetically Yours*, Hwa A. Lim).

Twenty three days into conception, the nervous system appears as a depression that folds onto itself to form a tube along the back of the embryo. By day thirty-two, still no larger than a small flying beetle, a primitive heart, eyes, blood vessels, flipper-like arms and legs begin to appear. Another week later, the human embryo resembles that of a chicken, fish, pig and in fact, most other creatures at this early stage of embryonic development: all have a tail, a yolk sac and rudimentary gills. Around this stage of development, the human embryo also begins to develop a sense of smell.[112]

By about day fifty-two, the embryo is still no larger than the size of a grape, but it already has nostrils and pigmented eyes, though the eyes will not be able to sense light for another four more months, pending the development of optic nerves. At two months old, the fetus has all the major organs: a brain without cognitive function, a heart pumping only 20% of the full capacity of an adult, a stomach doing no digestion, an umbilical cord to anchor to the mother's placenta, an esophagus connecting the mouth to the stomach, two kidneys which can already eliminate waste, two lungs with tubules, vertebrae to connect the brain with the rest of the body, and a liver pumping out only red and white blood cells. By about the third month, the rib cage begins to show.

---

[112] J. Madeleine Nash, "Inside the womb", *Time Magazine*, November 11, 2002.

Seven months into conception, the brain can control body temperature, rhythmic breathing, and intestinal contractions.[113]

A month before full term, the rapidly growing fetus gets ready for its grand entrance into the world: packing away nutrients and stocking up disease-preventing immune cells obtained from the mother through the umbilical cord, its stored fat insulates it and provides it with an energy source. Because of its larger size, the fetus begins to crouch into the classical well-known fetal position, with arms and legs drawn into the chest.

Table 1. Facts of fetal development.[114]

Day	
1	Fertilization takes place.
7	The tiny embryo implants in mother's uterus.
10	Mother's menstrual cycle stops.
18	The heart begins to beat.
21	It starts to pump its own blood through a separate closed circulatory system with its own blood type.
28	The eye, ear and respiratory system begin to form.
42	Brain waves can be detected; reflexes are present; the skeleton is complete.
49	Thumb sucking can be seen.
56	All the body systems are present.
63	It squints, swallows, moves tongue, and makes fist.
77	It shows sign of spontaneous breathing movements; it has fingernails, all body systems are working.
84	It weighs about an ounce at this point of development.
112	Its genital organ is clearly differentiated; it can grasp with hands, swim, kick, turn, somersault, but is not felt by the mother.
126	The vocal cord is working; it can cry.
140	It has hair on the head, weighs about a pound, and is about a foot long.
161	It has 15% of viability outside the womb if premature birth should occur.
168	It has 56% of viability outside the womb if premature birth should occur.
175	It has 79% of viability outside the womb if premature birth should occur.

► Birth Defects

The crafting of a human from a single fertilized egg is a vastly complicated affair. Given the number of steps in fetal development, it is a miracle that life ever comes into being without a hitch. Sometimes, though, the fidelity is compromised and a baby is born defective.

---

[113] Alexander Tsiaras, and Barry Werth, *From Conception to Birth: A life unfolds*, (Doubleday & Co., 2002).
[114] M. Allen et al., "The limits of viability", *New England Journal of Medicine*, 329(22), November 11, 1993, pp. 1597.

When the heart fails to develop properly, a baby may be born with a hole in the heart or even missing a valve or a chamber; when the neural tube fails to develop properly, a baby may be born with a brain not fully developed (anencephaly) or with an incompletely formed spine (spina bifida). Neural tube defects have been traced to insufficient levels of the water-soluble B vitamin folic acid since folic acid is essential to a dividing cell's ability to replicate its DNA. Vitamin A is another nutrient critical to the development of the nervous system, but Amy Ogle and Lisa Mazzullo caution would-be mothers to limit foods that contain lots of vitamin A. Too much vitamin A can be toxic and cause damage to the skull, eyes, brain and spinal cord.[115]

Vitamins and other nutrients reach the developing embryo by crossing the placenta, a temporary organ developed from the outer layer of cells of the blastocyst. The outer layer of cells is very aggressive, behaving like tumor cells as they burrow into the uterine wall. They even replace the maternal cells that form the lining of the uterine arteries by tricking the pregnant woman's immune system into tolerating the embryo's presence rather than rejecting it as a foreign tissue. In essence, the placenta is a one-way traffic cop that keeps good things in and bad things out, it does so by marshalling platoons of natural killer cells to patrol its perimeters and engaging millions of tiny molecular pumps that expel poison before they begin to harm the vulnerable embryo.

Nothing is infallible and the placenta's defense can be breached. Microbes such as rubella and cytomegalovirus, drugs such as thalidomide and alcohol, heavy metals such as lead and mercury, organic pollutants such as dioxin and polychlorinated biphenyls (PCBs) are potential culprits. When development goes off track, the cause can often times be traced to these teratogenic (birth-defect inducing) factors that disrupt certain aspect of the developmental program: when a human embryo is deprived of essential nutrients or exposed to a toxin such as alcohol, tobacco nicotine, or crack cocaine, the consequences can range from visible abnormalities including spina bifida and fetal alcohol syndrome, to subtler metabolic defects that may not be obvious for many years.

Timing is a critical factor in the effects of toxins on the developing embryo. Air pollutants such as carbon monoxide and ozone have been linked to heart defects when the exposure coincides with the second month of pregnancy, the window of time during which the heart forms; similarly, the nervous system is particularly vulnerable to damage while neurons are migrating from the part of the brain where they have been made to the area where they will eventually reside. Heavy metals such as lead and mercury

---

[115] Amy Ogle, and Lisa Mazzullo, *Before Your Pregnancy*, (Ballantine Books, Inc., 2002).

interfere with the migration of neurons formed during the first trimester while PCBs block the activity of thyroid hormone, which conducts the formation of fetal brain. Though PCBs are no longer produced in the U.S., some 150 chemicals pose possible risks for fetal development.[116]

A mounting body of evidence suggests that a number of serious maladies: atherosclerosis, hypertension and diabetes, trace their origins to detrimental prenatal conditions. So what goes on in the womb before a baby is born is just as important as the genes in determining who the baby is going to grow to be.

Not only what the expecting mother ingests and inhales can affect the development of the fetus, but also the hormones that surge in her body. Pregnant rats with high levels of blood glucose give births to female offspring that are unusually susceptible to developing gestational diabetes. These daughter rats are able to produce enough insulin to keep their blood glucose in check, until they become pregnant when their glucose levels soar because their pancreases have been damaged by prenatal exposure to their mother's sugar-spiked blood.[117]

Similarly, atherosclerosis may develop because of prenatal exposure to chronically high levels of cholesterol. There seems to be a kind of metabolic memory of prenatal life: genetically similar groups of rabbits and kittens born to mothers on fatty diets are far more likely to develop arterial plaques. Of all the long-term health threats, maternal undernourishments, which stunt growth, perhaps top the list. People who are small at birth have, for life, fewer kidney cells and so they are more likely to go into renal failure. The same is true of insulin-producing cells in the pancreas so low-birth-weight babies stand a higher chance of developing diabetes later in life because their pancreases have to work harder. Low birth weight has also been linked to heart diseases, and undernourishment can trigger lifelong metabolic changes since food scarcity in prenatal life causes the body to shift the rate at which calories are turned into glucose for immediate use or stored as reservoirs of fat.[118]

We now return to the special set of *homeobox* or *HOX* genes that is in charge of establishing the head-to-tail axis in a developing embryo. Their absence in insects has caused malformations including feet growing on the head or an extra pair of wings growing on the thoraxes. Many other genes interact with the *HOX* genes, including the *Hedgehog* and *Tinman* genes,

---

[116] Sandra Steingraber, *Having Faith: An ecologist's journey to motherhood*, (Perseus Publishing, 2001).

[117] Peter W. Nathanielsz, *The Prenatal Prescription: Your program for your child's lifelong health*, (Vermillion, 2001).

[118] David Barker, "Development plasticity: The biological origins of heart diseases", *Cornell Messenger Lecture Series*, October 2, 2000.

without which an insect grows a dense covering of bristles or fail to develop a heart.

Sonic Hedgehog plays a role in making spinal cord neurons. Cells at different places in the neural tube are exposed to different levels of the proteins encoded by the *Hedgehog* gene: cells drenched in significant quantities of the protein mature into one type of neuron, and those receiving the barest mature into another.[119] Indeed, in 2002, it was by using a particular concentration of *Sonic Hedgehog* that neurobiologist Thomas M. Jessell and his research team at University of Columbia coaxed stem cells from a mouse embryo to mature into seemingly functional motor neurons.

At the University of California at San Francisco, biologist Didier Stainier is working on genes important in cardiovascular formation. Removing one of the genes (*Miles Apart*), the study subject zebra-fish embryos result in a mutant with two non-viable hearts. In all vertebrate embryos, the heart forms as twin buds. In order to form a viable heart, the buds must join. The *Miles Apart* gene works by detecting a chemical attractant that entices the pieces to move toward each other.

Our expanding knowledge about the interplay between genes and prenatal environment will help reduce the incidence of both birth defects and serious adult diseases. While there have been several breakthroughs in model organisms such as rats and fish, there is still much to be done before regenerative medicine can be applied on humans because fundamentally, a rat is a rat and a human is a human. The rat and the human are different. The challenges in transferring the technology are significant, but they are likely not insurmountable.

As a sidebar to show that animal studies may not be directly applied to humans, the incident of thalidomide of the late 1950s is an excellent example. Thalidomide was introduced in 1957 in Europe to cure insomnia with none of the side effects of barbiturates. However the safety data was virtually all based on animal studies.

To the delight of pregnant women, the tranquilizer seemed ideal for the treatment of morning sickness. The drug's sleep-inducing effect, which was so evident in people, had not been evident in many of the test animals. More troubling was some patients on thalidomide experienced a tingling sensation in the fingers.

Though the effects of the drug had been studied on pregnant rats and even pregnant women, the manufacturer did not study the effects on pregnant women in their first trimesters. This is exactly when thalidomide wreaked

---

[119] J. Madeleine Nash, "Inside the womb", *Time Magazine*, November 11, 2002.

havoc with the developing fetus. The rats were not susceptible to thalidomide's effect; the human liver, unlike that of a rat, produces an enzyme that converts thalidomide into its dangerous form. This dangerous form of thalidomide is teratogenic, causing phocomelia – a terrible condition whose name derives from the Greek word for limb and seal – in newborns. A newborn with phocomelia has little flippers instead of limbs.

By the time thalidomide was withdrawn from the market as a treatment for morning sickness, at least 8,000 babies had been disfigured!

▶ The Invisible Hand in Human Body Commerce

We thus see that the development of a human – from the union of two sex cells all the way to a fully developed human of trillions of cells – is indeed a very complicated but wondrous event. Given the possibilities of glitches in each step of the development, it is a miracle that life ever comes into being. Some people believe that there is a higher being in guiding the development; others believe this is all a natural process. For the latter, whenever a glitch should occur, there is no reason why we should not intervene and fix the glitch if the glitch is fixable using existing knowledge and technologies. Hence arises a new aspect of life – the selling and engineering of life – the human shop, and the head-on collision of modern mores and traditions.

The history of selling and engineering of life began with the Enlightenment, also known as the Age of Reason. This is the name given to the period in Europe and America during the 1700s when mankind was emerging from centuries of ignorance into a new age enlightened by reason, science, and respect for humanity. People of the Enlightenment were convinced that human reason could discover the natural laws of the universe; determine the natural rights of mankind; thereby unending progress in knowledge, technical achievement, and moral values would be realized.

Thinkers of that time, including Galileo Galilei (1564-1642), Johannes Kepler (1571-1630), Rene Descartes (1596-1650), John Locke (1632-1704), Isaac Newton (1642-1727) and others, changed forever the way we would look at the world and ourselves. They also helped develop the view that life forms are no more than sophisticated machines, a doctrine known as mechanism. In fact, the French scientist Julien Offray de La Mettrie (1709-1751) called humans "perpendicularly crawling machines."[120]

---

[120] Julien Offray de La Mettrie, *"L'homme Machine"*, 1748.

The great thinkers of the seventeenth and eighteenth centuries also began to design a new system of human behavior, a system that is not based on traditional mode of social control through fear of God or church, and duty for community or king. Rather, it was based on the principle that each individual would always act in his or her own self-interest and Adam Smith (1723-1790) created the free-market system with self-interest as the invisible hand that would lead to social justice and endless wealth.[121]

The dogmas of mechanism and the free market are the twin tenets of human shops: mechanism is the reductionist view of our body and other life forms, justifying us to view them as biological technology products available for sale; the free market is the basic ordering principle of capitalist social life, granting the rationale and ethical foundation for selling body products. Our daily vernacular is good evidence of how much these dogmas have ingrained in our society: we speak of the heart as a ticker, the brain as a computer, thoughts as feedbacks, the digestive and sex organs as plumbing. The body is no longer seen as divine, but rather as similar to the engines of industry.[122]

From Alchemy to Algeny

Let us recapitulate the goals of the alchemists, from whom we may be able to learn about what the current human body shop is trying to achieve.

Alchemy, derived from an Arabic word meaning "perfection," is said to have originated as a formal philosophy and transformation process in Egypt during the 4th century BC. Many historians believe its root dates further back in antiquity to the great Egyptian king named by the Greeks Hermes Trismegistus, reputed to have lived in 1900 BC. To the alchemists, all metals are in the process of becoming gold or are gold *in potentia*.[123,124] Their goal was to accelerate the natural process of transforming base metals into gold by elaborate laboratory transmutations: the alchemic process began by fusing together several metals to form an alloy – a universal base material from which various transmutations could be made. For the alchemists, elements are just convenient labels for identifying metals, but the element boundaries are in no way impenetrable walls separating various metals. In alchemy, fire was an indispensable tool because the alchemists had to melt, purify, distill and fuse

---

[121] Adam Smith, *An Inquiry into the Nature and Causes of the Wealth of Nations*, two volumes, (Dent & Sons, London, 1904), Vol. I.

[122] Andrew Kimbrell, *The Human Body Shop: The engineering and marketing of life*, (HarperCollins Publishers, New York, 1993).

[123] Morris Berman, *The Reenchantment of the World*, (Cornell University Press, Ithaca, New York, 1981).

[124] Titus Burckhardt, *Alchemy: Science of the cosmos, science of the soul*, (Stuart & Watkins, London, 1967).

the base material to create new combinations and forms, each successive one closer to the ideal golden state. Because of the extensive use of fire, we also call the alchemical technology pyrotechnology.[125]

As we cross the disciplinary border from chemistry to biology, a new philosophy corresponding to alchemy emerges. Algeny, attributed to the Nobel laureate biologist Joshua Lederberg, states that the living world is perfection *in potentia*. The algenists are dedicated to the improvement and enhancement of existing organisms, and perfecting their performance. Their goal is to accelerate natural processes such as the evolutionary process by biofacturing new creations that they believe are more efficient than those that exist in the same natural state: the algenic process begins with slicing out or splicing in genetic materials to get genetic chimeras or transgenics from which more perfect chimeras or transgenics may be made. The repertoire of recombinant DNA tools, selective reproduction procedures and cloning techniques is indispensable to the entire process. Corresponding to the pyrotechnology in alchemy, in algeny we have biotechnology.[126]

Like alchemists, algenists do not regard an organism as a discrete entity, but rather as a temporary relationship existing in an ephemeral context, on the way to becoming something more perfect. For the algenists, species boundaries are just imaginary boundaries for identifying living organisms, but are in no way impenetrable walls separating various plants and animals.

Table 2. Comparison of pyrotechnology and biotechnology.

Pyrotechnology – Alchemy	**Biotechnology – Algeny**
– All metals are gold *in potential*	– Living world is perfection *in potential*
– It is possible to accelerate the natural process of transformation by elaborate laboratory procedures	– It is possible to accelerate the natural process by genetic means
– The process begins with fusing several metals into an alloy, a kind of universal base material from which various transmutations can be made	– The process begins with slicing out or splicing in genetic materials to get transgenics from which other chimeras may be created
– Fire is indispensable to the entire transmutation process	– Genetic tool is indispensable to the entire process
– Gold is the ultimate state	– Perfection is a function of the prevailing environment, which changes with time

[125] Jeremy Rifkin, *The Biotech Century: Harnessing the gene and remaking the world*, (Jeremy P. Tarcher/Putnam, New York, 1999).
[126] Hwa A. Lim, *Genetically Yours: Bioinforming, biopharming, and biofarming*, (World Scientific Publishing, New Jersey, 2002).

We should also make another distinction between alchemy and algeny. In alchemy, the ultimate goal is to get gold from base metals; in algeny, the goal is perfection, which is harder to define and is a function of the prevailing environment. Since environment changes, the definition of perfection may also change.

▶ *In Vitro* Fertilization

We saw earlier that hormones play a key role in female menstrual cycle. This provides a handle to control human reproduction.

Administered at given points in the menstrual cycle, not necessarily in sync with the normal bodily rhythm, gonadotropic hormones can inhibit or promote ovulation. The realization that ovulation could be controlled from outside prompted an exploration from the 1940s of ways in which hormones might be used as contraceptives. Margaret Sanger (1879-1966) who crusaded to legalize birth control, Gregory Pincus (1903-1967) and Min Chueh Chang (1908-1991) known for the development of the pill, intensified their efforts of contraception in this vein.

Going to the other extreme of controlling fertility, infertility is defined as the inability of a couple to conceive after twelve months of intercourse. It affects 2.3 million American couples with wives aging between fifteen and forty-four years old. The figure represents about 8% of married couples in the childbearing age group, or about one in twelve.[127]

In the U.S., about 40% of couple infertility is due to problems with women: 5 million women, or 8.5% of the 58 million women who are between fifteen and forty-four years old have an impaired ability to have children. Slightly more than 40% of these infertile women have primary infertility or never having been able to have a child; the remainder of the women have become infertile after having one or more children. Among the causes are ovaries that do not release eggs properly or scarred fallopian tubes. Damage to a woman's reproductive capability can be due to factors including sexually transmitted disease, infection after surgical procedures, surgical sterilization, endometriosis, unnecessary hysterectomies, cancer treatment and damage from contraception devices, smoking, and environmental toxins.[128]

---

[127] William D. Mosher, and William Pratt, "Fecundity and infertility in the United States, 1965-88", *Vital Health Statistics of the National Center for Health Statistics*, 192, December 4, 1990.
[128] "Infertility: Medical and social choices", Office of Technology Assessment, *OTA-BA-358*, (U.S. Government Printing Office, Washington, D.C., 1988).

As much as men do not like to admit it, men are just as often infertile as women. Up to 40% of couple infertility cases are due to male infertility problems: mainly low sperm counts, impotence, genetic disorders, exposure to environmental mutagens and sexually transmitted diseases.[129]

A way to overcome infertility is *in vitro* fertilization (IVF). As far as the chronology of IVF goes, it started in 1878 with the work of S.L. Schenk in Vienna. Schenk drew sperm from a male rabbit and added it to a flat glass containing unfertilized eggs. Though this first attempt failed, Schenk ultimately succeeded in fertilizing rabbits and guinea pigs and watched they spontaneously divide.

The acknowledged father of embryo transfer is Walter Heape (1855-1929) of Cambridge University. Heape wanted to elucidate if uterine structure and biochemistry shaped the phenotype of offspring. One way of getting to the core of the questions was to incubate embryos from one breed of animal in a female of another type. If the offspring emerged substantially altered, it would be an indication that the uterine environment conveyed something essential. Heape worked with Angora rabbits and Belgian hares. In the spring of 1890, he mated two Angoras, anesthetized the female (doe) with chloroform, and using a technique called embryo flushing, he flushed her ovaries out with saline solution to retrieve a clutch of fertilized eggs. The embryos were then implanted in a Belgian doe. About four weeks later, the doe produced a litter of which two out of six were Angoras, their phenotype unaltered by the foreign uterus they had grown in.

Using hormones for an opposite purpose of abortion, in early 1970, Patrick Steptoe (1913-1988) corralled a group of forty-nine infertile women, hoping to one day conceive with the aid of his unorthodox methods. He subjected them to the gonadotropin regime and was able to retrieve eggs from twenty-nine of them.

He and Jean Purdy incubated the women's eggs for up to four hours, washed them twice in one of the culture media, and placed them in droplets of medium containing follicular fluid and sperm gathered from the women's respective husbands. Within a few hours, the cumulus cells surrounding the eggs drifted away. After another ten to thirteen hours, they placed the fertilized eggs in one of the several test media with formula suitable for meeting the metabolic requirements of cleaving eggs. Altogether, thirty-eight eggs underwent cleavage: eleven to eight-cell stage, eight contained eight to sixteen cells, and three had more than sixteen cells. The achievement brought

---

[129] Annette Baran, and Reuben Pannor, *Lethal Secrets: The shocking consequences and unsolved problems of artificial insemination*, (Warner Books, New York, 1989).

Patrick Steptoe and Bob Edwards (1925-) one step closer to their goal of *in vitro* fertilization (IVF).

Then throughout 1971, opposition to IVF and embryo transfer grew. Ethicists, lawyers, theologians, and some physicians raised objections. At the October panel discussion, James Watson stared straight into Edward's face and declared the inevitable loss of embryos in the experiments tantamount to infanticide. Watson quite rightly pointed out that the procedure was not therapeutic: infertile women might have a child, but they would remain infertile. Thus physicians pursuing this line of work could be said to engage in experimentation.

Paul Ramsey of Princeton University echoed IVF should properly be considered a manufacturing process, with its goal an end product, subject to quality control. Ramsey wondered, "If medicine turns out to doctoring desires instead of medical conditions… is there any reason for doctors to be reluctant to accede to parents' desire to have a girl rather than a boy, blonde hair rather than brown, a genius rather than a tout, a Horowitz in the family rather than a tone-deaf child…?" Ramsey demanded that IVF be prohibited, for then and anon, on moral grounds.[130]

Undeterred, Steptoe and Edward continued their research. By 1977, the two decided to jettison the fertility drugs. They would let the woman's natural cycle unfold, settle for one egg, and trust that they could produce one viable embryo which would take hold.

This is the beginning of the world's first IVF baby. Lesley Brown came from a broken home. She met John Brown at sixteen at a sailor's café in Bristol. John was estranged from his first wife, with whom he had two children. Lesley lived with John for six years before they got married. At twenty-two, depressed because she had not conceived after years of trying, she visited a fertility specialist. The diagnosis was tube blockage and the chance of conceiving was one in a million. On a referral, she traveled to Manchester to see Patrick Steptoe. Steptoe did not inform Lesley that IVF had never worked yet, and Lesley thought that hundreds of children had already been born through being conceived outside the mothers' wombs!

Two weeks after implantation, Lesley's urine and blood tests were encouraging. A week later, there was a faint sign that the embryo had taken. On December 6, 1977, the results were unequivocal and Lesley received a letter reporting the good news. Under Steptoe's advice, Lesley cut down smoking. At week sixteen, the results of amniocentesis revealed no

---

[130] Paul Ramsey, "Shall we 'reproduce'? I. The medical ethics of *in vitro* fertilization", *JAMA*, 220, 1972, pp. 1348.

chromosomal abnormalities and the level of alpha-fetoprotein was normal, a sign that the fetus neural development was proceeding without a hitch. Had any anomalies been evident, Steptoe was ready to abort the fetus.

During this period, secrecy was the theme of the hour. A newspaper article had alerted the press and the press had descended on the hospital offering cash for insider information. To prevent leaks by staffers, Steptoe had Lesley Brown checked in and out under the assumed name "Rita Ferguson." He also falsified her official medical docket and kept the legitimate note on the case in a private diary.

At half past eleven in the evening, July 25, 1978, Steptoe hoisted a five-pound twelve-ounce baby girl out of the Caesarean incision in Lesley's abdomen. They named the baby Louise and the couple chose Joy as her middle name, and of course Brown as her last name.

IVF the Aussie Style

A team in Melbourne, Australia, headed by Carl Wood of Monash University, came close to besting their British rivals in 1973. Wood, having been intrigued by accounts of failed turn-of-the-century attempts to overcome tubal blockage by grafting ovaries into the uterus, had approached infertility from surgery. Wood in 1969 had attempted to replace faulty fallopian tubes with plastic prostheses. The operation, like other attempts, led to infection and had to be scrapped. His encounter with Neil Moore of Sydney University who wanted to perfect IVF and embryo transfer for the sheep industry, set Wood to try to do the same in humans. To form a team, he drafted physicians from Monash University's Queen Victoria Medical Centre and University of Melbourne's Royal Women's Hospital. One of Wood's team members, Alex Lapota, soon made a research foray to Germany, England, and the United States, gathering information and experience on IVF.

In 1973, the Melbourne team, which had by then swelled to about a dozen physicians and reproductive biologists, experienced some beginner's luck: two patients conceived but the pregnancies were terminated within weeks. They then had no similar luck for the next seven years. Despondent at the dry spell, the team at one point tried to induce human sperm and eggs to fertilize in a sheep oviduct, but to no avail. Dissatisfied, Wood found Alan Trounson, an embryology from Sydney University who gained experience in freezing and thawing cattle embryos at Cambridge University. In 1977, Trounson joined Wood's team. This experience would prove useful for Trounson who would later play an important role in stem cell research.

A little too late may be. In July 1978, Patrick Steptoe and Bob Edward went down in the annals as having achieved the first IVF. Wood and Trounson set to revamp their quality control, bringing greater systemization to the harvesting of sperm and eggs and the production of embryos. Meanwhile, Alex Lopata of Royal Women's Hospital was becoming proficient at growing embryos in culture and became more reluctant to share his findings with others. In June 1980, Candice Reed – Australia's first, and by general reckoning, the world's third IVF baby – was born at Royal Women's Hospital. A schism soon developed: Lopata and others broke off and formed their own team at Royal Women's Hospital; Wood and Trounson stayed at the Queen Victoria embryology laboratory. Trounson refined sperm, eggs and embryo handling and Wood developed a protocol for drug stimulation. The two led to a series of pregnancies in 1980.

The Australian press gave the births a lot of coverage, tinged with nationalistic pride. Of 1,000 men and women over fourteen polled: 77% approved of the procedure; 11% disapproved. Two similar polls in the U.S. in 1978 found that the Americans were somewhat less enthusiastic. A Gallup poll showed that 93% had heard or read about the British test-tube baby, 42% could correctly explain how IVF and embryo transfer were done. 53% said they would resort to IVF if infertile, 36% said no; 60% said IVF should be available to everyone.[131] A Harris poll of 1,501 women if they would choose IVF or adopt if faced with infertility yielded: 57% would adopt, 21% would choose IVF. Asked if they believed IVF should be banned until scientists had proof that it would not contribute to birth defect, a resounding 63% voted yes.

A key difference between the Wood's modern version of embryo flushing and Heape's version of embryo flushing is the addition of hormones to synchronize and direct the cycles of the two females involved – the donor from whom embryos are obtained and the recipient into whose womb one or more embryos will be implanted. Wood's version inseminates the donor with sperm of an infertile women's husband, waits a few days, floods her uterus with a saline solution and then drains the fluid out with a catheter to capture the embryo.

This procedure particularly appeals to women with X-linked diseases such as color blindness or hemophilia, which are caused by faulty genes residing on the X chromosome. These diseases manifest themselves in sons, who have only one X chromosome. Daughters possess two X chromosomes, generally inherit one good X chromosome from their fathers and thus have a spare, do

---

[131] Clifford Grobstein, *From Chance to Purpose*, (Addison-Wesley Publishing, Co., Reading, Massachusetts, 1981).

not fall victim but are carriers of the defective gene, unless both X chromosomes are defective. Obtaining an egg from a woman known to be free from such a genetic defect would not only eliminate the threat to male offspring but also break the chain of malign inheritance!

The Business of In Vitro Fertilization

Around 1980, two Chicago entrepreneurs Richard and Randolph Seed began contemplating the $32 million market for embryo flushing in cattle and got the idea that by transferring the technology into the medical arena, there might be more market – an over $2 billion spent each year by couples attempting to overcome infertility in adoptive pregnancy for humans.[132] A selling point of artificially inseminating a third party and then receiving embryo is that it does not require surgery. The Seed brothers thought that the whole venture would require only raising a small seed capital, lining up suitable physicians, doing some promotion and public relations, drafting some women willing to be inseminated by strangers' sperm and then have their wombs flushed out and acquiring some space for clinics throughout the nations. The Seed brothers incorporated Fertility and Genetics Research, but unbeknown to them, they had overestimated the altruistic urges of women. They had a hard time recruiting women willing to serve as stand-ins for infertile women and to chance pregnancy or abortion should an embryo implant itself rather than float out during the lavage with saline solution.

Assiduous application of the technique finally yielded two pregnancies that were reported in *Lancet* in July 1983. Encouraged by the news, Richard Seed attempted to patent the procedure because he was beginning to think of IVF in terms of an industry, where large numbers of process patents have been granted. Many in the medical field viewed Seed's pecuniary ambitions as distasteful for in this field the long-standing protocol demands that advances in surgery and treatment be shared freely among all for the good of humanity.[133] Besides, the prestige and popularity of IVF militated against a wholesale embrace of a common stock-breeding technique.

Other IVF programs are also extremely careful to avoid giving away their aggressive campaign to purchase oocytes. Donor oocytes are being regarded as the solution to the infertility of older mothers. For those over forty years old, even with the help of IVF using their own eggs, the chances of conceiving

---

[132] Glenn Kramon, "The infertility chain: The good and bad in medicine", *The New York Times*, June 19, 1992, pp. D1.
[133] Beverly Merz, "Stock breeding technique applied to human infertility", *Medical News*, 250, (1983), 1257.

are virtually nil. Younger donors are increasingly persuaded as egg donors for these older would-be mothers. Between 1990 and 1993, the number of IVF births through egg donors rose from 120 to 800 annually; the average age of donors also dropped from around thirty to early twenties.

A few infertility experts, including Mark Sauer of Columbia University and Richard Paulson of University of Southern California,[134] and Severino Antinori in Rome, are well known for helping women advanced in age to give births thanks to the vigorous eggs of young women. Some of these names in the commerce of *in vitro* fertilization will reemerge when we discuss human cloning.

IVF Children

As of 1982, some 2,000 couples were waiting up to two years to undertake IVF and by August the following year, seventy-five IVF babies had been born. By July 25, 2003 the twenty-fifth birthday of Louise Brown, a million babies worldwide have been conceived through IVF. To celebrate the occasion, 1,000 IVF babies met on that Saturday at the Bourn Hall Clinic in rural eastern England where Steptoe and Edwards had perfected the IVF technique.[135]

Currently IVF and other assisted reproductive techniques can help 75% of the estimated one in six couples who has a fertility problem to become parents. Since Louise Brown's birth, a technique for male infertility called intracytoplasmic sperm injection (ICSI) and embryo screening to select the

Figure 4. Louise Brown with mom Lesley and dad John in 2003. (Photo: courtesy of Dr. Peter Brinsden, Bourn Hall Clinic).

---

[134] "Birthing at 50", *CNN News*, November 13, 2002.
[135] "Profile: Louise Brown: A quarter century on", *BBC News*, July 24, 2003.

best embryos have pushed the success rate to about 25%. ICSI involves inserting a single sperm into an egg.

Though refinements in the technology have increased pregnancy rates and it is estimated that in 2004 about 1.5 million children have been born by IVF, the IVF technique is still far from perfect. One of the most pressing issues is the rate of multiple pregnancies, which arise because more than one embryo is implanted in the womb. Indeed the first U.S. IVF success was quadruplets delivered in September 1985. Twins and triplets can be extremely bad for the health of the mother and the children.

Several recent studies have highlighted the possibility that IVF babies and other assisted reprotech (reproductive technology) babies might be at a greater risk of genetic diseases. This will only become clear as this first generation of IVF babies grow up.[136]

▶ Donor Insemination

One of the options for treating severe male factor infertility, or for achieving fertility where no male partner is involved, is artificial insemination using donor sperm or, more commonly, donor insemination (DI). Donor insemination is a procedure in which a fine catheter (tube) is inserted through the cervix (the natural opening of the uterus) into the uterus (the womb) to deposit a sperm sample from a man other than the woman's mate directly into the uterus. The purpose of this procedure is to achieve fertilization and pregnancy.

Donor insemination is also called artificial insemination by donor (AID) or heterologous insemination; it is to be distinguished from artificial insemination by husband (AIH) which is homologous insemination. The acronym "AID" is no longer used since the advent of AIDS, and sometimes the procedure is called therapeutic donor insemination (TDI). Therapeutic donor insemination using donor sperm may be the treatment of choice in the following cases:

1. The males may have low sperm count and there are no treatment options to improve sperm count and quality.
2. Immotile sperm that is incapable of fertilizing an egg. Some men are unable to ejaculate because of neurological disorders.
3. When it is not possible to recover sperm capable of fertilizing an egg – even by ICSI.
4. When the man is a carrier of an undesirable hereditary condition.

---

[136] Helen Pearson, "Meeting celebrates IVF birthday", *Nature*, July 25, 2003.

5. When ICSI/IVF is not financially possible. The average cost of an IVF/ICSI cycle is about $7000 to $9000.
6. For single women or same sex couples.

Before embarking on donor sperm insemination the prospective mother will be evaluated to rule out any obvious fertility problems. The menstrual cycles will be monitored with a basal body temperature chart to see if she is ovulating. This will give information about the length of the cycle, and at what time of the month she normally ovulates.

Certain tests are performed, including hormone tests, and blood tests to rule out infectious illnesses such as hepatitis B and C, HIV, as well as genital tract cultures. The exact timing of ovulation will be determined by checking urine each day leading up to the fertile period using an ovulation predictor kit. Approximately 24-36 hours prior to ovulation, luteinizing hormone (LH) appears in the urine. The ovulation predictor kit allows the prospective mother to monitor her urine for the presence of this hormone. When the test is positive, the egg will be released the following day. If the prospective mother does not ovulate regularly, or if her cycles are unpredictable, she may be asked to use a fertility enhancing or ovulation induction agent.

The actual insemination process is very like having a Pap smear done. A speculum is inserted into the vagina, and the thawed, washed sperm is injected through the cervix into the uterine cavity using a special thin catheter. The insemination procedure usually takes only a few minutes. Sometimes difficulty is experienced passing the catheter through the cervix, and the cervix will need to be held steady using an instrument called a tenaculum.

After the insemination the prospective mother will be asked to lie quietly for 5-10 minutes, after which she will be free to leave the office and resume normal activities. She is however discouraged from doing any major exercise, going into a hot tub or public swimming pool for 24 hours after the insemination.

The success rate of DI depends on many factors, the most important of which is the prospective mother's age. For women under the age of 35, with no other fertility-related health problems, the success rate is about 15-20 % per treatment cycle or about 60% after 6 months. The success rate decreases as the age of the prospective mother increases.

A child born through donor insemination is considered to be the legal child of the mother and her spouse or partner. The legal obligations of the mother and her spouse to such a child are no different to that of any other couple. Although adoption can be an attractive alternative as well, there has been a growing scarcity of adoptable babies and prospective parents may have to wait a long time before adoption becomes possible. Artificial insemination, in selected cases, may have certain advantages. The child conceived by artificial insemination is biologically closer to the parents both physically and, in some cases, emotionally. The mother has an opportunity to experience pregnancy, birth, nursing, and all the other roles of motherhood she wishes.

DI is a highly confidential process. It is not necessary for the mother to disclose her participation to anyone. The mother will also have no access to the identity of the donor, nor will the donor have access to the mother's identity. Or is it?

DI Half-Siblings

Sperm banks typically pay men $50 to $100 per sample, and customers pay about $150 to $600 per vial, plus shipping and handling. Because of the confidentiality clause, an anonymous sperm donor will probably never meet any of the offspring he fathers through sperm bank donations.

But for children who often feel severed from half of their biological identity, finding a sibling – or in some cases, a few – can feel like coming home. It can also make them even more curious about the anonymous father whose genes they carry. While many donor-conceived children prefer to call their genetic father "donor," to differentiate the biological function of fatherhood from the social one, they often feel no need to distance themselves, linguistically or emotionally, from their siblings (more like half-siblings).

The half-sibling hunt is driven in part by the growing number of donor-conceived children who know the truth about their origins. As more single women and lesbian couples use sperm donors to conceive, children's questions about their fathers' whereabouts often prompt an explanation at an early age, even if all the information about the father that is known is his code number (such as Donor 150 of the California Cryobank) used by the bank for identification purposes and the fragments of personal information provided in his donor profile, such as medical history, ethnic background, education, hobbies and a wide range of physical characteristics.

Donor-conceived siblings, who sometimes describe themselves as "lopsided" or "half-adopted," can provide clues to make each other feel more whole, even if only in the form of physical details. It is precisely for this reason that Wendy Kramer and her donor-conceived son, Ryan, founded Donor Sibling Registry in 2002, when Ryan was only 12. Many of the site's registrants hope that the donor himself will get in touch. But others are just happy to settle for contacting half-siblings, who actually want to be found. As they do, they are redefining the meaning of family that both rests on biology and transcends it.

Many mothers seek out each other on the registry, eager to create a patchwork family for themselves and their children. As of January 1, 2006, the registry had a membership of 6,647 with matches between more than 1,242 half-siblings (and/or donors) facilitated. One group of seven feels bonded by the half-blood relations of their children, and perhaps by the vaguely biological urge that led them all to choose a particular donor. Several who have met describe a sense of familiarity that seems largely irrational, given the absence of a father, unrelated mothers and most of the time divergent interests.

Even as the Internet makes it easier for donor-conceived children to find one other, some of siblings are questioning the system of anonymity under which they were born. They argue that sperm banks should be required to accept only donors who agree that their children can contact them when they turn 18, as is now mandated in some European countries.

This makes good sense from the perspective of accountability. Sperm bank officials estimate the number of children born to donors at about 30,000 a year. In this largely unregulated industry, no one really knows the number for sure. As half-siblings find one another, it is becoming clear that the banks do not know how many children are born to each donor, and where they are.

Popular donors may have several dozen children or more, and there is always that risk of unwitting incest between half-siblings. Moreover, critics argue, no one should be able to decide for children before they are born that they can never learn their father's identity. Typically, women can learn about a donor only from the information provided in his donor profile at the sperm bank.

There have been some changes in the sperm bank confidentiality policy. More recently sperm banks have begun to charge more for the sperm from donors who agree to be contacted by their offspring when they turn 18. But these banks say far fewer men would choose to donate if they were required to release their identity.[137]

▶ Hello Dolly, Good-Bye Dolly

The new reproductive technology or reprotech – embryology, reproductive technology, clonetechnology – did not happen in the vacuum. In some cases, scientists involved in developing techniques for handling sperm, eggs,

---

[137] Amy Harmon, "Hello, I'm your sister. Our father is donor 150", *The New York Times*, November 20, 2005.

embryos and cloning held eugenics convictions. In other cases, those who promulgated reprotech aspired to further the cause of biological engineering. And couples seeking to use any of the reproductive procedures may not begin with the conscious intent to fabricate a better baby, but soon found themselves caught up in making decisions. With each new fillip to reprotech, we move further along into a continuum of human engineering in which parents will attempt not only to eliminate diseases, but also to determine which of many of the characteristics their children should be endowed with.

The manipulations going on in reprotech today can be seen as the culmination of a long sweep of events beginning some ten thousand years ago, when one of our hominid forebears abandoned a newborn, probably deformed, somewhere in the wild and left to die. In reference to genetic screening, genetic manipulation, reprotech and cloning, Clifford Grobstein (1916-1998) holds that humanity is on the verge of a revolutionary transition from chance to purpose, from genetic roulette to genetic determinism.[138]

On July 25, 1978, this transition culminated in the birth through Caesarean section of a five-pound twelve ounce Louise Brown, the world's first *in vitro* fertilization baby. The success of Louise Brown launched a new era, the era of human reproductive technology or reprotech. This is an era during which a range of technologies for manipulating human reproductive processes at the level of gametes (eggs and sperm) and zygotes (embryos) appeared with stunning rapidity.

Then July 5, 1997, likely the world's most famous lamb, Dolly, healthy at 14.5 pounds, entered the world. Dolly was the first mammal cloned from an adult cell of a six-year-old ewe. In January 2002, Dolly was diagnosed as having arthritis, a condition usually expected in older animals. It is not clear if the cloning process led to the arthritis, but research in 1999 suggested that Dolly might be susceptible to premature aging.

On December 27, 2002, Clonaid claimed to have produced the first human clone, Eve, a 7-pound girl born the day after Christmas to an American couple. The claim has still yet to be verified.

And on February 15, 2003, Dolly was euthanized at the age of six after being diagnosed with a progressive lung disease. The cause of the health problem is still under studies...

•••••   •••••   •••••

---

[138] Clifford Grobstein, *From Chance to Purpose: An Appraisal of External Human Fertilization*, (Addison-Wesley Publishing Co., 1981).

The first mammal cloned from an adult cell, Dolly the sheep, had a lot of health problems and had been euthanzied. To understand the potential sources of these health problems, let us now turn to the chronology of cloning in the barnyard.

7 Clonology

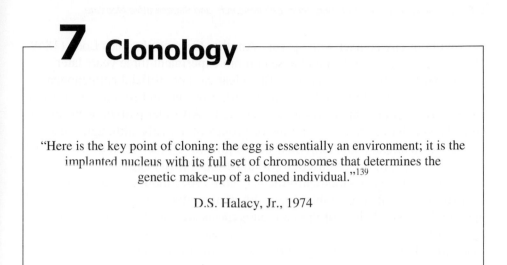

"Here is the key point of cloning: the egg is essentially an environment; it is the implanted nucleus with its full set of chromosomes that determines the genetic make-up of a cloned individual."[139]

D.S. Halacy, Jr., 1974

▶ Parthenogenesis – Virgin Birth

A natural form of asexual reproduction called parthenogenesis is an unusual form of reproduction that occasionally occurs in plants and animals. The word "parthenogenesis" is derived from a Greek word "parthenos" meaning "virgin" to signify that in parthenogenesis there is no mixing of parental genes because all the genes come from one parental organism. Parthenogenesis produces offspring by certain actions within a single cell, such as cell division in bacteria, cell budding in yeasts, and vegetative duplication in plant cutting or grafting.

Higher animals reproduce by copulation, but some lower animal forms can reproduce by parthenogenesis. Aphids (plant lice), some ticks, water fleas, ants, wasps, bees, certain lizards and snakes can all develop without male fertilization. Some species of the whiptail lizards in the western United States have only female populations.

Charles Bonnet (1720-1793) discovered the phenomenon of parthenogenesis in the 18th century. Later, pioneer experimenters with parthenogenesis use chemicals on the eggs of various species to artificially induce parthenogenic creation of offspring. In 1896, German embryologist Oskar Hertwig (1849-1922) added strychnine or chloroform to seawater containing sea urchin eggs. Remarkably, the process fertilized the urchin

---

[139] D.S. Halacy, Jr., *Genetic Revolution, Shaping Life for Tomorrow*, (Harper Row, New York, 1974).

119

eggs without any contact with sperm. Another German, Jacques Loeb (1859-1927), successfully duplicated the sea urchin experiment three years later.

In 1900, Loeb accomplished the first clear case of artificial parthenogenesis when he pricked unfertilized frog eggs with a needle and found that in some cases normal embryonic development ensued. Artificial parthenogenesis has since been achieved in almost all major groups of animals, although it usually results in incomplete and abnormal development. Numerous mechanical and chemical agents have been used to stimulate unfertilized eggs.

In 1936, Gregory Goodwin Pincus (1903-1967) induced parthenogenesis in mammalian, that is, rabbit eggs by temperature change and chemical agents. The postal child of the parthenogenesis was on the cover of the 1939 *Life* magazine. Despite its normal appearance, the rabbit had been created inadvertently by administering thermal shock treatment to a female rabbit.

Interestingly enough, in a complete career switch, Pincus would later conduct pioneering birth control pill work and became internationally known for his role in the development of the pill. He developed the first practical oral contraceptive birth-control pill with M.C. Chang after being persuaded to do so by Margaret Sanger and Katherine Dexter McCormik.[140]

No successful experiments with human parthenogenesis have been reported, even though in January 2002, Jose Cibelli of Advanced Cell Technology reported a minor breakthrough to perform parthenogenesis of a macaque monkey named Buttercup. The phenomenon is more rare among plants – where it is called parthenocarpy – than among animals. Unusual patterns of heredity can occur in parthenogenetic organisms. For example, offspring produced by some types are identical in all inherited respects to the mother. Note "mother" since males do not produce eggs.

Since parthenogenesis reproduces offspring identical in all inherited aspects to the mother, this implies that parthenogenesis is a form of reproductive cloning.

► Cloning

Cloning is not really a scientific word. Rather unfortunately, it has many meanings.

In this computer age, we hear friends talk about a computer clone and a compatible. In a nutshell, a clone is both software and hardware compatible; a compatible is only software compatible. If we are concerned with just

---

[140] James Le Fanu, *The Rise & Fall of Modern Medicine*, (Little, Brown and Company, London, 1999).

running programs, designed for the IBM computer for example, we need buy a compatible. But if we are concerned about future hardware upgrading, we better buy a clone.

In the office, our colleagues photocopy. To photocopy is to make a photographic reproduction of printed or graphic material, especially by xerography.[141] As part-time amateur authors, pirated books frustrate us. These pirated books are clones of our books; photocopies are clones of parts or whole of our books. They are selling at a much lower price and we get no royalties from these clones.

Photocopying of life forms, or cloning as it is usually called, is subtler. Cloning means creating an organism genetically identical to another organism. Cloning in reproduction can be of at least two forms:

❑ The synthesis of multiple copies of a particular DNA sequence using a bacterium or another organism as a host, and
❑ The creation of replicas of whole organisms.

The second, creation of replicas of whole organisms, has captured the popular imagination since the notion was first introduced in science fiction. Now cloning is almost a scientific reality, the populace is clamoring for news about the procedure that fascinates them as much as it repulses them.

There is a critical difference, however, between photocopying graphic materials and cloning (genetic copying) organisms. In photocopying a graphic material, the photocopy is an exact duplicate of the original, barring scale differences and tones; photocopying of a life form is performed at the genetic level. Cloning involves the production of genetically duplicate organisms from the biological information contained in the nucleus of a single cell. When first created, the copy or clone may not look like the original at all. It may involve development and growth processes, and hopefully the clone will develop into a duplicate of the original. But the development of a clone involves not only nature (the genetic makeup), but also nurture (the effects of environmental factors). Thus an exact duplicate of the original is almost impossible.

Clonology

Cloning is a very young science, being entirely a product of the twentieth century. The word "gene" first appeared in 1909. The chronology of cloning – clonology – started just a little before then with parthenogenesis. Parthenogenesis was the first significant step toward artificially reproducing

---

[141] *The American Heritage ® Dictionary of the English Language*, Fourth Edition, 2000.

some of the life forms. Cloning of life forms has to wait for more significant advances in the understanding of cells and genes until the 1950s.

Herbert Webber first used the word "clone" at the beginning of the 20th century (1903). Webber from the U.S. Department of Agriculture was looking for a unique word to describe the little sections of a plant that can be cut off and transplanted. At that time, there was a whole range of words in use, including bud, bulb, graft, runner and cutting. Instead, he invented the new word "clon," which soon evolved into "clone."[142] This new word was used to describe crops, such as strawberries, that can all come from a single parent plant and which are genetically identical to the parent plant. These identical crops are quite different from other crops such as wheat, which come from seed. In the latter case, the next generation of wheat can be different from the parent wheat.

By the 1950s, the word "clone" had picked up a slightly different meaning. By then, "clone" meant to replicate some sort of living creature by keeping the DNA the same from one generation to the next. And in these early days, the word "clone" mainly referred to cells instead of the original intended reference to sections of plants.

In historical hindsight, as early as in 1885 August Weissman (1834-1914) stated that genetic information of a cell diminishes with each cell division. Walter Sutton (1877-1916) proved that chromosomes hold genetic information in 1902. In that same year, the famous German embryologist Hans Spemann (1869-1941) divided a Salamander embryo in two and showed early embryo cells retain all the genetic information necessary to create a new organism. In 1928, Spemann performed the first nuclear transfer experiment, using again Salamander embryos. By then, Spemann and contemporaries knew that all animal cells contain a nucleus – the membrane-bound structure that houses a cell's genes. Phoebus Levene (1869-1940) discovered in 1929 a previously unknown sugar, deoxyribose, in nucleic acids that do not contain ribose; those nucleic acids are now known as deoxyribonucleic acids, or DNA.

Nine years later and a few years before his death, Spemann pondered over the idea called nuclear equivalency – whether a nucleus from a differentiated cell retain the know-how to develop into a complete organism. Scientists of that time generally dismissed nuclear equivalency, suggesting that as cells specialized, they irreversibly altered their DNA or even discarded whole genes. Spemann was unsure and theorized that by fusing an embryo with an egg cell animals could be cloned. This "fantastical experiment," as Spemann called it, was ahead of the technology of his time.

---

[142] Karl S. Kruszelnicki, "Cloning around", Great Moments in Science, Australian Broadcasting Corporation, 2000.

Oswald Avery (1877-1955) discovered genetic information is carried by the nucleic acids of cells in 1944. In 1952 Robert Briggs (1911-1983) and Thomas J. King performed a landmark experiment replacing the nuclei of freshly fertilized northern leopard frog eggs with foreign nuclei from an advanced frog embryo of the same species. They created tadpoles identical to the original donor. Briggs and King, biologists in Philadelphia, used the phrase "nuclear clones." The word "nuclear" has nothing to do with radioactivity – it comes, rather, from the fact that they were working with the nucleus of the frog cell. Although the experiment succeeded, scientists believed for more than 40 years thereafter that adult cells could not be used for cloning higher animals.

The word "clone" has since been used in other publications For example, in 1958, the Australian Nobel Prize-winning immunologist, Sir Frank Macfarlane Burnet (1899-1985) used the word in the title of his famous book, *The Clonal Selection Theory of Acquired Immunity*.

In 1953, biologist Francis Crick and biochemist James Watson discovered the structure of DNA at Cambridge's Cavendish Laboratory, advancing the field of genetics and creating the new field of molecular biology. For this path-breaking work, they later won the 1962 Nobel Prize in Physiology and Medicine. Biologist F. E. Steward grew, in 1958, a complete carrot plant from a fully differentiated carrot root cell at the Laboratory of Cell Physiology, Growth, and Development at Cornell University, which encouraged the belief that cloning from adult cells might be possible.

Following Briggs and King's pioneering work, other successful experiments in creating frog clones followed. In 1962, British molecular biologist John B. Gurdon announced he had cloned frogs using the nuclei of fully differentiated adult intestinal cells, though the tadpoles died before maturing.

Despite the successes, researchers would struggle for decades to achieve the far more difficult feat of cloning mammals. The successes of experiments in creating frog clone bridged the chasm from parthenogenesis to cloning. In all these experiments, the egg has been proven to be essentially an environment and that it is the implanted nucleus with its full set of chromosomes that determines the genetic makeup of a cloned individual.

Marshall Niremberg, Heinrich Mathaei, and Severo Ochoa (1905-1993), in 1966, determined which codon sequences specify each of the twenty amino acids, thereby "cracking the genetic code" and opening the door to advances in genetic engineering. James Shapiero of Harvard University, working with Jonathan Beckwith, isolated the first gene in 1969.

By 1972, Paul Berg invented the techniques of genetic manipulations. Four scientists, Stanley Cohen and Annie Chang of Stanford University,

Herbert Boyer and R.B. Helling of University of California at San Francisco, constructed a novel form of DNA by splicing together two pieces of DNA from different sources. They then implanted this new DNA molecule into a bacteria cell. At the time of the first successful splicing of different pieces of DNA together, cloning also stepped down to the level of cloning of a gene. In cloning a gene, scientists first isolate the gene, bound it to an organism – in this instance a yeast – that incorporates the gene into its own DNA. The multiplication of the host organism produces many copies of the desired gene.

In 1976 Rudolf Jaenisch of the Salk Institute for Biological Studies in La Jolla, California injected human DNA into newly fertilized mouse eggs to produce mice that were part human. When these founder mice reproduced, they passed the human genetic material to their offspring, creating a slew of so-called transgenic mice. Transgenic animal models are useful for studying different human diseases by creating for example, mice with the appropriate genetic composition.

Fred Sanger invented in 1977 a method for sequencing DNA, which later enables researchers to map the genomes of various species.

A year later, just before midnight (Greenwich time) on Tuesday, July 25, 1978, the world clamored for a glimpse of Baby Louise Joy Brown. Baby Brown, weighed 5 lb 12 oz at birth, was the first child conceived through *in vitro* fertilization (IVF). Fertilization "*in vitro*" means "in a glass tube" to distinguish it from fertilization "*in vivo*" which means in the living body. Using the husband's sperm, British doctors Bob Edwards and Patrick Steptoe fertilized an egg in a petri dish, and then implanted the embryo in the uterus of the wife.

By a simple analogy, IVF is like a sophisticated plumbing process for women with blocked fallopian tubes, whose eggs cannot pass from the ovary down into the uterus to be fertilized by the husband's sperm. The solution to overcoming the blockage, at least in theory, is obvious: obtain an egg from the ovary, add the husband's sperm, then pop the fertilized conceptus back into the uterus. With luck, it will stick. Nature does the rest of the hard work in which the tiny fertilized egg grows and multiplies to form a fetus made up of billion of cells, each with its own specialized function.

Though the contribution of human agency through the procedure of IVF in initiating the process is important enough, it cannot bear comparison with the real miracle – the ineffable mysteries of fetal development itself. IVF may seem to have little to do with cloning, but it does point out the fact that an application of cloning can be yet another plumbing process to overcome infertility. In fact, as we will see in a later chapter, almost all current cloning experts were at one point IVF experts. More importantly, whether in IVF or in cloning, the miracle of fetal development is still left to nature.

On the legal front, in a court case that first started in 1971 when General Electric in the U.S. sought a patent for its oil-digesting microbe, the U.S. Supreme Court finally ruled in 1980 that live, human made organisms are patentable materials.

By 1981, Karl Illmensee of the University of Geneva and Peter Hoppe of the Jackson Laboratory in Bar Harbor, Maine claimed to have cloned mice from mouse embryo cells. However, James McGrath and Davor Stolter of the Wistar Institute in Philadelphia immediately disproved the claim the following year. McGrath and Stolter reported they could not repeat the cloning experiment and concluded that once mouse embryo reached the two-cell stage they could not be used for cloning. Other seemed to confirm McGrath and Stolter's findings.

In another important development, while working at Chiron Corporation in 1983, Kary Mullis invented polymerase chain reaction (PCR), which revolutionizes biotechnology by allowing a fragment of genetic code to be identified and reproduced (cloned) indefinitely.

Steen Willadsen of the British Agricultural Research Council cloned a live lamb from immature sheep embryo cells in 1984. This was the first verified cloning of a mammal via nuclear transfer. In the same year, Sir Alec Jeffreys accidentally invented DNA fingerprinting in Leicester, England while studying how genes evolved. A year later Willadsen joined Grenada Genetics to commercially clone cattle. Other scientists later replicated his experiment using cattle, pigs, goats, rabbits and rhesus monkeys, and thus barnyard cloning picked up pace. For example, in 1986, Neal First, Randall Prather and Willard Eyestone cloned a cow from an embryo cell.

University of Utah researcher Mario Capecchi developed a way to create specifically targeted mutations in mice in 1987. These "knockout" mice help researchers understand gene function. In February 1988, a front-page newspaper story featured Grenada's newfound ability to bring factory-like efficiency to animal reproduction – successfully used a modified version of the Briggs and King frog experiment to clone cattle embryos.[143] But animals cloned from embryonic cells contain the genetic material of both parents because the embryos are sexually fertilized. Clones from embryonic cells from the same parents fertilized at different times are as different as brothers and sisters.

In 1994, Neal First succeeded in cloning calves from embryos that had grown to at least 120 cells. In 1995, Ian Wilmut and Keith Campbell repeated First's experiment. They worked with sheep by putting the embryo cell into a

---

[143] Keith Schneider, "Better farm animals duplicated by cloning", *The New York Times*, February 17, 1988.

resting state before transferring their nuclei to sheep eggs. In 1996, Dolly the sheep, the first animal cloned from adult cells, was born. Infigen, Inc. produced Gene, the first cloned cow, from a fetal cell.[144] President Bill Clinton proposed a five-year moratorium on cloning in 1997, while Richard Seed announced his plan to clone a human.

Teruhiko Wakayama and researchers at the University of Hawaii Medical School, in 1997, produced Cumulina, the first mouse cloned from an adult cell. He also developed the so-called Honolulu technique and created generations of genetically identical mice. Researchers at Japan's Kinki University cloned eight calves from a single cow, but only four survived to their first birthday.

In 1998, several notable events took place. The Missyplicity Project, an effort to clone a beloved mutt named Missy, was founded. The project was backed by entrepreneur John Sperling and was based initially at Texas A&M University. Across the Pacific Ocean, the Ishikawa Prefectural Livestock Research Center produced Noto and Kaga, the first cows cloned from adult cells. Genzyme Transgenics Corporation and Tufts University produced Mira, the first goat cloned from an embryonic cell.

The same group of researchers at the University of Hawaii who had cloned Cumulina produced Fibro, the first male clone, in 1999. The mouse was named after the type of cell – a fibroblast, or connective tissue cell – that was taken from the genetic donor. All previous clones of adult mammals had been female. Medical researchers at Kyunghee University Hospital in Seoul succeeded in cloning a human cell from an infertile woman, creating a four-celled embryo that theoretically could have grown into a genetically identical replica of the woman. Because of legal and ethical implications of the work, they ended the experiment without implanting the embryo into a surrogate mother. At the same time that the Korean experiment was going on, scientists at Texas A&M University cloned a calf, Second Chance, from the cell of a 21-year-old Braham. This is believed to be the oldest animal ever cloned.

The new millennium started with a bang in cloning. Researchers at PPL Therapeutics produced Millie, Christa, Alexis, Carrel, and Dotcom, the first pigs cloned from adult cells.[145] A female rhesus monkey – the closest human genetic relative named Tetra, was cloned for the first time. Teams of researchers in Japan and Scotland announced they had cloned pigs. Porcine tissue is likely to be the most compatible with humans and thus porcine is an excellent candidate for growing organs for transplants. Researchers at China's Northwest University of Agriculture, Forestry Science and Technology

---

[144] Infigen, a pioneering cloning company in Wisconsin, laid off all of its employees in 2004.
[145] PPL Therapeutics, the Scottish company that helped clone Dolly the sheep, was dismantled in 2004.

produced "Yuanyuan," the first goat cloned from an adult cell. Researchers at the University of Teramo in Italy cloned the first mouflon – a rare kind of sheep – from an adult cell.

In 2001, Advanced Cell Technologies produced "Noah," making gaurs the first endangered species to be cloned. Researchers at Advanced Cell Technologies produced the first human clone embryo. The plan was to use the embryo to produce embryonic stem cells. However, the embryo stopped dividing before this could be accomplished.

Summer of 2001, Steve Stice pioneered a technique that tripled the success rate for calf cloning, from one in twenty to one in seven. The technique produced a series of cloning successes in the collaboration between University of Georgia at Athens and ProLinia: December 2001, seven Angus breed calves, cloned from skin tissue, were registered; April 22, 2002, an Angus-Hereford cross female calf, KC (for kidney cell), was cloned. In the latter case, genetic material for the clone was taken from a cow's kidney area – a part routinely left with the side of beef in processing – approximately 48 hours after the cow had been slaughtered in a local commercial facility. This was the first calf ever cloned from cells of a slaughtered cow.

As of January 2003, Stice had produced 30 cows.

The first cloned kitten, Cc: (for "copycat") was born on Dec. 22, 2001, but the announcement of her birth was delayed until genetic tests proved Cc: was a clone and it was certain she would survive. Since, there have been a series of other cat clones: Tabouli and Baba Ganoush were born in June 2004, Peaches was born in August 2004; as the first clone delivered to a paying client, Little Nicky was born in December 2004, Little Gizmo was also born in December 2004. Thus begins an era of commercial petory (pet cloning factory).[146]

In 2002, Scott L. Pratt produced three cloned piglets from the skin cell of a commercial hog on June 22. Researchers at France's National Institute for Agricultural Research produced the first rabbits cloned from adult cells. In November 2002 the first rat, Ralph, was cloned. Since then, the cloned rat has produced healthy pups.[147]

The year 2003 was another good cloning year. Researchers at Seoul National University are the first to derive stem cells from human clone embryos. On May 8, Morne de la Rey, a South African scientist, announced South African and Danish scientists had successfully cloned the first animal on the African continent, a Holstein heifer calf born on April 19 at the Embryo

---

[146] "At play with firm's clone kittens", *BBC News*, August 9, 2004.

[147] Zhou, Q., et al., "Generation of fertile cloned rats using controlled timing of oocyte activation", Sciencexpress, published online, doi:10.1126/science.1088313 (2003).

Plus Centre. The calf, weighing 32 kg at birth, has been named Fut, meaning "replica" or "repeat" in Zulu. Fut is a super cow of sort because the nine-year-old DNA donor cow is a former South African milk production record holder, producing some 78 liters of milk per day.[148]

Cloning efforts at Trans Ova Genetics and Advanced Cell Technologies produced the first bantengs – an endangered species – from adult cells. The bantengs were born to cows. The genetic donor had died 23 years earlier, and his skin cells had been preserved in the "Frozen Zoo" at the San Diego Zoo's Center for the Reproduction of Endangered Species. Idaho Gem, the first mule cloned from a mule fetus, was cloned at the University of Idaho.

There was also a breakthrough in cloning technique. Researchers at Aurox LLC developed chromatin transfer (CT) technology, a new cloning technology that involves pre-treating the donor cell to remove molecules associated with cell differentiation. The technology was first used to produce cattle. CT will be the main technique used in pet cloning (cat) later.

Italy's Consortium for Zootechnical Improvement cloned Prometea, the first cloned horse, from a few skin cells. It was born on May 28 in Italy. The name was so given as a form of defiance to the Italian government, which discouraged cloning. Prometea is the mortal of Greek mythology who stole fire from the Gods and gave it to man.[149]

In a collaborative effort between Texas A&M University and ViaGen Inc., Dewey, the first deer (*Odocoileus virginianus*) was cloned from an adult cell. Researchers at France's National Institute of Agricultural Research produced the first rats cloned from adult cells. At the Audubon Center for Research of Endangered Species, researchers produced Ditteaux, the first African wildcat (*Felis silvestris*) cloned from an adult cell.

The year 2004 was a year during which several cats were cloned using the chromatin transfer technology. The cats were cloned for commercial purposes, starting a line of business of pet cloning service or petory.

On August 3, 2005, Snuppy made a splash when it was introduced to the world. Sixteen weeks old at the time of the announcement, Snuppy (for Seoul National University puppy) was the first canine clone. The validity of the claim is still under investigation.[150]

Continuing their cloning work on wildcats, researchers at the Audubon Center for Research of Endangered Species naturally bred unrelated African

---

[148] "South Africa clones 'super cow'", *The Star*, May 8, 2003.
[149] Helen R. Pilcher, "First cloned horse born", *Nature*, August 7, 2003.
[150] "South Korea unveil first dog clone", *BBC News*, August 3, 2005.

wildcat clones, resulting in the birth of African wildcat kittens. This is the first time unrelated clones of a wild species have produced offspring.

The cloning quests continue in 2006, and beyond…

▶ Cloning Hollywood Style

There have been no shortages of novels and movies on cloning.

In Aldous Huxley's *The Brave New World* (1931) universal human happiness had been achieved: control of reproduction, genetic engineering, conditioning, and a perfect pleasure drug called Soma are the cornerstones of the new society.[151] Reproduction has been removed from the womb and performed on the conveyor belt, where reproductive workers tinker with the embryos to produce various forms of human beings: from the super-intelligent Alpha Pluses to the dwarfed semi-moron Epsilons. Each class is conditioned to love "its" type of work and its place in the society. Outside of work, the people spend their lives in constant pleasure: continually buying new things, participating in elaborate sports, and free-floating sex. While uninhibited sex is universal and considered socially constructive, love, marriage and parenthood are obscene…

David Rorvik's book, *In His Image: The cloning of a man*, was published in 1978.[152] The book sparked a worldwide debate on cloning ethics. In the book, J.B. Lippincott Company published as fact a story of an eccentric, wealthy businessman in his twilight years longing for a clone of himself as a son. At the time of publication, readers were immediately intrigued by the book not only because of its subject matter, but also because it was authored by someone seemingly in the know – a former science and medicine reporter of the *Time* magazine.

There are about thirty movies that deal with cloning. We shall only cite a few more popular ones:

The Human Duplicators, also known as *Jaws of the Alien* (1965), is a low-budget flick about an alien who came to Earth and began a sinister plot involving replacements of several world leaders with robot duplicates. It has primitive special effects, but is hilarious nonetheless. *The Boys from Brazil* (1978) is a movie about what would happen if cloning technology fell into the wrong hands? The thriller is based on the premise that the Nazis had relocated to South America and had continued their evil quest for a pure race through cloning. *Blade Runner* (1982) is a flick about human replicants, created by

---

[151] Aldous Huxley, *The Brave New World*, (Harper Perennial, New York, 1989).

[152] David Rorvik, *In His Image: The cloning of a man*, (J.B. Lippincott Company, 1978).

humans to do the dangerous work of colonizing outer space. The replicants later sought to expand beyond their status as tools and ultimately resort to violence against humans. Dinosaur clones appeared in *Jurassic Park* (1993) from a few strains of fossilized prehistoric DNA. The special-effect movie is about an ambitious theme-park owner who hired scientists to clone raptors. *All About Eve* (1993) is an episode in *The X-Files*. Television's best-loved FBI agents (who) were assigned to a mysterious case in which identical girls on opposite coasts each wound up with a dead parent. *Multiplicity* (1996) is a comedy about a contractor unable to juggle the demands of his busy life. A mad scientist made a few extra copies of the contractor to go around helping him with his daily chores...

The premises on which these movies are based can be very flawed. Popular films such as *Multiplicity* feed these public fears by obscuring the fact that cloning cannot instantaneously yield a copy of an existing adult human being. What cloning technology produces are cloned human embryos. The embryos still require the labor- and time-intensive processes of gestation and child rearing to reach adulthood. The now-deposed Saddam Hussein would have to wait 20 years to realize his dream of a perfect army, so would have a basketball franchiser to wait as long for a team of six Michael Jordans. And the Donald Trumps of the world would also have to enlist thousands of women to be the mothers of their clones since the rate of success is still very low.

▶ The Celebrated Dolly, But Second to Computer Chips

With news that Scottish scientists cloned a mammal – Dolly the sheep, the world's first mammal cloned from a cell of an adult animal – the very community that shunned Rorvik's claim is now embracing the new technology and delighting in its simplicity. Dolly the sheep was born on a soft summer night, July 5, 1996, at 5 p.m., but her existence was not revealed to the world until February 1997.

Dolly the sheep is the first viable offspring ever derived from adult mammalian cells. The researchers, Ian Wilmut, Keith Campbell and colleagues of the Roslin Institute, Edinburgh, Scotland removed an unfertilized oocyte (egg cell) from an adult ewe, replaced its nucleus with the nucleus of a mammary gland cell of an adult sheep (the donor). This egg was then implanted in another ewe as surrogate mother. Dolly was the result after months' of development in the womb of the surrogate mother. By February

1997 when Dolly was announced to the world, she was already seven months old and appeared normal in every respect.[153,154,155]

The adult cells with which Dolly was cloned came from the mammary gland of the donor. Thus the clone is appropriately named after the well-endowed country singer Dolly Parton. The surrogate mother was intentionally selected to be a blackface sheep so that it would look different from Dolly. Otherwise, Dolly could very well be its offspring.

Figure 1. Left: Dolly the cloned sheep (left) and her surrogate mother. The surrogate mother is a Scottish blackface sheep. The mother and the baby look different. Right: Ian Wilmut, whose team cloned the celebrated Dolly, was runners-up *Times Magazine* Man of the Year, 1997, after Andy Grove, the co-founder of Intel. (Photo: Courtesy PPL Therapeutics. Photo of Wilmut: adapted from *Times*, December 29, 1997, Man of the Year, Runners up).[156]

Prior to Dolly the sheep, researchers had succeeded in cloning tadpoles and cattle from embryo tissue. Indeed, Ian Wilmut, Keith Campbell and colleagues at the Roslin Institute had only a year earlier before Dolly the sheep raised seven other sheep from oocytes whose nuclei had been replaced with nuclei from either fetal or embryonic tissue.[157] Using a similar process that had proved useful in the earlier experiments with fetal cells, the researchers built on the knowledge to work with adult cells.

The significance of Dolly's birth was evident – it has provided the answer to a sixty-year-old question, the nuclear equivalency and the fantastical experiment proposed by Hans Spemann of whether the genetic material of differentiated cells from adult animals undergo some form of irreversible modification, thus rendering it useless for cloning.

---

[153] Sean Henahan, "Send in the clones", *Access Excellence*, February 24, 1997.

[154] Gina Kolata, Clone: *The road to Dolly and the path ahead*, (Allen Lane, The Penguin Press, London, 1997).

[155] Ian Wilmut, Keith Campbell, and Colin Tudge, *The Second Creations*, (Farrar, Straus and Giroux, 2000).

[156] Hwa A. Lim, *Genetically Yours: Bioinforming, biopharming, and biofarming*, (World Scientific Publishing Co., New Jersey, 2002).

[157] K.H.S. Campbell, et al., *Nature* 380, 1996, pp. 64-66.

Cloning with adult cells had not been successful before the Dolly experiment because the process depends on an experimental protocol which forces the donor cell – the cell providing the nucleus, into a "quiescent" state so that it is not replicating its DNA. This makes the nucleus more susceptible to re-programming by the recipient egg cell.

Incidentally, *In His Image* of 1978, David Rorvik also described fusion techniques whereby the DNA from an adult cell is implanted in the cell of an enucleated egg, and the synchrony problem – fusing the donor cell and an ovum at a key stage of development when the implanted adult DNA can manage the production of all cells.

Cloning with adult cells is still not a well-understood process. It still has a high failure rate. In the Dolly experiment, the researchers produced live births derived from three cell populations: 9-day old embryo, 26-day fetal tissue and 6-year old mammary gland. Two-thirds of the reconstructed embryos did not survive, but eight surrogate ewes did give birth. Only one of these births, Dolly, resulted from adult mammary cell nucleus cloning.

The technique used to clone Dolly the sheep involves several steps. First, the donor cells are grown under special conditions in cultures. This increases the number of cells by several orders of magnitude. Though not performed in the Dolly experiment, it is also possible to make genetic modifications and to select just those cells in which the desired modifications have occurred. The selected cells are then fused with unfertilized eggs. In principle, each egg and the introduced nucleus can lead to the formation of an embryo. The embryos are then transplanted into surrogate female sheep and the lambs are born naturally.

Even months after Dolly's debut, not everyone accepted Dolly's uniqueness and there were still skeptics. In fact, at the 9th International Congress on Isozymes, April 17-19, 1997, San Antonio, Texas for which HAL [158] was a speaker, one of the six Nobel laureates claimed that the experiment was not doable and suspected that contamination could have occurred. The Congress was an occasion to celebrate the eightieth birthday of Clement Markert, a co-inventor of the isozyme concept in 1959. [159,160]

In 1998 some researchers suggested that Dolly's DNA could have come from an adult stem cell or from a fetal stem cell since Dolly's mother was

---

[158] HAL are the initials of the author, Hwa A. Lim. Go to hal_lim@yahoo.com for comments.

[159] Clement Markert, and Freddy Moller, "Multiple forms of enzymes: Tissue, ontogenetic and species-specific patterns", *Proc. Natl. Acad. Sci., USA*, 45, 1959, pp. 753-763.

[160] Guoxiong Xue, Yongbiao Xue, Zhihong Xu, Roger Holmes, Graeme Hammond, and Hwa A. Lim, *Gene Families: Studies of DNA, RNA, enzymes and proteins*, (World Scientific Publishing Co., New Jersey, 2001), 305 pages.

pregnant when the cell for the clone was extracted. But DNA fingerprinting of frozen tissue cleared the argument.[161] The implications of the Dolly experiment are far-reaching. The same method can be used to clone other mammals, including barnyard animals, pets and humans.

Cloning barnyard animals would allow farmers to clone disease-resistant cows and goats. On the negative side, too many clones in a herd can potentially reduce essential genetic diversity to a dangerous level, leaving the animals vulnerable to diseases and unforeseen problems (see the Red Queen hypothesis). On the positive side, cloning livestock can allow the creation of animal herds that can be farmed for milk, blood and organs.

The breakthrough in cloning has excited interest in cloning pets, and extinct species in which a compatible living host can be found. For example, elephants and mastodons are compatible. Mastodons might be cloned using elephants as surrogates to create a real life Jurassic Park. It is also not too farfetched to imagine that one could clone one's dead relatives, or even one's self. Aldous Huxley described such a world in his novel *The Brave New World*.

As significant as the Dolly experiment, in the 1997 Man of the Year of *Time* magazine, Ian Wilmut and Dolly came second after Andrew Steven Grove, the co-founder of Intel Corporation:

> "Fifty years ago this week – shortly after lunch on Dec. 23, 1947 – the Digital Revolution was born...
>
> That Digital Revolution is now transforming the end of this century the way the Industrial Revolution transformed the end of the last one... And in 1997, as the U.S. completed nearly seven years of growth, the microchip has become the dynamo of a new economy marked by low unemployment, negligible inflation and a rationally exuberant stock market...
>
> This has been a year of big stories. The death of Princess Diana tapped a wellspring of modern emotions and highlighted a change in the way we define news. The cloning of an adult sheep raised the specter of science outpacing our moral processing power and had a historic significance that will ripple through the next century. But the story that had the most impact on 1997 was the one that had the most impact throughout this decade: the growth of a new economy, global in scope but brought home in the glad tidings of personal portfolios, that has been propelled by the power of the microchip...

---

[161] John Whitfield, "Obituary: Dolly the sheep", *Nature*, February 18, 2003.

And so *TIME* chooses as its 1997 Man of the Year Andrew Steven Grove, chairman and CEO of Intel, the person most responsible for the amazing growth in the power and innovative potential of microchips..."[162]

▶ Clones of Clones of a Clone

On July 23, 1998 the first reproducible cloning of a mammal from adult cells, which has successfully yielded more than three generations and more than 50 identical cloned mice, was reported by an international team of scientists led by Ryuzo Yanagimachi of the University of Hawaii at Manoa.[163]

The distinctive cloning technology, described as the Honolulu technique, can be more viable for the production of drugs using transgenic animals than earlier techniques because of its efficiency of reproducibility. When used in genetic and embryonic development studies, the clones will shed new light on the cellular and molecular activities involved in aging and diseases. The technology has been licensed to the Hawaii-based biotechnology company ProBio America, Inc., a venture capital company, for commercialization and to test it for expanded uses.

Due to similarities between the developments in mammals, the technique will be applicable to larger animals. For example, efficient and accurate cloning can improve the reliability and safety of reproducing transgenic mammals, such as cattle, pigs and sheep that can be used in the economical production of lower cost protein-based pharmaceuticals. The technique may also be useful for cloning wild or endangered species in a controlled environment.

Figure 2. Scientists at the University of Hawaii cloned more than 50 mice from adult cells, creating multiple generations of identical laboratory animals. (Photo: Courtesy of Peter Morgan).[164]

[162] *Time Magazine*, vol. 150, no. 28, December 29, 1997 / January 5, 1998.
[163] *Nature*, July 23, 1998.
[164] Hwa A. Lim, *Genetically Yours: Bioinforming, biopharming, and biofarming*, (World Scientific Publishing Co., New Jersey, 2002).

Teruhiko Wakayama, then a postdoctoral researcher working in Yanagimachi's laboratory, pioneered the Honolulu cloning technique.[165] The technique uses an injection method and adult cells, making it substantially different from other techniques that generate clones either by injection or fusion of embryonic or fetal cells, or by the fusion of adult cells. Fusion of adult cells is how Dolly the sheep was cloned.

In brevity, the technique involves using a special pipette to microinject a donor nucleus into an egg whose nucleus has been previously removed. The resulting cell is then cultured, placed in a surrogate mouse and allowed to develop. By repeating the procedure, multiple generations of cloned mice that genetically match their sister/parent, sister/grandparents and sister/great grandparent... – essentially identical mice born a generation or more apart, can be created.[166]

In the original groundbreaking experiments, the donor nuclei came from cumulus cells, which are cells surrounding developing ovarian follicle in female mice, and are shed with eggs on ovulation. Within five minutes of the removal of donor nuclei, the researchers inserted the nuclei into developing egg cells (or oocytes) using injection pipettes. The oocytes removed from adult female mice had already undergone the first part of their two-step maturation process.

A relatively high proportion of the oocytes developed into blastocysts and then further developed when a delay of anywhere from one to six hours was introduced between nuclear injection and oocyte activation of the second part of maturation. The exposure after injection of donor nuclei into oocyte cytoplasm, which are rich in factors that promote cell divisions, appears to facilitate the nuclear changes essential for normal development.

The team performed a series of four sets of experiments. In the first set of experiments, they transferred 142 blastocysts into 16 surrogates. Between 8.5 and 11.5 days after nuclear insertion (dpc), the scientists found five live and five dead fetuses.

In the second set of experiments, the team transferred 800 blastocysts into 54 foster mothers. Caesarean sections at 18.5 and 19.5 dpc revealed 17 live fetuses. Of these, 10 survived, six died after delivery and one died a week later. All of the surviving offspring, including the first born named Cumulina on October 3, 1997, grew into adults, were mated and raised normal offspring. Several of the offspring went on to produce the next generation of offspring.

---

[165] Dr. Teruhiko Wakayama is now Head of Laboratory, Laboratory for Genomic Programming Center for Developmental Biology, RIKEN, Kobe, Japan.
[166] "First reproducible cloning of mammals from adult cells reported in July 23 issue of journal Nature", *University of Hawaii press release*, July 22, 1998.

In the third set of experiments, the team attempted to prove genetically that the offspring produced in the studies were clones. By injecting nuclei from agouti mice, whose coats were coffee-colored, into the oocytes of genetically black mice, the team produced mice with the agouti coloring. This experiment verified that the clones were pure agouti mice after further genetic analysis of their placental tissue.

Clones produced in this third set of experiments were themselves used as nucleus donors in a fourth set of experiments. This set of experiments showed, perhaps for the first time in any species, that clones could be cloned from clones.

Overall, Yanagimachi and his team found a high blastocyst implantation rate of up to 71% yielding development rates of fetuses at 5-16% and of full term mice at 2-3%. The experiments clearly confirmed that mammals could be reproducibly cloned from adult cells.

▶ Jerky Bulls and Yang Cows

When it comes to cloning of cows and bulls, Xiangzhong (Jerry) Yang of University of Connecticut Animal Science Department has a number of firsts.

Jerry Yang (nothing to do with the co-founder of Yahoo!, who happens to have the same name), working in collaboration with Japanese scientist Chikara Kubota of the Kogashima Cattle Breeding Development Institute, produced six genetically identical calves using cells extracted from the ear of a prize bull in Japan. The genetic elite was 17 years old and is a Japanese black cattle bull named Kamitakafuku, known for its superior meat quality.

In the experiment, four calves were born in December 1998 from cells cultured for two months, but only two calves survived from this group. Two more calves were born in February 1999 from cell cultured for three months. The four surviving bulls are named Kamitakafuku-1, -2, -3 and -4 in Japanese. Their corresponding anglicized names are Tommy, Andy, Timothy and Anthony (or TATA – a regulatory region of DNA).

The work is significant for it shows that cells of aged animals remain competent for cloning, and that prolonged culture does not affect the cloning competence.

Then on June 10, 1999 a Holstein heifer named Amy was delivered by Caesarean section at the university's Kellogg Dairy Center. She was cloned using an ear skin cell from her genetic mother Aspen. Aspen was about 14 years old and was too old to reproduce naturally. Amy's birth is significant for a few reasons: first, she is the first large animal cloned from genetic material extracted from an adult farm animal in the U.S.; second, she was cloned from the ear skin fibroblast cells of an adult cow, not from reproductive cells; third, extraction of skin cell is less invasive and skin cells may be taken from either sex of the animal.[167]

On July 7, 1999, Daisy was born by cloning cells from, again, the 14-year-old high-merit cow Aspen. Thus Daisy was born "sister" to Amy. On June 3, 2001, Daisy gave birth to Norm. What makes Norm's birth special is that Daisy was cloned from an aged animal (14 years old and passed menopause). The fact that Daisy could reproduce normally demonstrates that cloning of an aged individual does not produce an aged copy.

Judging from the successes, it is expected that clones will be produced with regularity at Yang's facility.

▶ Second Chance – Clone of the Oldest Bull

Mark Westhusin and colleagues at the Texas A&M University successfully cloned an adult bull in September 1999. The 21-year-old Brahman, Chance, died just before his DNA was used to produce Second Chance. It is the oldest mammal ever cloned.[168]

Chance was unable to reproduce naturally because of the removal of both diseased testicles two years before its death. Cloning Chance was the only option for preserving his genetics. The owners of Chance wanted to have their prized bull cloned because of his unusually gentle nature, and they considered the cloning effort a good opportunity to see if an identical copy of Chance might also have such an easy-going disposition.

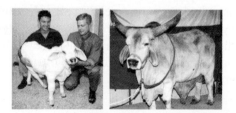

Figure 3. Left: Second Chance, clone of the oldest bull, Jonathan Hill (left) and Mark Westhusin (right). Right: A 1998-photo showing the late bull Chance, a Brahman who died at about 21 years of age. He is the oldest animal ever cloned. (Photo: Courtesy of Texas A&M).[169]

---

[167] "Amy, the first cow cloned in United States", *Bovine Telomere Length Reprogrammed News Release*, June 10, 1999.

[168] "Texas A&M scientists clone first-ever bull", *Texas A&M Press Release*, September 2, 1999.

[169] Hwa A. Lim, *Genetically Yours: Bioinforming, biopharming, and biofarming*, (World Scientific Publishing Co., New Jersey, 2002).

Second Chance is the product of a yearlong project involving 189 attempts, or transferring 189 cells into 189 different eggs, before a pregnancy ended in the delivery of the clone. He displays identical markings as his father and has identical DNA. Because Second Chance came from the oldest animal cloned to date, he has received intensive monitoring and treatment since birth by a team of veterinarians and intensive care technicians at the Texas A&M Large Animal Hospital. Like many previously cloned calves, at birth he displayed some symptoms that resembled those seen in premature human babies.

In the spring of 1999, scientists revealed that the DNA of Dolly, the first cloned sheep, had some characteristics of the older cells that were used to clone her. The caps at her chromosome ends, called telomeres, which get shorter each time a cell divides, were 20% shorter than was normal for a sheep her age. There is therefore considerable interest in keeping track of Second Chance throughout his lifetime because he came from the oldest animal cloned to date.

▶ Artificial Monkey Twins

The first success at creating identical monkey twins was achieved by Gerald Schatten and his colleagues at the Oregon Regional Primate Center in January 2000.[170] Twinning is an alternative method of cloning.

In the artificial twinning process, Schatten and colleagues used a procedure very similar to *in vitro* fertilization (IVF): an egg is taken from the mother monkey and the sperm from the father monkey is used to fertilize the egg. Once the embryo had grown into eight cells, the embryo is divided into four identical embryos consisting of two cells each. Repeating the process numerous times, they created 368 tiny embryos. They did 13 embryo transfers into surrogate mothers, got four pregnancies and one live birth, Tetra, 157 days after implantation.

---

[170] Tom Abate, "First monkey born using new method of cloning", *The San Francisco Chronicle*, January 14, 2000.

Figure 4. Tetra, the singleton rhesus macaque, was cloned using artificial twinning: Tetra was the result of splitting a very early embryo, consisting of only eight cells, into four pieces. These were then nurtured into new embryos. Of the 13 embryos implanted, only one survived. So unlike Dolly, Tetra has both a mother and father, is a clone of neither, but is rather an artificial quadruplet. (Photo; Courtesy of *Science* magazine).[171]

Tetra, from the Greek word for four, is not the first time this technique has been used to create mammalian twins. The same technique has already been used in cattle, a physician also reported using the technique to clone human embryos as far back as 1993. Tetra is not the first monkey cloned either, researchers from the same Oregon research group, Donald Wolf, director of IVF Laboratory at Oregon Health Sciences and colleagues, reported cloning monkeys in March of 1997 using the nuclear transfer method. Unlike the Dolly experiment that produced a genetically identical animal from adult cells, the two monkeys born in August of 1996 and reported in March of 1997 were cloned from embryos.[172]

What makes Tetra unique is that it is the first monkey to have a perfect genetic copy with its sibling, none of which survived unfortunately. Unlike earlier monkey clone experiments, Tetra is the first to possess both identical nuclear and cytoplasmic components. This should be contrasted with the technique used to clone Dolly. Dolly the cloned sheep was created by a method in which the nucleus is scooped out of an adult cell and placed into an egg cell whose nucleus has been previously removed. In that method, a small set of auxiliary genes were left behind in the egg cell, leading to Dolly not quite an exact copy of her cell-donor parent.

Exact genetic copies such as Tetra and sibling offer researchers for the first time the opportunity to produce a line of identical primates for medical research. In other words, the control and the test groups will be exactly the same.

---

[171] Hwa A. Lim, *Genetically Yours: Bioinforming, biopharming, and biofarming*, (World Scientific Publishing Co., New Jersey, 2002).

[172] Sean Henahan, "Tetra the singleton twin monkey", *Access Excellence*, January 14, 2000.

In addition, most medical therapies are currently first tested in mice. Monkeys would be a more reliable animal model in testing new techniques such as gene therapies or growing new organs from stem cells. Since research using human embryonic stem cells is controversial because to produce the cells requires the death of an embryo, genetically identical monkeys may be used as alternatives to develop treatments.

▶ Clone of a Clone – More Bulls

On January 24, 2000, Japanese scientists at the Kagoshima Prefectural Cattle Breeding Development Institute (KPCBDI) reported that they had succeeded in cloning clones of a bull. Though researchers in Hawaii have successfully bred clones of cloned mice, this experiment marks the first time a large cloned animal has itself been cloned.[173]

To create the new calves, skin tissue was taken from the ear of a cloned bull in April 1999 when the animal was four months old. Those cells were fused with an unfertilized egg that had been stripped of its nucleus. The resulting bull calves weighed nearly 100 pounds at birth.

The primary objective of KPCBDI is to produce good cattle consistency. In other words, the Japanese are interested in producing a herd to provide tasty beef. Cloning also reduces the amount of time needed for breeding. The tissues of an animal as young as three months can be used for cloning, while cows do not mate naturally until they are about 14 months old.

Figure 5. These four calves were cloned from the ear cells of the bull standing next to them. These clones are part of an experiment to answer the question of life expectancy of clones and acceptance by diners. (Photo: Courtesy of *The Associated Press*).[174]

---

[173] Shihoko Goto, "Clone of cloned steer bred in Japan", *The Associated Press*, January 24, 2000.
[174] Hwa A. Lim, *Genetically Yours: Bioinforming, biopharming, and biofarming*, (World Scientific Publishing Co., New Jersey, 2002).

▶ Cloned Piglets

In March 2000, PPL Therapeutics, the British company that helped clone Dolly the sheep, announced that it had created the first cloned pigs – animals that can eventually be used to harvest for organs for transplant to human beings.[175] The five cloned piglets were born on March 5, 2000, ushering what can be a new era in cell and organ transplant and an end to the chronic shortages of donors worldwide.[176]

Like Dolly, the five piglets – Millie, Christa, Alexis, Carrel and Dotcom – were cloned from adult cells. The first, Millie, was named after the millennium; Christa, after Barnard Christiaan,[177] the surgeon who performed the first human heart transplant in 1967; Alexis and Carrel after transplant pioneer and Nobel prize winner Alexis Carrel; and Dotcom to reflect the growing use of the Internet.

In performing this cloning project, PPL Therapeutics was eyeing the $6-billion organ market and a similar market for cellular therapies to treat different diseases. Pigs are preferred over other species for xenotransplantation (or animal-to-human transplant) because they can be bred quickly and their organs are the same size as those of humans.

Figure 6. Millie, Christa, Alexis, Carrel, and Dotcom, the five piglets born on March 5, 2000. (Photo: Courtesy of Mike Theiler, *Reuter*).[178]

---

[175] PPL Therapeutics, the Scottish company that helped clone Dolly the sheep, was dismantled in 2004.
[176] Marjorie Miller, "New breed of cloned pigs – organs wanted for humans", *Los Angeles Times*, March 15, 2000.
[177] Dr. Barnard Christiaan died a year and a half after the naming, on September 2, 2001, at age 78, suffering from a fatal asthma attack after going for a swim at a resort in Paphos on the southwest of Cyprus.
[178] Hwa A. Lim, *Genetically Yours: Bioinforming, biopharming, and biofarming*, (World Scientific Publishing Co., New Jersey, 2002).

▶ Reversing Aging

Figure 7. Five of the six Holsteins that started life younger than the cells from which they were cloned. (Photo: Courtesy of Advanced Cell Technology).

In the spring of 2000, news of six calves cloned using a new technique was announced.[179,180] The technique allows the animals to start life biologically younger than the aged cells from which they were derived.

The DNA used to produce the calves was taken from cells of a calf fetus. The donor cells were molecularly aged in the laboratory by allowing them to replicate to the point of exhaustion. We recall that telomeres are caps at the ends of DNA strands that ordinarily wear away each time cells divide. When no further replication is possible, the cell dies. This cellular senescence is linked to age-related disorders, including problems affecting the skin, eyes and internal organs.

In the cloning process, nuclei of the biologically aged cells were then implanted into bovine egg cells from which the nuclei had been removed. Despite having been aged, the cells of the cloned calves appear to be as young as they look: the calves have lengthy telomeres, even though the donor genetic material had been aged to exhaustion.

In other words, although the calves look ordinary, they have cells that appear to be younger than their chronological age, thus raising the tantalizing possibility of a new era in regenerative medicine – any worn-out body part may be replaced as easily as in replacing automobile parts! In regenerative medicine, the goal is to fashion replacement parts from a patient's own rejuvenated cells to overcome the problem of organ shortages and transplant rejection.

---

[179] Carl T. Hall, "New way to clone raises hope for medical miracles", *The San Francisco Chronicle*, April 28, 2000.
[180] *Science*, April 27, 2000.

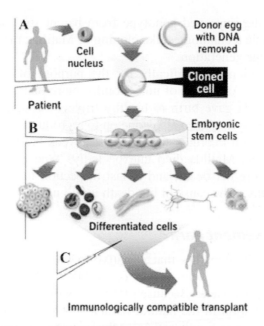

Figure 8. In regenerative medicine, replacement parts of aging tissues come from the patient's own rejuvenated cells: (A) A cell nucleus from a patient is transferred into an enucleated egg cell. (B) Stem cells, the ancestral cells from which all organs and tissues develop, taken from the resulting clone are genetically identical to the patient, but have youthful traits. (C) Stem cells differentiated in the laboratory morph into various genetically matched human cells and tissues for transplantation. (Figure adapted from Advanced Cell Technology, and from Todd Trumbull, *The San Francisco Chronicle*).[181]

▶ Other Notable Cloning Efforts

In April of 2000, the South Australian Research and Development Institute and the University of Adelaide successfully delivered by Caesarean section the first cloned merino sheep named Matilda. Scientists cloned Matilda using the same technique similar to that used to clone Dolly.

The Australian wool industry had been in a perilous state with the production of less marketable lines of wool exceeding demand. Fine wool tinier than 19 micron had been attracting premiums relative to the medium merino wool of 21-23 microns. In the current industry structure based on parent breeding studs, daughter studs and production flocks, it may take up to

---

[181] Hwa A. Lim, *Genetically Yours: Bioinforming, biopharming, and biofarming*, (World Scientific Publishing Co., New Jersey, 2002).

13 years to get the improved genotype from parent stud down to the flock level. With genetic technologies like cloning of Matilda, the time needed may be shortened to one generation.[182]

Matilda was cloned specifically for the purpose of creating a herd to produce fine wool. At the age of nine months, more than a year younger than most sheep, Matilda gave birth to healthy triplets through a speed-breeding technique. The offspring have also gone on to reproduce.

Unfortunately, on February 8, 2003, just a week before the euthanization of Dolly the sheep, Matilda died unexpectedly. She had been particularly sprightly just before her death, and at last inspection that Saturday, she was remarkably healthy. The cause of her death is still unknown.[183]

Jurassic Amusement Park?

The gaur, an ox-like creature that is native to India and Southeast Asia, is usually brown or black with white or yellow socks on each leg and a humplike ridge in its back. Because of excessive hunting and poaching, and because of degradation in the natural habitat – forests, bamboo jungles, and grasslands in India and Southeast Asia, the gaur population has dwindled to about 36,000. The gaur population in Malaysia (seladang in the Malaysian language), for example, has dropped to about 500 individuals and are scattered across several reserves.[184] The gaur is thus an endangered species.

In a new way to save endangered species, in October 2000, scientists at the Massachusetts-based biotechnology company, Advanced Cell Technology (ACT), cloned an Asian gaur and implanted the resulting embryo into a cow. In the cloning process known as cross-species cloning, the scientists took a frozen skin cell from a recently deceased gaur and fused it with an enucleated cow's egg. The DNA of the gaur commandeered the egg to grow into a gaur embryo. The embryo was then implanted into the womb of a domestic cow serving as the surrogate mother. Previously, many scientists thought such cross-species cloning was impossible because the DNA of the animal to be cloned would not be able to interact properly with the rest of the egg cell.[185]

As in other cloning attempts, the technique currently has a low rate of success. Of the several hundred embryos created, only 81 developed to the stage suitable for implantation. Some 42 embryos were implanted into 32

---

[182] "Matilda: Australia's first cloned merino", *SARDI Media Release*, May 3, 2000.

[183] Kim Arlington, "Clone death a mystery", *News.com.au.*

[184] *National Policy on Biological Diversity*, (The Ministry of Science and the Environment, Kuala Lumpur, Malaysia, 1998).

[185] Andrew Pollack, "Cow pregnant with cloned ox", *The New York Times*, October 9, 2000.

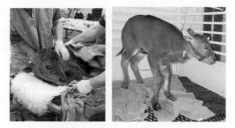

Figure 9. Noah weighs 36 kilos (80 lbs) at birth. Noah exploring his new environment shortly after birth.

surrogate mother cows, but only eight cows became pregnant. Fetuses from two of the eight pregnant cows were extracted for examination, while five other pregnant cows suffered spontaneous abortions, leaving only one viable pregnant cow.

On January 15, 2001, ACT reported that the 80-pound baby gaur had been delivered at 7:30 p.m. on Monday, January 8, 2001. The birth of the baby bull gaur, named Noah, was the first successful birth of a cloned animal that is a member of an endangered species. Collaborating with ACT were Trans Ova Genetics, Genetics Advancement Center in Iowa, Jonathan Hill of Cornell University, and The San Diego Frozen Zoo and the Reproduction of Endangered Species of Captive Ungulates (RESCU) International that supplied the gaur cells.[186]

Unfortunately, while healthy at birth, Noah died within 48 hours from a common type of dysentery not related to cloning. Robert Lanza, vice president of medical and scientific development at ACT, explained Noah died from clostridial enteritis, a bacterial infection that is almost universally fatal in newborn animals.

Jonathan Hill, who was also involved in the 1999 cloning of Second Chance at the University of Texas A&M and now at Cornell University, explained that the first 48 hours is a particularly vulnerable time for newborn animals similar to a premature human baby. Careful observations are needed to determine how well the animal has made the transition from the womb to the outside world. Immediately after Noah's birth, veterinarians and technicians under the direction of Hill intensively monitored the newborn and administered a variety of treatments. Within 12 hours of birth, Noah was able to stand unaided and begin an inquisitive exploration of his new surroundings; at 1 day old, Noah began to exhibit symptoms of a common infection, and

---

[186] Laura DeFrancesco, "Clone of first endangered animal is born and dies", *Bioresearch Online*, January 15, 2001.

succumbed to it another day later despite intensive treatment efforts. His surrogate mother, Bessie the domestic cow, remains healthy.

Philip Damiani, a researcher with ACT, explained that the data collected clearly indicate that cross-species cloning worked. Despite the unexpected loss, Noah's birth brightens the prospects of cloning endangered species on the verge of extinction. Kurt Benirschke, former president of the Zoological Society of San Diego and founder of the Frozen Zoo, expressed optimism that science has advanced to the point of being able to successfully create a healthy trans-species gaur clone.

"We are gratified by the hard work and vision of many people on an effort carried out almost flawlessly," Michael West, president and CEO of ACT, summed it up nicely, "While we set the 'bar' at long-term survival, this was clearly a huge step forward." Though saddened by the news of Noah's death, the Zoological Society is encouraged that scientists are learning to perfect the cloning process and has continued hope for the inevitable role cloning will play in the conservation of endangered species. In fact, the government of Spain has granted ACT the permission to clone the already extinct bucardo mountain goat, using cells collected from the last goat before she died earlier in 2000.

More Calves and Piglets

In the summer of 2001, ProLinia, Inc. and the University of Georgia pioneered a technique that virtually tripled the success rate for calf cloning, and in April of 2002 their scientists became the first to clone a calf from carcass cells after the carcass had been graded. K.C. is the first calf cloned from the cells of a slaughtered adult cow rather than a live one. It was delivered by Caesarian section, April 22, 2002 at 11 a.m.

But K.C. is only the second time any mammal had been created from cells of a dead one. Italian scientists recovered cells from two endangered wild sheep dead for 18 to 24 hours and used them to produce a healthy clone. When two ewes died at a wildlife rescue center in Sardina, the staff sent tissue to Pasqualino Loi of the University of Teramo, Italy. The team substituted the nuclei from cells of the mouflon ewes for nuclei in egg cells from a domestic sheep. Out of the 23 substituted eggs, seven developed enough for transfer into surrogate domestic sheep mothers. In the summer of 2000, one ewe delivered a mouflon lamb. The surviving female clone of the endangered European mouflon sheep, a rare breed of sheep found on Sardinia, Corsica and

Figure 10. The surviving clone of an endangered sheep, grazing by the side of her domestic surrogate mother.

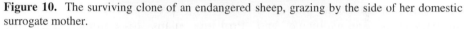

Cyprus, was already seven months old when the news was announced on October 1, 2001.[187,188]

The DNA, or genetic material, used to create K.C. (short for kidney cell), an Angus-Hereford cross sometimes called Black Baldy, was extracted from the kidney area of an adult cow.[189] The cow carcass from which K.C. was cloned went through a cooling process for 48 hours, so the kidney cells were still alive. This was the same cooling process that any side of beef goes through after an animal has been slaughtered.

The live kidney cells were then permitted to grow in a petri dish, and cell nuclei were placed in unfertilized eggs. The next steps were incubating the eggs, culturing them for seven days and transferring them to surrogate mother cows. In total, seven cloned embryos were transferred and one developed to term. The 14% success rate is nearly three times better than the average 5% survival rate for cloned cattle embryos.

The cloning technique will permit cattle producers to select and clone choice beef from their stock. But Randall Prather, a cloning specialist at the University of Missouri, believes it would be difficult to guarantee cloned cow meat quality would be identical to that of a genetic parent. Differences in the quality of feed and other environmental factors could cause variations. Others believe the differences can be overcome and the technology offers the opportunity to produce a more valuable product.

---

[187] Susan Millus, "Rare sheep cloned from dead donor", *Science* 160(16), October 20, 2001..

[188] Loi, P., et al, "Genetic rescue of an endangered mammal by cross-species nuclear transfer using post-mortem somatic cells", *Nature Biotechnology*, 19, 2001, pp. 962-964.

[189] Joyce Howard Price, "Cloned calf means second helping of good steak", *The Washington Times*, April 28, 2002.

Two months later, the University of Georgia (UGA) made news again by producing three healthy cloned piglets using skin cells from a commercial hog.[190] The birth of the piglets is the third in a series of recent cloning successes for UGA and ProLinia, Inc.

Cloning technology promises to provide significant improvements once commercialized within the hog and cattle production industries. Some industry experts estimate that the implementation of pig cloning will save $5 to $15 per pig while providing consumers a consistently superior product.

Smithfield Foods, the world's largest pork producer, has already partnered with ProLinia to implement cloning within a large-scale hog production operation as part of a technology development agreement. The agreement is non-exclusive and ProLinia plans to commercialize the technology with other large-scale producers as well.

More Endangered Species Cloned

In the 1970s, the San Diego Zoo started an effort to put cells and genetic materials from animals in plastic vials for cryo-preservation at −196°C. One of the cryo-preserved tissue samples came from a captive male banteng before it died in 1980. The banteng is a wild species of cow of which there are only about 8,000 individuals in existence, most of them on the island of Java, and some in Malaysia and Myanmar.[191]

The San Diego Zoo sent tissue samples to Advanced Cell Technology. ACT fused skin cells of banteng into cow eggs. Trans Ova Genetics then implanted 30 such eggs into cows. Of the 30 implanted eggs, 16 pregnancies resulted but only two pregnancies led to live births after a normal gestation period of nine months. One was delivered on April 1 at a normal weight of 20 kg and a second two days later with a large calf syndrome at 36 kg. Both were delivered via Caesarean section seven to ten days before term. The second had to be euthanized shortly after birth because of poor health. The first animal is currently thriving and is a proof-of-concept that a cross-species cloning is possible, after the disappointment with Noah in January 2001.

---

[190] Kim Carlyle, "Cloned pigs demonstrate continued commitment to advances in food biotechnology", *ProLinia Press Release*, June 25, 2002.
[191] "Scientists clone long-dead animal", *CBS News*, April 8, 2003.

Figure 11. The banteng clone, delivered in the U.S.A. Banteng is an endangered wild species of cow found in Java.

► Multi-Legged Bio-X

Why have Ian Wilmut and colleagues persisted for so many years in attempts to create a cloned sheep from embryonic, fetal and adult cells? Like in most other cloning efforts of other animals, to a large extent, commercial interests motivate the effort. PPL Therapeutics, a Scottish biotechnology firm interested in genetically altering female animals so they secrete valuable drugs in their milk funded Wilmut's cloning effort.

Multi-Legged Bio-Reactors

Following closely on Dolly's success came Gene, a cloned calf. Perhaps of greater commercial interest, though, was Polly, the first cloned transgenic lamb. Polly was cloned from fetal sheep fibroblast cells that had been genetically engineered to contain the human gene for factor IX protein, a blood-clotting agent used to treat hemophilia. Although not derived from an adult cell, Polly is proof of principle that a viable animal can arise following genetic manipulation of the donor's DNA.

The advent of recombinant DNA technology in the early 1970s gave scientists the means to insert a foreign gene into the genome of a host organism to clone the gene. In 1982, the commercial potential of this technology was first realized when the biotechnology company Genentech brought to market recombinant human insulin. Recombinant human growth hormone, β-interferon and the blockbuster drug erythropoietin (epo) soon followed. These drugs are all produced by the transfer of human genes into bacterial or mammalian cells, which are then grown in large-scale culture systems to produce commercial quantities of the therapeutic proteins. The bacteria function as bioreactors in this large-scale culture system.

Some human proteins that might have value as pharmaceuticals or nutritional supplements are not amenable to economical production in bioreactors, as they require complex modifications that cannot be carried out by bacterial cells. Alternatively, foreign genes can be inserted into mammalian embryos, typically by microinjection, yet it is an inefficient, hit-or-miss technique. This traditional method is to inject multiple copies of a human gene into a newly fertilized egg, but the injected genes rarely integrate properly into the egg's genetic material, producing few drug-producing offspring. This way of creating a transgenic "founder" animal that carries the acquired trait in its germ line so it can pass the trait on to its progeny requires a great deal of time and expense.

Wilmut and colleagues have laid the foundation for a more efficient approach to genetically alter animals to produce milk for pharmaceutical purposes. In cloning, there is a brief period during which researchers can grow donor cells in the laboratory before transplanting the nuclei into eggs. During that time, researchers can manipulate the cells by adding or deleting genes. Only nuclei of cells that have been effectively altered are transplanted to drastically increase the odds of creating a commercially useful animal.[192]

The ability to clone such founder animals to create transgenic herds for large-scale drug production offers the industry a more efficient alternative. This attractive alternative, dubbed pharming, is to transfer these genes into livestock. By expressing the target genes selectively in the mammary glands of lactating goats, sheep or cows, for example, researchers can then purify the desired protein products from the large quantities of milk produced by herds of these transgenic animals.

Whether it is chicken, rabbit, cattle, goat, sheep or pig, the most striking feature of the new animals being concocted is to pluck a gene from one species (in this case, a human) and insert it into the DNA of another species (in this case, one of these animals). These new animals are thus transgenics, and if they are used as a bioreactor for producing milk of pharmaceutical values, they are also called pharm animals.

Pharming is a combination of the old with the new: the barnyard animals and their milk are the old; genetic engineering is the new.

---

[192] Hwa A. Lim, *Genetically Yours: Bioinforming, biopharming, and biofarming*, (World Scientific Publishing Co., New Jersey, 2002).

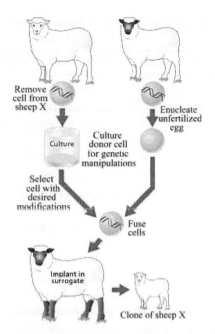

Figure 12. In pharming, the donor cell is multiplied under special conditions in culture. In this way the number of cells can be increased by several orders of magnitude. Genetic manipulations are made and only cells with desired modifications are selected to fuse with an enucleated egg from another sheep with a distinctive phenotype (black-faced). Viable embryos are implanted into surrogate mothers to carry them to terms. Lambs are born naturally or by Caesarean birth. The distinctive phenotype is just a way to assure that the clone is really an offspring of the donor cell, and not of the egg or surrogate mother.

Depending on the purpose and proteins desired, different pharm animals offer different advantages. Different proteins are expressible in different pharm animals, for example:

- Chicken: monoclonal antibodies, lysozyme, growth hormone, insulin, human serum albumin;
- Rabbit: calcitonin, extracellular superoxide dismutase, erythropoietin, growth hormone, insulin-like growth factor 1, interleukin 2, α-glucosidase, glucagons-like peptide;
- Cattle: lactoferrin, α-lactalbumin;
- Goat: antithrombin III, tissue plasminogen activator, monoclonal antibody, α1-antitrypsin, growth hormone;
- Sheep: α1-antitrypsin, factor VIII, factor IX, fibrinogen, insulin-like growth factor 1;
- Pig: Factor VIII, protein C, hemoglobin.

Table 1. A table to compare and contrast different pharm animals. BELE is short for "breed early, lactate early."

	Gestation time (months)	Number of offspring	Time to sexual maturity (months)	First lactation of founder (months)	Annual milk yield (liters)	Raw protein per female (kg/year)
Chicken	0.6	250	6			Up to 0.25
Rabbit	1	8	5	7	4-5 (2-3 lactations)	0.02
Cattle	9	1	16	33	8000-9000	40-80
Goat	5	1-2	8	3-6 in BELE; 18	365 in BELE; 800-1000	4
Sheep	5	1-2	8	18	500	2.5
Pig	4	10	6	16	300 (2 lactations)	1.5

Besides expressing proteins, there are other attributes of pharm animals to consider. For example, the rabbit is gaining a growing recognition as an ideal animal to produce molecules of pharmaceutical interest. Rabbits could be particularly suitable for the production of recombinant proteins in their milk due to their short generation times. A female rabbit produces up to 10 liters of milk a year. Although 20-40 rabbits will be needed to produce the same volume of milk as one sheep, the rabbit gestation time is only one month. It takes only four months for rabbits to reach sexual maturity, giving the first milk production at around six months of age.

Multi-Legged Bio-Facturers

Transgenic animals can also be used to manufacture materials. When manufacturing is achieved biologically, it is called biofacturing.[193]

Spider silk is an excellent example of a material that can be biofactured. Spider silk has some of the best of desirable properties: enormous tensile strength and elasticity, and ultra-lightweight. The tensile is 400,000 pounds per square inch (2,800 MPa), or 10 times stronger than steel or 3.5 times the strength of widely used para aramid fibers. An inch thick of the silk can stop

---

[193] Hwa A. Lim, *Genetically Yours: Bioinforming, biopharming, and biofarming*, (World Scientific Publishing, New Jersey, 2002).

a fighter jet landing on an aircraft carrier. Despite the superior mechanical properties, spider silk is not used commercially because of a shortage of supply.

The basic art of silk production dates back to the earliest weavers that evolved 125 million years ago and the recipe leaves little to be improved upon. The spider has learned to be an excellent weaver, but seems not to have learned to be sociable. Silkworm farming is possible because the silkworm is harmless; the spider is territorial and aggressive, making intensive cultivation impractical. They are cannibals, ruling out large-scale spider farming.

Genetic engineering comes to the aid: if we can engineer the good genes into more sociable animals, we can overcome the hostility of the spider. In other words, by isolating the spider genes that code for silk protein and genetically engineering them into a founder herd or plant, silk can be mass-produced. This is perfectly doable when several spider genes have already been isolated and characterized.[194,195] Bacterial and other fermentation systems work well with certain recombinant proteins; but they are inadequate in producing authentic silk protein. Transgenic animals provide an elegant production method.[196]

A serendipitous discovery indicates that the silk gland of the spider and the milk gland of the goat are almost identical. The goat is prized for its ability to convert plant biomass into high quality proteins in its milk. Nigerian dwarf goats offer certain competitive advantages: they mature and lactate sooner, in about 16 months or in 10-20% less time than other goats and sheep, and twice as fast as cows. In the trade, they are said to breed early and lactate early (BELE). Thus the spider silk proteins can be expressed in Nigerian dwarf goat milk by genetically engineering spider genes into the 70,000 or so goat genes. During lactation, the goat is milked as in conventional dairy farms, and silk proteins are then extracted from the milk.[197]

Biofactured silk finds application in lightweight and flexible body armors, in military and civilian aviation and space vehicles, in high performance sports gear and apparel, and in biodegradable fishing line, medical devices such as suturing and artificial tendons. To put things in proper perspective, each liter

---

[194] R.V. Lewis, "Spider silk: unraveling of a mystery", *Acc. Chem. Res.*, 25, 1992, pp. 392-398.
[195] P.A. Guerette, D.G. Ginzinger, B.H.F. Weber, and J.M. Gosline, "Silk properties determined by gland-specific expression of a spider fibrion gene family", *Science* 272, 1996, pp. 112-115.
[196] J. Scheller, K.H. Gührs, F. Grosse, and U. Conrad, "Production of spider silk proteins in tobacco and potato", *Nature Biotechnology* 19, 2001, pp. 573-577.
[197] F. Vollrath and D.P. Knight, "Liquid crystalline spinning of spider silk", *Nature* 410, 2001, pp. 541-548.

of milk can contain 2 to 10 grams of the protein. It takes a goat a month to produce enough silk for making a lightweight vest of about a pound.

Comparatively speaking, cows are more prolific producers yielding 7,000 to 8,000 liters of milk per year; dwarf goats produce 250 to 500 liters per year. So for large-scale production, longer-to-production but higher-production cows are more ideal.

Nature Versus Nano

Nanotechnology, technology in the nano-scale dimension, has challenged the monopoly enjoyed by the spider. Researchers at the University of Texas at Dallas and Trinity College in Dublin, led by Ray Baughman, have spun nanotubes into fibers as thick as human hair and more than 100 meters long. Carbon nanotubes, discovered in 1991 by NEC Research Fellow Sumio Iijima, are hollow, cylindrical, hexagonal tubes of molecules of carbons that resemble rolled up chicken wire, about 20 billionth of an inch wide and 20 thousandth of inch long.[198]

To make something larger, these nanotubes have to be connected together. Making fibers turns out to be a challenging problem. Spider silk glands contain a solution of silk proteins dissolved in water. As each thread of silk is pulled out, the proteins line up in tandem and interlock, producing nature's toughest material.

To spin fibers from nanotubes, the researchers used an improvement of a technique developed by Philippe Poulin of the French National Center for Scientific Research. The team started with a mixture of nanotubes, water and a special soap to prevent nanotubes from clumping. When the team injected the mixture into a pipe with a flow of polyvinyl alcohol, the soap reacted with the alcohol and the mixture coagulated into gelatinous fibers. The polyvinyl alcohol acted as a glue to hold the nanotubes together as they were spun into fiber and dried. The resulting fiber was 60% nanotubes and is four times as tough as spider silk or 20 times as tough as steel, 17 times tougher than Kevlar, the material used in bulletproof vest.

The researchers also made energy-storing capacitors out of the nanotube fibers to weave them into fabrics. Because electrical resistance changes when the fibers stretch, these fabrics could potentially be used for clothes that could track body movements of athletes. At $500 a gram, the clothes are currently still unaffordable, even if it were bulletproof and could play MP3s.[199]

---

[198] Kenneth Chang, "Those intriguing nanotubes create the toughest fibers known", *The New York Times*, June 17, 2003.
[199] "Nanofiber fabric Unveiled", *Science Magazine*, June 11, 2003.

Multi-Legged Bio-Models

In the laboratory, transgenic animals, such as mice and rats, are invaluable to medical researchers. Scientists can study the function of uncharacterized genes and can evaluate their role in biochemical pathways and in disease processes. By inserting a known pathogenic gene into a laboratory animal, researchers can create animal models of human diseases. These models are powerful and preferred tools for understanding diseases, for identifying novel therapeutic targets and for evaluating the effectiveness of experimental therapies. Based on the work in mice and some of the novel features of the Honolulu cloning technique, it should be possible to rapidly advance the method for use in large animals.

The Honolulu technique was first published in the July 23, 1998 issue of *Nature* by a scientific team led by Ryuzo Yanagimachi of the John A. Burns School of Medicine at the University of Hawaii at Manoa.[200] Yanagimachi's collaborators in the effort included co-authors from the University of Hawaii at Manoa (in Honolulu), the University of Tokyo (Japan), the University of Pavia (Italy), the Jackson Laboratory (Bar Harbor, Maine) and the Babraham Institute (Cambridge, UK).

The technique, which has been licensed to the biotechnology company ProBio America, Inc., provides scientists a valuable tool for the creation of model systems useful for evaluating and controlling the molecular mechanisms that influence the genome and for the study of embryo formation and cell differentiation. The technique may be more useful than earlier techniques because it affords researchers greater ability to manipulate the adult donor nucleus. This ability will have applications industry-wide such as the study of the role genes play in aging and disease processes.

In some cases, rabbits can offer several advantages over other laboratory species such as mice in the study of several human physiological disorders. Because of its larger size, physiological manipulations can be more easily carried out in rabbits than in mice. Furthermore, in evolutionary terms, lagomorphs (rabbits and hares) are closer to primates than rodents (mice).[201]

Rabbits are used extensively in heart-disease research. They are used to make monoclonal antibodies, which are immune system proteins used in diagnostics and occasionally as medical treatments. The rabbit immune system resembles that of the human in many ways, and a large amount of organ-rejection research has also been done in the species. In addition, much

---

[200] T. Wakayama, A.C.F. Perry, M. Zuccotti, K.R. Johnson and R. Yanagimachi, "Full-term development of mice from enucleated oocytes injected with cumulus cell nuclei", *Nature*, 394, 1998, pp. 369-374.

[201] "Animal cloning, rabbits join cloned menagerie", *AgBiotech News*, April 5, 2002.

of the basic research on reproduction that led to successful cloning was done in rabbits.

Multi-Legged Bio-Factories

The prospects for whole organ transplants are improving. When a regular pig organ is attached to a primate or a person, the host immune system will destroy it within hours. In recent years, scientists have genetically engineered pigs whose cells display antigens, a genetic flag that the human immune system recognizes as belonging to the body. Organs from these partially humanized animals when transplanted into baboons have survived up to eight weeks. Alternatively, they have been used externally to sustain transplant patients for brief periods of time while they are waiting for organ transplant.[202]

In 2002 alone, in the U.S., there were:

- ☐ 12,800 organ donor (deceased or living),
- ☐ 24,900 life-saving transplants,
- ☐ 88,242 patients still on the waiting list at the end of the year,
- ☐ 6,439 people who died while waiting for a transplant.

These numbers do not take into account the estimated 100,000 potential candidates who die before being placed on a waiting list.[203] In total, the cost of all organ replacement therapies in the U.S. is estimated to exceed $100 billion per year.[204]

Organ demand is a major health care issue that is growing in magnitude. Over the past 10 years, while organ donations have increased, the waiting list has grown even more:

- ☐ In 1992, there were 28,952 patients on the transplant list and 7,092 donors,
- ☐ In 1996, there were 49,381 patients on the transplant list and 9,172 donors,
- ☐ In 2001, there were 81,528 patients on the transplant list and 12,607 donors.

Despite organ donation education campaigns, the rate of donations has been greatly outstripped by the increase in need. Tissue and organ failure is clearly a serious problem that will only increase as the population grows and ages.

---

[202] Geoffrey Cowley, "The new animal farm", *Newsweek*, April 2, 2001.

[203] Bruno Gridelli, and Guiseppe Remuzzi, "Strategies for making more organs available for transplantation", *The New England Journal of Medicine*, 343, 2000, pp. 404-410.

[204] Robert Langer, and Joseph P. Vacanti, "Tissue engineering", *Science*, 260, 1993, pp. 920-926.

When animals are used for organ harvesting, they function as organ shops or tissue shops and are biofactories, just like car spare parts are manufactured in auto factories.[205] Pigs are thought to be the best for xenotransplant – cross-species transplant – because of the similarity between their organs to human organs in size and function. Not surprisingly, pigs are sometimes called "horizontal humans." So do not feel too insulted when someone calls you a pig!

•••••　　•••••　　•••••

If cloning techniques have been successfully used in the barnyard, there is no reason why they should not be transferred to clone pets, and humans for that matter.

---

[205] Hwa A. Lim, *Change: In business, corporate governance, education, scandals, technology, and warfare*, (EN Publishing, Santa Clara, 2003).

8 Petory and Biofactory

"I'll be back."

- Arnold Schwarzenegger, speaking to a pet clone shop salesman
in the movie *The Sixth Day*

▶ PerPETually Yours

By refining the cloning techniques that have successfully cloned sheep, mice, pigs, and monkeys, it is conceivable that the same techniques may be applied to the cloning of other species, particularly, pets. This has created a potentially huge market for petory (or pet factory) or pet clone shop. This being the case, pets will be genetically yours and perpetually yours (or perpetually owners).[206]

Pet owners started preserving genetic materials soon after Dolly the sheep. The method to store genetic materials is commonly known as "gene banking." It is a rather straightforward procedure: a cloning kit is sent to the pet's veterinarian to conduct a simple biopsy on the animal's belly or neck. Samples from live animals are preferred, but at Genetic Savings and Clone, about 40% of the samples are from animals that have been dead for less than a week. The extracted tissue, about the size of a pencil eraser, is sent to a laboratory to be grown in culture dishes. The results are cryo-preserved in liquid nitrogen to wait for science to catch up to fantasy – from an owner's perspective, or science to catch up with fiction – from a science writer's perspective.[207]

---

[206] Hwa A. Lim, *Genetically Yours: Bioinforming, biopharming, and biofarming*, (World Scientific Publishing Co., New Jersey, 2002).

[207] David W. Chen, "Pet cloning is a boon for some, raises hackles for others", *The New York Times*, December 28, 2000.

Owners' waits have been rewarded: the first cloned pet, a feline – Copy Cat or Cc: – was born on December 22, 2001 at Texas A&M University, and on March 29, 2002, successful clones of rabbits were announced by a French group. It may not be long before canines and other family pets are cloned with regularity.

Pet cloning still requires relatively deep pockets: the cost to extract the genetic material of a pet is about $1,000 to $3,000, and it costs $100 per annum for storage fees. It is also expected to initially cost $200,000 per animal clone before dropping to a more affordable $10,000 in several years. Despite the costs, pet owners have already frozen samples from an estimated 500 to 1,000 animals, not only dogs and cats, but also rabbits and gerbils, at more than a half-dozen companies and clinics around the U.S.

K-9 Cloning

Canine sperm banks, once used chiefly to breed show dogs, are freezing more samples from the more ordinary family purebreds and mutts. Some pet owners claim cloning or artificial insemination is the best, most heartfelt tribute to a beloved family member.

Mark E. Westhusin is a veterinarian at University of Texas A&M involved in the Missyplicity Project. The Missyplicity Project is a multi-year effort to produce the first cloned dog. The anonymous sponsors – who for years were rumored and now confirmed to be John Sperling and his spouse – invested $2.3 million to partially fund the project to produce a clone of their pet dog, Missy, a mixed breed border collie. Sperling made his millions as the founder of University of Phoenix and chairman of the Apollo Group, Inc., the parent company of University of Phoenix. According to *The Wall Street Journal*, the 81-year-old financier Sperling wants to charge wealthy pet owners to clone their animals. He was quoted by *Reuter* news agency as saying that he would also like to see cloning used for socially useful animals such as rescue dogs.[208]

The Missyplicity Project has a team of about 20 researchers who are using the knowledge gained from cloning Second Chance to advance that research.[209] In attempting to clone Missy, Westhusin's team uses the same protocol used to create Dolly the sheep. But before the protocol can be applied, there is a set of species-specific problems that need to be resolved. A dog has a bursa, or pouch, encasing their ovaries making egg harvesting more difficult. Dogs release ova randomly, once every 6 to 12 months. In

---

[208] "First pet clone is a cat", *BBC News*, Friday, February 15, 2002.
[209] Charles Graeber, "How much is that doggy in the vitro?" *Wired* Magazine, March, 2000.

Figure 1. Missy, a mixed breed border collie, the dog that the Missyplicity Project is trying to clone.

comparison, sheep ovulate regularly, once every 19 days. Species-specific issues also explain why cats are easier to clone than dogs.

To improve the chances of success of the Missyplicity Project, sixty bitches are serving as egg machines. Collected eggs are retrofitted with Missy's DNA, cultured *in vitro*, and if any embryo is viable, it will be surgically implanted into the oviduct of a surrogate mother dog for gestation. If everything goes perfectly, Missy II will debut in 63 days. To date, the project has yet to lead to a perceptible heartbeat.

Genetic Savings & Clone (GSC) of Sausalito, California started funding the Texas A&M project in response to interest generated by the Missyplicity Project. The company had spent years since it was founded in 2000 and more than $19 million in its attempts to clone a dog. Mark Westhusin, after years of trying, realized that the reproductive biology of dogs makes them a nightmare to clone and just "quit." In 2004, GSC opened a lab in Madison, Wisconsin, with 50 employees. But, so far, no dogs have been cloned.

Dogs present a number of challenges to researchers. They have such an unusual reproductive biology, far more so than humans that the methods that allowed cloning of sheep, mice, cows, goats, pigs, rabbits, cats, a mule, a horse and rats simply do not work with them. Ovulation in dogs is once or twice a year, but not predictable, and no one has found a way to induce ovulation by giving dogs hormones.

With other animals, scientists collect mature eggs from ovaries, but the eggs dogs ovulate are immature. They mature in the oviduct and so far it has proved impossible to extract eggs from a dog's ovary and mature them in the laboratory. So researchers will need to pinpoint when to pluck a mature egg from the oviduct, and will need surgery to retrieve it, instead of the kind of needle suctioning used in other animals.

Snuppy

Through trial and error, a South Korean team, headed by Woo-suk Hwang, had discovered a signature spike in the hormone progesterone that signaled ovulation and they had cloned a dog.[210]

The group worked for nearly three years, seven days a week, 365 days a year and used 1,095 eggs from 123 dogs before finally succeeding with the birth of a cloned male Afghan hound on April 24, 2005. The surrogate mother was a yellow Labrador retriever.

Woo-suk Hwang, the principal author of the dog cloning paper, wrote that the puppy, an identical twin of the adult Afghan but born years later, was delivered by Caesarean section. The pregnancy lasted a normal 60 days and the newborn puppy weighed 1 pound 3.4 ounces. The Korean scientists named the puppy Snuppy, for Seoul National University puppy.

In cloning Snuppy, Hwang's team first plucked a mature egg from the oviduct of an ovulating bitch. The next step in cloning of any other animal is to replace the egg's nucleus with that of an adult and let the cloned embryo grow in the lab for several days.

But no one has been able to grow dog embryos in the lab. So the South Koreans quickly started the cloning. They removed the genetic material from the eggs and replaced it with skin cells from the ears of Afghan hounds. When the altered eggs started to develop into embryos, the researchers anesthetized a female dog, slipped the eggs into the animal's oviduct, and hoped the eggs would grow into early embryos, drift into the uterus, and survive. They found they had less than four hours after starting the process to get the eggs into the female dogs.

Figure 2. Snuppy (middle) is an identical twin of the adult male Afghan hound (left). He was born on April 24, 2005 to a surrogate mother, a Labrador retriever.

---

[210] "S Korea unveils first dog clone", *BBC News*, August 3, 2005.

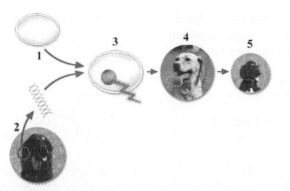

Figure 3. Cloning of Snuppy: 1) A mature egg is enucleated; 2) DNA from the car skin cell of the 3-year-old donor is extracted; 3) the enucleated egg and the DNA are fused together and cultured to trigger embryo growth; 4) the growing embryo is implanted into the Labrador surrogate mother; 5) the Afghan puppy was born after a normal gestation of 60 days.

Ordinarily, researchers give hormones to female animals that are to serve as surrogate mothers, preparing them to become pregnant with a cloned embryo. This is not so with dogs since no one knows how to prepare a dog for pregnancy. To overcome this, the researchers used the same dogs for egg donors and for surrogate mothers, 123 dogs in all.

In the end, three pregnancies resulted. One ended in a miscarriage, one was carried to term but the puppy died a few weeks later of respiratory failure, and one resulted in Snuppy, a clone of the three-year-old male Afghan hound. The Afghan Snuppy appeared on the cover of *Time* magazine, which declared the feat of Hwang's team as the most amazing invention of the year 2005.

Until dog cloning becomes a lot more efficient, few people will be able to afford to clone their pets. Lou Hawthorne, CEO of GSC, estimated that it would cost more than $1 million to repeat what the South Koreans have done.

Snuppy is the second coup in 2005 for the Seoul researchers. In May 2005, Dr. Hwang's lab announced that it had created cloned human embryos and extracted stem cells from them. The dog project is separate. Hwang explained that the canine cloning pursuit is not for pet cloning. Rather, the goal is to use dogs to study the causes and treatment of human diseases. Dogs have long been used to study human diseases. Rabies, in fact, was first discovered in dogs; insulin was discovered in dogs; and the first open heart surgery was performed on dogs. Eventually, the team hopes to make dog embryonic stem cells and test them in the animals as treatments.

In December 2005, Dr. Hwang's stem cell research breakthroughs were called into question. Subsequently, the dog clone is now under validation. If the dog clone turns out to be a fake, then Snuppy will just become a big snoopy investigation into how real science is being conducted. As of date, as far as we know, Snuppy is a clone.

Dupli-Cat

In an advance that takes cloning out of the barnyard and into the living room, researchers have cloned a cat. The female domestic shorthair is called "Cc:" for "copycat" or "carbon copy." It was born on December 22, 2001 and is healthy and frisky. This is the first reported success in cloning cats.[211]

The work was an offshoot of the Missyplicity Project, the $3.7-million (original funding level) effort to clone a mixed-breed pet dog named Missy mentioned in the preceding subchapter. The cloning attempt used two types of cells from adult cats. In a first, failed experiment, the researchers attempted to insert DNA derived from the cheek cells of a male cat into donor eggs. When that failed, they tried using cumulus cells from a female cat. Cumulus cells are found near the ova of female cats. This approach was much more efficient.

The kitty clone was the only success in 188 tries to produce viable embryos, and transfer of 87 cloned embryos into eight female cats, including Allie, Cc:'s surrogate mother. Sixty-six days later, Cc: was delivered by Caesarean section. This makes the overall success rate comparable to that seen in other cloned species such as sheep, cattle, goats, pigs and mice. If these odds can be improved and Cc: remains in good health, the possibility of pet cloning and concomitant ethical problems could be just a whisker away.[212]

Figure 4. Cc: the Copy Cat, first cat ever cloned, was born on December 22, 2001, and her tabby surrogate mother. (Photo: Courtesy of *Nature*).

---

[211] Malcolm Ritter, "Big payoff for a pet project", *The Associated Press*, February 14, 2002.
[212] "Scientist clone 'carbon copy' cat", *CBC News*, February 15, 2002.

Cc:'s coat color suggests that she is a clone. A genetic match between Cc: and the donor mother confirms that. She does not, however, look identical to her DNA donor. The calico kitten differs from its genetic donor in its color pattern. The pattern on cats' coats is only partly genetically determined. It also depends on other factors during development. Apart from difference in appearance, pet owners should realize a new clone does not come with a ready-made bond to the owner or carry other memories.

Cc: Genetic Anomaly

The cloning of Cc: was greeted with joy and criticisms, but the public was mystified. Cc: is a white and tiger-tabby domestic shorthair, whereas her genetic donor Rainbow is a calico domestic cat. If Cc: is a clone of Rainbow, shouldn't Cc: be a calico too? The answer has to do with a genetic anomaly unique to calico cats, a fascinating and less-than-fully-understood issue called X-linked inactivation.

Calicos are almost always female, which means they have two X-chromosomes, versus the male's XY. One of these X chromosomes contains a gene for orange coat color and the other contains a gene for black coat color; white patches are specified by a different set of genes that are not relevant for our discussion.

For reasons that are still not fully understood, as the embryo develops, a phenomenon called X-linked inactivation occurs, in which one or the other X-chromosome in every cell in the Calico embryo becomes randomly inactivated. If the specific X-chromosome containing the gene for orange coat color becomes inactivated, that cell, if it eventually becomes a coat follicle cell, will go on to produce black coat color; the inverse is true if the X-chromosome containing the gene for black coat color becomes inactivated.[213]

Figure 5. Cc: (left) is not an exact "carbon copy" of her genetic donor, Rainbow (right). (Photo: courtesy of *Nature*).

---

[213] Laura Carrel, and Huntington F. Willard, "X-inactivation profile reveals extensive variability in X-linked gene expression in females", *Nature*, 434, March 17, 2005, pp. 400-404.

The question now is if the inactivation is random, why do we not get a very fine distribution of orange and black hairs within the coat? For reasons beyond the scope of our discussion here, the inactivation occurs in larger patches of orange and black.

"Mosaicism" is the term for distribution of different cell types – of which the X-linked inactivated cells are one – within a single organism. Mosaicism is three-dimensional, meaning that the inactivation of orange or black-producing genes occurs within cells throughout the calico's body regardless of whether the cells have anything to do with production of the animal's coat. Thus, even the specific cumulus cell used to clone Cc: would have been inactivated for either orange or black coat color.

If the nuclear transfer process were able to reset the inactivated X-chromosome the way it resets the nuclear differentiation, then one might expect to see a calico clone with a calico coat; if nuclear transfer does not reset X-inactivation then one would expect to see a clone with a black coat if the donor cell used had an orange coat gene on the inactivated X-chromosome, and conversely one would expect a clone with an orange coat if the donor cell used had an black coat gene on the inactivated X-chromosome.

The fact that Cc: has no orange in her coat means that nuclear transfer does not reset X-inactivation, and that the cumulus cell used had an orange coat gene on the inactivated X-chromosome.

Commercial Grade Pet Cloning

Genetic Savings & Clone (GSC), based in Sausalito, just north of San Francisco, is the world's first firm to go commercial and offer the public the chance to clone their cats and dogs. In October 2003, GSC announced that it had obtained an exclusive license to use the chromatin transfer (CT) technique, developed by the Connecticut-based Aurox LLC, for pet cloning. Briefly, CT involves pre-treating the cell of the animal to be cloned to remove molecules associated with cell differentiation. CT has been shown to have a higher success rate than the conventional nuclear transfer (NT) technique, and by October 2003, Aurox LLC had used it to produce over 50 healthy calves.

Tabouli and Baba Ganoush, born in June 2004, were the first kittens cloned using the CT technique. They are clones of Tahini, then a one-year-old female Bengal cat, and live in the San Francisco Bay Area.

Figure 6. Tabouli and Baba Ganoush (left), the first kittens cloned using the chromatin transfer technique, are clones of Tahini (right). (Photos: Courtesy of Genetic Savings & Clone).

Following Tabouli and Baba Ganoush is Peaches, who was born in August 2004. The donor is Mango, two years older than Peaches. Peaches was the last cat GSC cloned before shifting to the production of cat clones for paying clients.

After the success of Tabouli and Baba Ganoush, five cat owners signed up at a cost of $50,000 to have their pets copied. The first pet cat clone for a paying client is Little Nicky who was born on October 17, 2004 in Austin, Texas. The genetic donor is a Maine Coon named Nicky.

Figure 7. Peaches (left) is a clone of Mango (right), who is two years older. (Photos: Courtesy of Genetic Savings & Clone).

Figure 8. The first cat cloned for a paying client. Little Nicky (left) was cloned from Nicky's genetic material. Nicky had died a year earlier in November 2003, at age seventeen. (Photos: Courtesy of Genetic Savings & Clone).

Figure 9. The second cat cloned for a paying client. Little Gizmo (left) was cloned from Gizmo (right). Gizmo died at age thirteen. (Photos: Courtesy of Genetic Savings & Clone).

Little Gizmo, also born in Austin, Texas, but was produced two months later than Little Nicky in December 2004. She is a clone of Gizmo, a Siamese mixed breed cat who died at age 13 in March 2004.

Note that unlike Cc: who was cloned using the nuclear transfer (NT) technique, all these cats were cloned using the CT technique. Cc:'s coloring and disposition are different from those of her genetic mother; these cats show uncanny likeness to their genetic donors.

CT is safer and more efficient than traditional methods, including the NT technique. The efficiency and safety of the technique were verified when Aurox LLC used it to produce over 50 healthy calves. Now with the first commercially cloned cats – Little Nicky and Little Gizmo – coming off the production line, CT has been truly validated as a commercial grade (from a clone service provider's perspective) or a consumer grade (from the pet owner's perspective) cloning technology. The age of commercial petory has begun.

In a reversal of the Missyplicity Project, in which the original goal was to clone a dog (Missy) but a cat (Cc:) was cloned first, work is being perfected to clone the world's first dog using the CT technique, after having successfully cloned numerous cats using the technique. Several hundred customers are already on the books banking their dogs' DNA with the company. According to Lou Hawthorne's estimate, it would cost more than $1 million to repeat what the South Koreans have done to produce Snuppy using the NT technique. This makes dog cloning very unaffordable. Hawthorne is confident that, instead of days of raining cats and dogs, the days of cloning cats and dogs will come pretty soon.[214]

Cloning Rabbitly

Lagomorphs are rabbits and hares. Magicians may be able to pull rabbits out of hats, but it takes more than a hat and a magician to clone a rabbit. On

---

[214] "At play with firm's clone kittens", *BBC News*, August 9, 2004.

March 29, 2002, a team headed by Jean-Paul Renard of INRA (Institut National de la Recherché Agronomique), France, announced the successful cloning of the rabbit.[215]

Rabbits have been notoriously resistant to attempts at somatic cloning. The French team overcame the technical limitations by paying particular attention to the species-specific characteristics of oocyte (egg) physiology and early embryonic development. They had to shorten the time used for oocyte activation and delay the embryo transfer to improve implantation efficiencies.

The team began by using established somatic methods for cloning other animals. However, they found that the rapid kinetics of the cell cycle of rabbit embryos and the narrow window of time for their implantation into foster recipients meant that they had to devise an alternative cloning method. In order to combat the narrow time frame available for successful implantation, the team used foster mothers that were asynchronous with the donor animals. The best success rate was achieved in foster mothers that were mated with vasectomized males 22 hours behind the donor females.[216]

The efficiency of rabbit cloning was low. Of 371 reconstructed embryos implanted into 27 asynchronous foster mothers, 10 became pregnant, and only 4 females delivered a total of six kits. Of these six animals, only four survived to weaning. The remaining animals, all female New Zealand rabbits about a year old at the time of the announcement, appear to be healthy and normal and have been grown to maturity. Two of the rabbits, when mated naturally, have given birth to healthy offspring. The addition of rabbits to the cloned menagerie suggests that any mammalian species can probably be cloned if the physiology and oocyte cell cycle are carefully considered.

Although the main goal of the INRA research is pharming – using farm animals as drug factories – we group the successful cloning effort under pet cloning since rabbits are pets for many.

A Stubborn Clone

Mules result from breeding a male donkey (62 chromosomes) with a female horse (64 chromosomes) to produce an animal with 63 chromosomes. Thus mules are sterile, and because mules are sterile, prized mules cannot be bred.

A mule cloning effort, partly motivated and financed by Don Jacklin, president of the American Mule Racing Association, successfully produced the

---

[215] "Rabbits join the cloning club", *BBC News*, March 29, 2002.
[216] Chesné P, Adenot PG, Viglietta C, *et al.*, "Cloned rabbits produced by nuclear transfer from adult somatic cells", *Nature Biotechnology*, 20, April 2002, pp. 366-369.

first equine cloned foal, Idaho Gem. Idaho Gem was born at 3:05am on May 4, 2003 in the University of Idaho, Moscow (yes, the city of Moscow in the State of Idaho, USA). The cloning team consisted of Gordon Woods and Dirk Vanderwall, both animal and veterinary science experts at the University of Idaho, and Ken White, animal and cloning expert at Utah State University.[217]

In the experiment, a donor fibroblast (skin) cell line was established from a 45-day mule fetus. Of 334 manipulated oocytes, 305 were transferred to recipient mares, resulting in 21 14-day pregnancies. Of these, only the embryos maintained and activated in medium with elevated calcium concentrations established pregnancies through day 45, resulting in the birth of the live male mule after 346 days of gestation. The foal, born naturally, weighed 49 kg at birth. The surrogate mother is a paint mare named Idaho Syringa. Idaho Gem himself is a full brother of Don Jacklin's fleet-footed famous racing mule, Taz.

Despite several efforts around the world, equine have proved particularly tricky to clone using the cloning technique to clone Dolly. For example, cloning of equine has eluded Robert Godke of Louisiana State University in Baton Rouge. Wood's team eventually succeeded by boosting calcium levels inside the eggs and embryos, a trick that is believed to trigger the embryos to divide and grow.

Woods was gratified by the culmination of five years' research, but the success rate is still very low.

Figure 10. Idaho Gem, first equine cloned successfully. (Photo: Courtesy of P. Schofield, University of Idaho).

---

[217] Gordon L. Woods, Kenneth L. White, Dirk K. Vanderwall, Guang Peng Li, Kenneth I. Aston, Thomas D. Bunch, Lora N. Meerdo, and Barry J. Pate, "A mule cloned from fetal cells by nuclear transfer", *Science*, 301(5636), pp. 1063.

Cloned Equine Twin of Surrogate Mother

Previous celebrity clones including Dolly and Idaho Gem received their DNA from a donor but were raised in the wombs of unrelated surrogate mothers. The first cloned foal born identical to her surrogate mother was delivered on May 28, 2003. This makes the foal the first horse to be cloned and the first clone to be carried by her twin mother.

The Haflinger foal, Prometea, was cloned in the laboratory by fusing an adult skin cell and an empty egg from a female horse, then returning the resulting embryo to the same female's womb after a few days. Cesare Galli of the Laboratory of Reproductive Technologies in Cremona, Italy led the cloning effort.[218]

Cloning of horses has been tricky because the breeding seasons are short and eggs are hard to come by. Galli's team cultured skin cells from the Haflinger mare and fused them with enucleated oocytes. Of the more than 300 embryos created, only 14 were viable after being cultured for seven days. Two of the viable embryos were implanted into the same mare from which the cell line was derived. The remaining embryos were implanted into other surrogate mares. The only embryo that survived to term was Prometea. The team put eight-day-old embryos into the mare five days after ovulation. This is believed to have helped the embryos to develop.[219]

The approach used to clone Prometea is unusual because some researchers thought it was impossible for a mother to give birth to her identical twin since in normal pregnancies, the maternal immune system recognizes the fetus as foreign and produces certain immune proteins. These proteins are thought to sustain the pregnancy. Thus Prometea comes as a surprise because it challenges the idea that for an embryo to survive, it needs to be recognized as different by the mother's immune system.

Figure 11. Prometea, the first cloned horse, with her genetically identical surrogate mother. (Photo: Courtesy of C. Galli, Laboratory of Reproductive Technologies, Cremona, Italy).

---

[218] C. Galli had worked with Ian Wilmut, the Scottish scientist who created Dolly in 1997.
[219] C. Galli, et al, "A cloned horse born to its dam twin", *Nature*, 424, 2003, pp. 635.

► Ethically Yours

Regardless of whether the science works, ethicists and animal welfare groups are already wary of the concomitant moral implications and ethical issues. In November of 2000, a year before Cc (1 BCc) or 4 years after Dolly (4 AD), Arthur Caplan, director of the Center for Bioethics at the University of Pennsylvania, made a poll about cloning at Duke University and at a private high school in suburban Philadelphia. Although very few of the polled audience supported cloning humans, roughly 90% of them favored cloning pets.

Pet owners may not realize that an animal's personality is more a function of its environment and its experiences (nurture) than of its genes (nature): any clone, even if it is raised in the same family, will have a different life. There is no reason to believe that clones of pets will be the reincarnated version of the original. Identical twins – natural clones – may look very much alike and even act in very similar ways as each other, but they are not the same people. What cloning does is to make identical twins that live serially instead of "simultaneously in time." Cloning Rin Tin Tin will give us Rin Tin Tin II, rather than Rin Tin Tin again!

Many clients learn about cloning through the Internet. Companies promoting pet cloning may be preying on the vulnerability of pet owners who have lost a pet and are too attached to think rationally. They can barely contain their excitement because pets are emotional ballasts providing unconditional love in an uncertain world. Some people believe pet cloning companies are selling a fantasy that they cannot possibly fulfill: the notion that an old friend will rise from the dead.

There are many unknowns related to cloning. Critics fear it would demean the individuality of a pet to know that its DNA is in the freezer, ready to grow into a replacement, to be reproduced, whenever and wherever desired. Cloning is inefficient: it produces many stillborns or deformed animals for each live cow, sheep or goat, and now cats and rabbits. Given the enormously high rate of miscarriages and birth defects, would a pet owner who really loves the pet want to subject that pet's genetic twin to such travail?

Critics also argue that we already have too many cats and dogs. For example, the world's dog population numbers some 400 million, divided into about 400 breeds. Producing them through cloning "goes in the opposite direction of where we need to be." So for people who would like a new version of a deceased cat, many communities have too many cats for too few homes, and adoption is probably a better solution.

On a more positive side, cloning offers a variety of benefits. In the U.S. for example, 5 million or so cats and dogs are destroyed at shelters each year;

cloning would produce a single pup or kitten instead of a litter, owners could spay and neuter their pets, as vets forcefully recommend, and still breed their favorite animals.

In a small twist on a familiar argument, efforts at pet cloning will also bear fruit for human health. Pet cloning can serve as a test bed for human cloning. If proved safe in pets, the practice will likely accustom the public to cloning as a form of human reproduction. Pet cloning could help research that uses for example, cats for learning about human diseases; but others voice concern and suggest moving away from using animals in research.

As for cloning endangered or extinct species, Kent Redford, director of biodiversity analysis and coordination at the Wildlife Conservation Society in New York said the product would be more like an amusement park version of the species rather than the wild species. He believes species should be preserved in their natural environments. The biggest problem endangered species face is habitat destruction.

Odd-Inarily Yours

After Prometea's birth, two other mule clones were born in June and July of 2003.

When Galli and colleagues cloned the first bull in 1999, they named the bull Galileo.[220] As expected, the Italian health minister, Rosy Bindi, a fervent Catholic, opposed the effort and claimed the researchers had violated a 1998 decree forbidding cloning. Galileo and the father, a famous bull named Zoldo, were confiscated – a fate very reminiscent of Galileo Galilei, the 17th century astronomer who was imprisoned for his "heresy." Much to the chagrin of the health minister, Galli and colleagues eventually won their case because Galileo had been cloned before the new law was enacted.

Prometea was so named by the Italians as a sort of continuing joke of naming clones after mythological figures who defied authority. Prometea is named after Prometeo, the Titan in Greek mythology who rebelled against Zeus and stole fire from Mt. Olympus.

After Idaho Gem and Prometea, equine regulatory bodies are now seriously assessing the implications of their birth. Geldings (neutered males) comprise 50% of champion horses, and 90% of big champions are castrated horses. For example, in the 1998 Badminton show jumping competition, 117 of the 121 entries were geldings. Many were gelded young before their capacities were realized to calm their natural friskiness.

---

[220] C. Galli, R. Duchi, R.M. Moor, and G. Lazzari, "Mammalian leukocytes contain all the genetic information necessary for the development of a new individual", *Cloning*, 1, (1999), pp. 161-170.

Figure 12. Galileo, Italy's first cloned bull. Galileo was cloned using a white blood cell from Zoldo, in the background. (Photo: Courtesy of C. Galli, Laboratory of Reproductive Technologies, Cremona, Italy).

Though sterile, a champion gelding can be turned into a champion stallion using cloning. In fact, a French biotechnology company, Cryozootech, has already collected cells from about 30 champion horses and is freezing the cells for clients who want to clone champion geldings. The company charges $7,500 for the removal, culture and ten-year cryo-preserving of equine tissue samples.

Currently, cloned horses are automatically barred from horse racing because of an existing ban on horses bred using artificial insemination. John Maxse, director of public affairs at the Jockey Club, which regulates the racing industry, saw little prospect of lifting the ban because it would risk undermining the whole industry for the sake of satisfying curiosity, and the whole industry would end up dominated by clones of the top champions. He stressed the very essence of any sport was dependent on there being some doubt as to the outcome, the fact that against all odds an underdog can triumph.

Though clones are banned from horse racing, there are no corresponding restrictions in all other horse sports, including show jumping, dressage, eventing, and polo. It is conceivable that a polo team could consist of only identical clones of a champion horse!

●●●●●   ●●●●●   ●●●●●

To conclude this chapter, I will summarize some of the best-known clones. The list is by no means exhaustive:

Table 1. Clones produced from adult cells and the number of attempts to produce the clones.

Date	Name	Species	Attempts	Remarks
Jul 5, 1996	Dolly	Sheep	277 Scotland	Nuclear transfer from a cell of the mammary g and of a mature six-year-old ewe. Euthanized at age six on Feb 15, 2003 for progressive lung disease.
Dec 9, 1998	8 cows	Cow	99 150 Japan	Cumulus cells added to 99 cow eggs, 47 took up the cumulus cells. Eighteen of the resulting embryos survived in the laboratory for eight to nine days until they were ready to be transferred to surrogate mothers. With the Fallopian tube cells, out of 150 eggs, 94 took up the tube cells, and ended up with 20 embryos. 10 of the 38 embryos were transferred to surrogate cows. Four of the eight calves died at or soon after birth; the others survived and appeared normal.
Aug 9, 1999	Second Chance	Cow	189 USA	Cloned from oldest animal cell. 139 attempts (i.e., transferring 189 cells into 189 eggs), 25 embryo transfers, and 6 pregnancies.
Jun 3, 1999	Fibro	Mouse	274 Hawaii	Cloned from cells from the tip of a donor male tail using the Honolulu technique. Fibro was the first clone of male mice clones. Three male offspring from tail-tip cells. Two died shortly after birth. The survivor, dubbed Fibro because the cultured tail-tip cells resemble fibroblast.
Mar 14, 2000	Millie, Christa, Alexis, Carrel, and Dotcom	Pig	72 USA	Five piglets cloned from a body cell of an adult female pig after 72 double nuclear transfers. 3 are identical clones from one burse cell, and the other 2 are identical clones from another. All delivered by Caesarean section, weighed about 2.72 lbs, or 25% less than piglets from a natural mating.
Mar 2000	1 lamb	Mouflon sheep	Italy	Out of 23 substituted eggs, seven developed enough to be transferred into surrogate mother. One ewe delivered
Apr 2000	Matilda	Sheep	Australia	The first cloned sheep outside of the Roslin Institute.

Table 1. (*Continued*)

Date	Name	Species	Attempts	Remarks
Jun 22, 2000	Yangyang	Goat	China	Sister of the first goat ever cloned from adult cells. The first cloned goat, Yuanyuan, died within 36 hours after birth. Both were cloned from cell from adult goat.
Jan 8, 2001	Noah	Gaur	692 USA	692 attempts to create embryos from the cells of its "father." Only 30 were sufficiently robust to be transferred into surrogate mothers, giving 8 pregnancies and resulting in 1 live birth.
Dec 22, 2001	Cc:	Cat	188 USA	188 attempts to produce viable embryos from cumulus cells of adult female, 87 transfers of cloned embryos into eight female cats. Cc: was the only delivered by Caesarean section.
Mar 29, 2002	3 rabbits	Rabbits	371 France	371 reconstructed embryos implanted into 27 asynchronous foster mothers, 10 became pregnant, and only 4 females delivered a total of six kits.
Apr 1, 2003	1 wild cow	Banteng	30 U.S.	30 embryos implanted leading to two live births, of which one had to be euthanized.
May 4, 2003	Idaho Gem	Mule	334 U.S.	From a mule fetal skin cell. Of 334 manipulated oocytes, 305 were transferred to recipient mares, resulting in 21 14-day pregnancies. Only one was carried to the full term of 246 days and was born naturally.
May 28, 2003	Prometea	Horse	300 Italy	Out of more than 300 embryos created, only 14 were viable. Two were implanted into the same mare from which the cell line was derived. One survived to term.
June 2004	Tabouli, Baba Ganoush	Cats	USA	The first two cats cloned using the chromatin transfer (CT) technique.
October 2004	Little Nicky	Cat	USA	The first cat cloned for a paying client, using the CT technique.
April 24, 2005	Snuppy	Dog	1095 Korea	The first dog cloned, using nuclear transfer (NT) technique. 1,095 eggs were collected from 123 dogs.

I have gone into length to discuss the clonology, the cloning of notable barnyard animals, pets and other animals to bring out several important aspects of cloning:

First, these animals were cloned for their special attributes. In some cases they were cloned because of their potential market value such as consistency of meat (cattle), for organ extraction (pigs), or for pharming (cows); in some cases they were cloned because they were pets (cats), champions (horses); in other cases they were cloned because they were sterile (mules), extinct (mastodons) or endangered species (bantengs).

Second, cloning is still very inefficient and the rate of success is very low. It produces many stillborns or deformed animals for each live birth.

Third, species-specific characteristics of oocyte (egg) physiology and early embryonic development are different for each species. In cloning mules for example, a critical success factor is regulating the calcium level of the developing embryo.

Fourth, using a safer and more efficient "commercial" grade cloning technique, commercial cloning of pets using the CT technique began in the year 2004.

Let us keep these aspects: cloned for attributes, the low success rate and species-specific issues in mind when we begin to discuss human cloning.

9 Cloning Techniques

"If a superior individual – presumably, then, genotype – is identified, why not copy it directly, rather than suffer all of the risks, including [the] sex determination, involved in the disruptions of recombination (sexual procreation). Leave sexual reproduction for experimental purposes: when a suitable type is ascertained take care to maintain it by clonal propagation."

- Joshua Lederberg, 1958 Physiology or Medicine Nobel Laureate, 1996

► Twins – Twice the Fun

In many animals, each pregnancy produces a litter of offspring. In humans multiple births are rather uncommon: 1 in 90 births produces twins, 1 in 8,000 births produces triplets, and 1 in a million births produces quadruplets... The odds get more unfavorable as the number of multiplet increases. Though uncommon, multiplets have been a subject of human fascination for centuries. Recently, because of advances with genetics, they have also become ideal subjects for medical studies.

To understand multiplets better, let us take a closer look at twins. There are about 100 million twins in the world. Human twins come in two types:

Nonidentical Twins – Siblings of the Same Age

In a fertile woman, the ovary normally releases a single egg during menstruation. Occasionally, two eggs may be produced at the same time. If both eggs are fertilized, of course each by a different sperm, nonidentical twins result. Nonidentical twins are called dizygote twins, meaning from two zygotes (zygote is just a fancy name for embryo). They are also commonly called fraternal twins.

Genetically, nonidentical twins share 50% of the same genes, which is the same proportion of genes shared by all brothers and sisters. Nonidentical

179

twins are simply siblings that happen to be of the same age. As in other siblings, nonidentical twins may be of the same sex, or they can be of different sexes.

The frequency of nonidentical twins varies among human racial groups. In North American Caucasians, nonidentical twins are born at a rate of 7 in 1,000 pregnancies; in Asia, the rate is 3 per 1,000 pregnancies; the highest rate is among Nigerians, particularly the Yoruba tribe, for whom 40 in 1,000 pregnancies end in nonidentical twins. The mother's conditions influence the chances as well: from 18 to the age of 37, older women are more likely to produce nonidentical twins, but after the age of 37, the rate of twinning declines. Apparently, the age of the father has no effect on the probability of twinning.

Heredity also plays a role for producing nonidentical twins. The trait tends to run in families. For example, genealogical records show female twins produce twins at a rate of about 17 per 1,000 pregnancies, higher than the average of 7 per 1,000 pregnancies among Caucasians. Male twins however do not have more twin offspring than other fathers. Sisters of twins are more likely to have twins. These facts indicate that the tendency to produce nonidentical twins is genetic, which is only expressed in females. Producing twins is not inherited as a simple genetic trait, but rather appear to be multifactorial, involving multiple genes and environmental factors.

The statistics for having multiple births are now being influenced by work habits and intake of fertility enhancement therapies such as fertility drugs. For nonidentical twins to arise, the mother's ovaries must release two eggs in the same menstrual cycle. Thus twinning is a trait of the mother, not of the twins. Any factors, such as fertility drugs, that stimulate the release of more than one egg will increase a woman's chances of producing multiplets. In a recent report, overall the number of twin births rose more than 50% (from 68,339 to 104,137), and triplet and higher births rose 404% (from 1,377 to 6,737).[221]

Identical Twins – Natural Clones

Identical twins (or we may say twinplicity) arise from a single egg fertilized by a single sperm, initially producing one embryo. The embryo later divides producing two embryos that develop into two babies. Identical twins are called monozygote twins, meaning from a single zygote. Because they originate from a single egg and sperm, these twins are genetically identical,

---

[221] "Trends in twin and triple births: 1980-1997", vol. 47, no. 24.20, National Center for Health Statistics, Public Health Services, 99-1120, September 14, 1999.

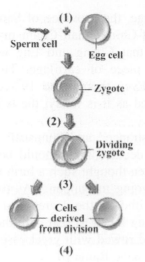

Figure 1. In natural identical twinning: (1) A sperm fertilizes an egg to form an embryo (zygote). (2) The embryo spontaneously divides into two embryos. (3). The two embryos separate. (4) Each embryo contains identical genetic material. The end result of this spontaneous embryo splitting is a pair of identical twins.

sharing 100% of the same genes. One consequence is that they are always the same sex. The splitting of the original embryo appears to be largely a chance event, which is not affected by the mother's age. In most ethnic groups this event occurs with a frequency of 4 twin pairs per 1,000 pregnancies.

Mirror Image and Conjoined Twins

Mirror image twins are also identical twins. They are identical twins created when a fertilized egg splits quite late, around days 9-12, any later the twins can be conjoined. Like identical twins, mirror image twins are genetically identical; but unlike identical twins, they have small mirror image differences: some parts of the body can be mirrored, but not the appendix or heart or other internal organs as far as we know. Some of the mirror image differences include: opposite handedness, mirror image dental problems, cowlicks (tuft of hair at fringe) are on opposite sides, or the opposite foot is larger.

When a fertilized egg splits rather late – later than 9-12 days after fertilization – the twins can be conjoined, that is, two human beings joined at birth by living tissue. Chang Bunker and Eng Bunker (May 11, 1811–January 17, 1874) were the twin brothers whose condition and birthplace became the basis for the term Siamese twins (conjoined twins). They were born in Siam (now Thailand), in Melange, the province of Samutsongkram, to a Chinese father (Ti-eye) and a half-Chinese/half-Cham mother (Nok). The unusual thing about the twins is that Chang and Eng were joined at the sternum (breastbone) by a small piece of cartilage. Their livers were fused but independently complete. Even though the 19th-century surgical technology was not nearly as advanced as it is today, the twins could have easily been separated.

When news of the "strange" and "unusual" birth reached the king of Siam, King Rama II, he decided they should be put to death. A lack of scientific understanding then thought such a birth was an evil sign – an omen that something bad was going to happen. As time passed and no disaster occurred, King Rama II withdrew the decree of death for the two boys. As young boys, Chang and Eng loved to fish with their fisherman father. They learned to use the oars and rowed with great ease. At the age of 16, Chang and Eng were presented to King Rama III.

In 1829, they were discovered in Siam by Captain Coffin and British merchant Robert Hunter, who made plans to exhibit them as a curiosity during a world tour. Upon termination of their contract with their "discoverers," they successfully went into business for themselves. In 1839, while visiting Wilkesboro, North Carolina with P.T. Barnum – of the Barnum American Museum on Broadway in New York City – the twins were attracted to the town and settled there, becoming naturalized United States citizens and adopting the name "Bunker."

The brothers settled on a plantation and were accepted as respected members of the community. On April 13, 1843, they married two Yates sisters: Chang to Adelaide and Eng to Sarah Anne. Chang and his wife had ten children; Eng and his wife had 11.

Figure 2. Chang and Eng, conjoined twins born in Siam, shown with their families.

As times grew harder and the number of children increased, the wives squabbled and eventually two separate households were set up, separated by less than a mile. The wives lived apart, with Chang and Eng sharing three days a week with each and the children. This arrangement would last for the rest of their lives.

After many childhood and adult illnesses, including a stroke suffered by Chang, the twins shared their lives together until one cold day in 1874, Eng woke up to find his brother cold. Chang was dead; Eng started to sweat and feel faint, dying a short time later on the same day.

Interest in Twins

The following is a list of names that would be linguistically logical for naming twins. This is not a list of names that sound alike (such as Ann and Andy), but a list of names that are connected through other properties such as palindromes, anagrams: Aidan and Nadia; Amy and May; Dolly and Lloyd; Ira and Ria; Johan and Jonah; Leon and Noel; Mary and Myra; Reva and Vera...[222]

The interest in studying twins is obvious. Twins can be used to estimate the importance of genes in determining the variation of traits. As an example, consider a complicated disease such as schizophrenia, which is a severe mental disorder. The study may begin by studying a number of twins for the presence of schizophrenia. If one member of the twin has schizophrenia, but the other member does not, the twins are discordant; if both of the twins have schizophrenia then they are concordant. From the study, we can calculate the percentages of identical and nonidentical twins that are concordant.

Since identical twins are genetically identical whereas nonidentical twins share only 50% of the same genes, if genes are important in determining the trait, we would expect greater similarity or higher concordance in the identical twins. Schizophrenia shows such a concordance and therefore the disease is genetic.

There is a caveat, however. The important thing about concordance is that it is the difference in the concordance of identical and nonidentical twins that indicate a genetic component to the trait. Many people misinterpret concordance studies, erroneously concluding that high concordance in identical twins alone signal the importance of genes. This is incorrect for the identical twins usually share the same environment – the same home, the same parents, the same friends, and so on – as well as the same genes. Therefore high concordance in identical twins might be due to their identical genes or it might result from their similarity in environment.

---

[222] The pairs of names are adapted from Mike Campbell, *Behind the Name: The etymology and history of first names, logical names for twins*, http://www.behindthename.com/twins.html.

Thus high concordance by itself in identical twins tells us nothing. If similar environments produce the high concordance observed in identical twins as the high concordance observed in nonidentical twins, then the trait is likely to be environmental.

To facilitate genetic investigations of twins, a number of twin registries, such as the Mid Atlantic Twin Registry, the Australian Twin Registry, and the Swedish Twin Registry, have been created.

Identical Twins are More Identical Than Clones

A clone and its progenitor have identical genes, so have a pair of identical twins. But clones are "serial twins" (twin *seriatim*) for they are born during different births; identical twins are "simultaneous clones" for they are born during the same birth.

According to Nancy L. Segal, director of the Twin Studies Center at University of California, Fullerton, as a group, identical twins are in fact more similar to each other in personality than fraternal twins or ordinary siblings: identical twins develop from an egg while fraternal twins develop from two eggs; clones and siblings are both serial. This observation shows a clear influence of genes.[223]

Identical twins are strikingly similar in many ways, but they are not completely alike. For all the differences, identical twins are more identical to each other than clones will ever be. But even identical twins are influenced by nongenetic factors – starting from the womb and extending to parental influence, influence of friends, opportunities in life, chance occurrence – that influence all of us. Since a clone and its progenitor would be born into different families at different times, these nongenetic factors could be expected to be more powerful.[224]

There are also surprising genetic differences in identical twins. A multigenic disease such as schizophrenia is clearly genetic, but if an identical twin is schizophrenic, the other only has a 45-50% chance of being schizophrenic.

Robert Plomin of the Institute of Psychiatry in London observed that in general, the correlation between identical twins for phenotypes from strong to weak are: height, IQ, weight, and personality. We emphasize that these findings are population averages and cannot predict similarities between individual clones and their progenitors.

---

[223] Nancy L. Segal, *Entwined Lives: Twins and what they tell us about human behavior*, (Dutton, 1999).
[224] "Expert: Twins more identical than clones", *CNN News*, December 30, 2002.

Figure 3A. HAL's identical twin nephews: Ooi Tec Ling and Ooi Tee Ping, delivered by Caesarean birth on March 23, 1994. Ling was delivered first. Shown here at three months old.

Figure 3B. The same identical twins at four months old.

Figure 3C. The identical twins at three years old. Note that they are beginning to show some observable differences.

Figure 3D. The identical twins at four years old. Note that the "younger" Ping is more extrovert while the "older" Ling is not only slightly smaller in size, but also more shy.

Figure 3E. The identical twins at nine years old. Note that the "younger" Ping is taller by about an inch and is about 2.5 pounds heavier.

The identical twins shown in the figures are HAL's nephews, from whom HAL learns quite a lot about twins. The names in Chinese are: Ooi Tee Ling (黃迪淋) and Ooi Tee Ping (黃迪彬). Note the family name (黃) is bilaterally symmetrical, the first given name is the same (迪), second given names (淋 and 彬) are bilaterally reversed.[225]

They do look alike, so alike that Uncle HAL, teachers and friends mix them up frequently. But there are clearly discernable differences:

---

[225] HAL are the initials of the author, Hwa A. Lim. Please go to hal_lim@yahoo.com or hal@dtrends.com for comments.

Table 1. Statistics of the identical twins OTL and OTP (date of birth: March 23, 1994) taken on March 25, 2003, and on March 2, 2006. Scholastically, the younger brother (OTP) is slightly stronger; the parents believe that the older (OTL) is actually slightly stronger, but slightly more careless. A cowlick is a knot in the hair: one has one and the other has two. There are no discernable differences in their fingerprints. (Published with permission from the parents).

Statistics	OTL	OTP	OTL	OTP
	2003		2006	
Height	121.5cm	124.2cm	141cm	145cm
Intelligence	-	+		
Weight	26kgs	27kgs	40kgs	42kgs
Cowlick	1	2		
Color Preference			Blue	Red
Meal Preference			Noodle	Rice

At age nine, Ping, "the younger," was about an inch taller and about 2 pounds heavier. Ping, being more outgoing, was also more popular among friends and girls! As they grew older, they also began to have different preferences in color and food. In early 2006, they were diagnosed to have a minor form of color blindness: they have problem recognizing a small patch of green color (with a little shade of brown), which they identify as red; but they have no problem telling a larger patch of green (with a little shade of brown). The good news is they have no problem telling the colors on traffic lights. In school, at age 12, they were both in extra active group and needed extra attention from teachers. Ping, however, was still more outgoing of the two.

So they are identical twins, but not exact replicas!

▶ En Route to Commercialization

A clone is almost an exact copy – or almost a genetic replica. Cloning refers to the generation of multiple copies of a gene, a cell or an entire organism. Whereas gene cloning and the manipulation and transfer of genetic material within and across species have been a scientific reality since the early 1970s with the pioneering work of Paul Berg, Stanley Cohen, and Annie C.Y. Chang of Stanford University, Herbert Boyer and Robert B. Helling of the University of California at San Francisco, cloning complete organisms, and in particular mammals, has long been a topic of science fiction thrillers, shrouded in futuristic mystique. Since the 1980s, experts such as Steen Willadsen and Neal First succeeded in cloning barnyard animals using embryonic cells.

With the arrival of Dolly, the first mammal cloned using an adult cell in 1996, cloning took a huge step forward. After Dolly were Polly and Gene – a lamb and a cloned calf. The future is here, and researchers have begun to explore the potential scientific and commercial applications of the technology. With each cloning success scientists seek to duplicate their findings and refine their techniques, further defining the molecular mechanisms of replication, gene activation and development involved in the cloning process. In cloning Dolly, Ian Wilmut, Keith Campbell and colleagues at the Roslin Institute, Scotland produced 277 genetically reprogrammed eggs, but only one survived and developed into a viable lamb. For cloning to realize its commercial potential, the technique had to become more efficient – veritably a standard laboratory procedure.

Not surprising, there are alternative ways to clone, each with its advantages and disadvantages. There are basically four ways to clone mammals: artificial twinning, the Roslin technique, the Honolulu technique and the chromatin transfer technique.

Artificial Twinning Technique

In the preceding subchapter we discussed natural twins as natural clones. Twins can also be obtained artificially, meaning with external intervention.

In this technique, an egg from a mother and a sperm from a father are used to create a fertilized embryo. After the embryo grows into eight cells, researchers split the embryo into four identical embryos, each containing two cells. The four embryos are then implanted into four different surrogate mothers. In effect, the single embryo becomes four embryos, all genetically identical.

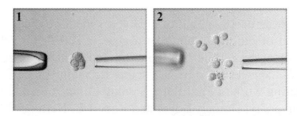

Figure 4A. Like in natural twinning, artificial twinning begins with a sperm fertilizing an egg. (1) The embryo has developed into an eight-cell embryo. (2). The embryo is split into individual cells.

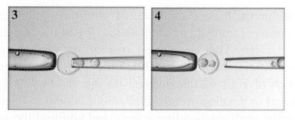

Figure 4B. (3) Two cells are injected into an empty egg. (4) One of the resultant four embryos, complete with two cells.

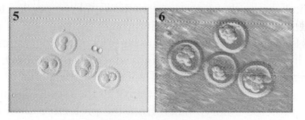

Figure 4C. (5) The resultant four embryos, each with two cells. (6) The resultant four embryos further develop, gaining more cells as they divide. If necessary, twinning may be repeated.

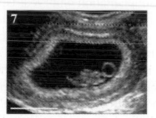

Figure 4D. (7) A sonogram showing a successful artificially twinned embryo growing into a fetus.[226]

As the name stipulates, this is a form of twinning, the only difference being the twinning is achieved not naturally, but rather artificially from external manipulation.

Clones obtained using this method are genetically identical. This is important because researchers will have identical twins for a variety of studies. Just like studies involving identical twins, the control and the test groups will be genetically identical.

As we will see below, in nuclear transfer technique, the nucleus is scooped out of an adult cell and placed into an enucleated (hollowed out) egg cell. In

---

[226] Hwa A. Lim, *Genetically Yours: Bioinforming, biopharming, and biofarming*, (World Scientific Publishing Co., New Jersey, 2002).

the nuclear transfer method, a small set of auxiliary genes are left behind, leading to the clone not quite an exact genetic copy of the nuclear donor!

In retrospect, Jerry Hall performed one of the first reported series of experiments on artificial twinning in 1991. Like most experiments of the time, the original work of Hall was performed on mice. The first task faced by them was the need to develop an artificial zona pellucida ("clear zone") – the acellular protective shell surrounding a fertilized egg and the subsequent pre-implantation embryo. The zona plays an important role during the processes of fertilization and embryo transport through the fallopian tube. Earlier experiments had shown that a pre-implantation embryo needed to be enclosed within a zona pellucida during its transit through the oviduct. If the zona were absent, by dissolving experimentally using enzymes for example, an early embryo were likely to come apart or might adhere to the oviduct, rendering the embryo failing to reach the uterus. Hence the zona pellucida functions to sequester a growing embryo during its transit to the uterus. Upon reaching the uterus, the sequestered embryo hatch from the protective shell and adhere to the uterine lining, implanting itself into the uterus to establish a pregnancy.[227]

Hall managed to create artificial zona using alginate (a jelly-like substance) to surround pre-implantation embryos. Upon successful experiments with mice, they turned their attention to human triploid embryos that had been fertilized by two sperm. Since these embryos contained an extra set of chromosomes (and thus triploid instead of diploid), they would not develop normally and would miscarry when returned to the uterus. In fact 25% of spontaneous abortions in pregnancies arise from triploid embryos.

Finally, in October 1993, Robert Stillman, a physician at the George Washington University Medical Center fertility clinic quietly announced at the American Fertility Society meeting in Montreal that a team headed by his colleague Jerry Hall had cloned human embryos using the technique of artificial twinning.[228] Hall worked with surplus embryos from an IVF clinic. The seventeen embryos from the IVF clinic were genetically abnormal and would not have survived because they have each been penetrated by more than one sperm. Hall divided the embryos and instead of inserting the cell clusters into emptied zonae of human embryos, he coated them with a gel to form an artificial sac. He then allowed the resulting forty-eight embryos to cleave for six days in culture dishes before terminating the experiment. Hall's eventual

---

[227] J.L. Hall, and S. Yee, "Implantation of zona-free mouse embryos encased in an artificial zona pellucida", 47th Annual Meeting of the American Fertility Society, Oct 21-24, 1991, Orlando, Florida.
[228] J.L. Hall, D. Engel, P.R. Gindoff, G.L. Mottla, R.J. Stillman, "Experimental cloning of human polyploid embryos using artificial zona pellucida", Conjoint Meeting of the American Fertility Society and the Canadian Fertility and Andrology Society, Oct 11-14, 1993, Montreal, Quebec, Canada.

goal was to perfect the cloning technique to give IVF practitioners a way of manufacturing more embryos, that is, identical ones, from a couple's zygotes. More embryos available for transfer would increase the success rate of IVF.[229]

Hall and Stillman were wrong in their reassurance that their experiment was a modest scientific advance that might someday prove useful for treating certain types of infertility. When the news broke, everyone focused on the one thing the scientists seemed willing to overlook: they were manipulating not with plants, pigs or rabbits, but with human beings!

Nuclear Transfer Technique

Hans Spemann first explored the nuclear transfer technique in 1928. By 1938 he theorized that by fusing an embryo with an egg cell, higher animals could be cloned. This technique is currently commonly used for cloning adult animals.

The nuclear transfer technique requires two cells: a donor and an egg cell. The egg cell works best if it is unfertilized. Otherwise, it may not accept the donor nucleus as its own. The donor cell must be forced into Gap Zero or G0 cell stage, a dormant phase. This dormant phase is critical because it causes the donor cell to shut down but not die. This is a form of synchronization process so that the donor cell is ready for transplanting into the egg cell.

The egg cell is enucleated or hollowed of its nucleus to eliminate most, not all, of its genetic material. This is because most of the DNA is in the nucleus of a cell, but not all. A small part of about 60 genes of the human cell's DNA is carried independently of the nucleus in structures called mitochondria. Mitochondria in cells are responsible for the energy metabolism that cells need to stay alive.

In the nuclear transfer technique, the nuclear DNA is transplanted from the donor cell to the hollowed out egg cell, thus substituting for the egg's nucleus with the donor nucleus. The egg is then prompted to begin to form an embryo. If an embryo results, it is implanted into a surrogate mother to carry to term.

In this new embryo, the mitochondrial DNA of the egg remains and very little, if any, mitochondrial DNA comes from the donor. This embryo thus contains most of its genetic information in the nucleus from the adult donor cell, and a small amount of possibly conflicting information in the egg's original mitochondrion. Thus, clones arising from nuclear transfer technique are not exact clones of their donor parent because of the residual egg mitochondrial DNA.

---

[229] Robert Cooke, "Experiments shows possibility of cloning human beings", *Newsday,* 113(52), October 26, 1993, pp. 3.

The Roslin technique, the Honolulu technique and the chromatin transfer technique are just refined variations of the nuclear transfer technique.

The Roslin Technique

Ian Wilmut, Keith Campbell and colleagues at the Roslin Institute pioneered this technique in 1996 to clone Dolly the sheep. In the Roslin technique, an appropriate adult cell is selected from the donor, the mammal to be cloned. The cell is then allowed to divide to form a culture *in vitro*. This produces multiple copies of the same nucleus, and this step is important only if the donor nucleus has been genetically modified. In this case, multiple copies are required so that a number of genetic modifications can be performed and studied to make sure the modifications have taken effect.

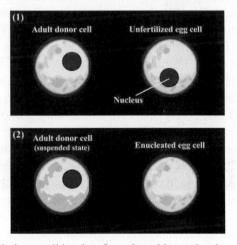

Figure 5A. (1) An adult donor cell is taken from the subject to be cloned. (2) The donor cell is starved into a dormant phase, and the egg cell is enucleated. The dormant phase synchronizes the donor cell with the egg cell.

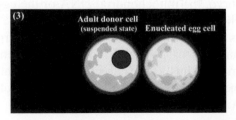

Figure 5B. (3) The donor cell and the egg cell are fused together using an electric current and to stimulate development. (4) Experience shows that embryos placed inside an oviduct are more likely to survive than those incubated in laboratory.

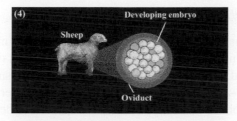

Figure 5B. (*Continued*)

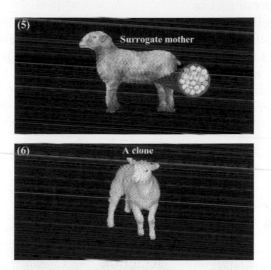

Figure 5C. (5) After developing in the oviduct for a few days, the embryo is implanted into a surrogate mother to develop to term. (6) After a normal pregnancy, a clone is born. Because of residual egg mitochondrial DNA, the clone is not an exact replica of the donor host.[230]

A donor cell is then taken from the culture and starved in a mixture, which has only enough nutrients to keep the cell alive. In this G0 phase, the donor cell is in a dormant phase, that is, all the genes shut down but the cell is not dead. An enucleated egg cell is placed next to the dormant donor cell when an electric pulse is applied to fuse the two cells and at the same time activate the development of an embryo. If an embryo develops, it is implanted into a surrogate mother to carry to term.

---

[230] Hwa A. Lim, *Genetically Yours: Bioinforming, biopharming, and biofarming*, (World Scientific Publishing Co., New Jersey, 2002).

The Honolulu Technique

Teruhiko Wakayama and Ryuzo Yanagimachi of the University of Hawaii pioneered this technique in July 1998 to clone generations of mice.

For this method, the donor cells extracted from the subject to be cloned are of special type – those that remain in the G0 phase: sertoli and brain cells remain in the G0 phase naturally; cumulus cells are either in the G0 or G1 phase. The donor nucleus is taken from the donor cell immediately after the cell extraction. An unfertilized, enucleated egg is prepared for use as the recipient of the donor nucleus.

The donor cell nucleus is then inserted into the enucleated egg cell. The egg cell is then placed in a chemical culture to jumpstart the cell growth. The culture contains a chemical, cytochalasin B for example, to prevent formation of a polar body, a second cell that normally forms before fertilization. If a polar body is formed, it will take half of the genes of the cell, preparing the other cell to receive genes from sperm, just like in natural fertilization of an egg by a sperm.

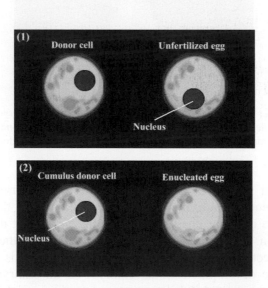

Figure 6A. (1) The donor cell from the subject to be cloned and the egg cell to function as donor cell nucleus receptacle. (2). The egg cell is enucleated; the donor cell is naturally in a dormant state.

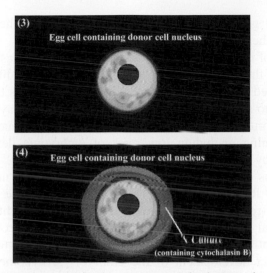

Figure 6B. (3) It takes a few hours for the egg to accept the donor nucleus. The fused cell is then allowed to sit for a few more hours. During this period, the fused cell undergoes no development. (4) The cell is then placed in to culture to jumpstart. The culture functions in the same way as an electric shock but is less strenuous on the cell.

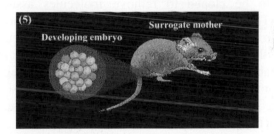

Figure 6C. (5) The egg cell develops into an embryo that is then implanted into a surrogate mother. After a normal gestation period, a clone is born.[231]

After jumpstarting, the cell develops into an embryo that can be implanted into a surrogate mother to carry to term.

The Chromatin Transfer Technique

In the conventional nuclear transfer technique, an intact nucleus from an adult or somatic donor cell, such as a skin cell, is placed into an oocyte (egg cell)

---

[231] Hwa A. Lim, *Genetically Yours: Bioinforming, biopharming, and biofarming,* (World Scientific Publishing Co., New Jersey, 2002).

from which the genetic material has been removed. The oocyte is then activated and it then divides to become a cloned embryo.

As a result of differentiation into their specific cell types (skin cell in the example above), the nuclei of donor somatic cells exhibit a different pattern of markers compared to the nuclei of normal embryonic cells that have not yet differentiated into specific cell types. These differences may lead to some of the widely reported limitations of the application of nuclear transfer in mammalian cloning, including low rates of embryonic development, high rates of pregnancy loss and low survival of cloned offspring.

Examinations of the behavior of the donor nucleus in nuclear transplant embryos have identified several nuclear defects due to incomplete reprogramming (remodeling), including assembly of the differentiated cell-specific structural protein, lamin A, enhanced content of pronuclei in TBP, a general transcription factor, and high resistance of DNA to DNAse, all of which may affect development of the embryo. A hypothesis is that these abnormalities may result from incomplete remodeling of the donor cell nuclei and/or from mis-regulation of expression of differentiated cell-specific genes.

The goal of the new chromatin transfer technique is to produce a cloned embryo that more closely resembles a normal embryo by pre-treating the cell of the animal to be cloned to remove molecules associated with cell differentiation. The first step is to create holes in the donor cell membrane (permeabilize) to provide direct access to the cell nucleus. Soaking the permeabilized cell in a special media causes the membrane of the nucleus to dissolve and enables removal of specific somatic cell regulatory proteins (remodeling) from the chromatin (genetic material and associated proteins). The permeabilized cell, containing the remodeled chromatin, is then fused to an egg cell and activated to create a cloned embryo.

This technique, led by Philippe Collas, professor at the Institute of Medical Biochemistry, University of Oslo, Norway and championed by Aurox LLC (based in Connecticut, USA) researchers, is a system for reversibly permeabilizing the donor cell to allow direct remodeling of donor nuclei *in vitro* (i.e., in a test tube). Transplantation of the resulting permeabilized cell, containing remodeled chromatin into recipient oocytes alleviates the defects of incomplete remodeling and yields nuclei that more closely resemble the nuclei of normal embryos. The DNA in chromatin transfer embryos showed increased sensitivity to DNAse treatment, suggesting that it is in a more "open" configuration (like embryonic chromatin). In addition, the nuclear remodeling lowered TBP content in nuclei and induced repression of lamin A gene expression in the cloned embryos. Implications of the repression of lamin A expression may include enhanced availability of factors required to turn on proper sets of embryonic genes and/or necessary to help turn off additional donor cell-specific genes. By permeabilizing the donor cell to allow condensation of the donor chromatin *in vitro* and inserting the condensed chromatin into the recipient oocyte, the researchers eliminated many DNA-bound components from the cloned embryos such as transcription factors or other potentially inhibitory somatic components.[232]

This technique has been exclusively licensed to Genetic Savings & Clone (GSC) for pet cloning. There have been several successes in cloning cats by GSC for commercial clients.

▶ Twinning Roslin in Honolulu

It is interesting to compare and contrast the four cloning techniques.

The artificial twinning technique involves an egg from a mother and a sperm from a father to fertilize the egg. Thus it still involves the stochastic process of sperm-egg union. The clones (or offspring twins) are clones of neither the father nor the mother, but they are the exact genetic replicas of each other.

The nuclear transfer technique, Roslin and Honolulu techniques included, involves a donor cell from the subject to be cloned and an egg cell to be enucleated to function as a receptacle for the donor nucleus. The clone is not an exact genetic replica of the donor because of residual mitochondrial DNA from the egg. Nonetheless, the nuclear transfer technique provides a better handle over the attributes of clones than the fertilization process of artificial twinning.

The Honolulu technique might work better for producing genetically engineered clones because it is easier to do. Researchers at PPL Therapeutics that collaborated with the Roslin Institute to produce Dolly had licensed the Honolulu cloning technology for research on cloning pigs and other animals.[233] The Honolulu technique has a higher success rate (3 clones out of 100 attempts) compared to the Roslin technique (1 clone in 277 attempts); the Honolulu technique is also highly reproducible; and it gives researchers precise control over the cloning outcome, unlike the Roslin technique, which involves fusing cells and mixing their contents.[234]

---

[232] Eddie J. Sullivan, Sriranjani Kasinathan, Poothapillai Kasinathan, James M. Tobl, and Philippe Collas, "Cloned calves from chromatin remodeled in vitro", *Biology of Reproduction*, 70(1), 2004, pp. 146-153.

[233] Kristie Coale, "A clone is a clone is a clone", *Wired News*, July 22, 1998.

[234] PPL Therapeutics, the Scottish company that helped clone Dolly the sheep, was dismantled in 2004.

The chromatin transfer technique is an improved version of the nuclear transfer technique. Remodeling of the donor genetic material is better controlled in the improved version, thus resulting in higher rates of success.

As with any cloning method, questions remain whether the clone will inherit the genetic baggage carried by the adult whose cell is fused in the process. By the time an animal becomes an adult, its cells will have divided thousands of times, differentiating along the way. The differentiation is cumulative as certain genes over time are turned off, causing cells to age. This is why there are arguments for using embryonic material or fetal material for cloning: by using embryonic or fetal cells instead of adult cells for cloning, researchers hope to clean the adult cell's slate.

Another unanswered issue pertains to the residual mitochondrial DNA. Research is still undergoing to find out what this will mean to the newborn clone.

Honolulu Versus Dolly

The Honolulu cloning technique describes the first successful mammalian cloning from adult mouse somatic cells using a technique called nuclear transfer by microinjection. In contrast, Dolly, like most other clones, was created via a different method called nuclear transfer by electrofusion.

Dolly, the first mammal cloned from an adult cell, stunned the scientific community when she was introduced on February 27, 1997 as a 7-month-old lamb. Animal cloning was, however, not something completely new. For decades prior to Dolly, animal breeders have divided young embryos and coaxed clumps of these embryonic cells to give rise to fully formed animals, a process called embryo splitting or artificial twinning (see above). Each of the two, three or more offspring that result from the splitting of a single embryo share the same genetic makeup, as do naturally formed identical twins. Embryo splitting is possible because young embryonic cells share an attribute called totipotency – the cells have not yet begun to differentiate into a specialized cell types (kidney, skin or eye cells, for example); they still have the potential to become any and every type of cell in the adult animal.

Before Dolly, nuclear transfer was only successful when the nucleus came from an embryonic or fetal cell. Attempts to clone animals, including mice, from adult cells failed. Scientists assumed that adult cells, because they had already differentiated and lost their totipotency, would require some sort of "molecular reprogramming" before they could regain the developmental capabilities of embryonic cells.

The Roslin group appears to have achieved this genetic reprogramming by starving the adult sheep mammary cells before harvesting their nuclei. Depriving the cells of nutrients induces their genes to enter an inactive stage, ceasing growth and cell division. Once the adult nucleus is inserted into the egg, reactivation of the genetic material leads to a healthy, dividing embryo. The embryo is then implanted and carried to term in a surrogate mother. Alternatively, the embryo can be used to repeat the cloning process and produce a second generation of clones.

The cloning of adult mice, reported by Wakayama, Yanagimachi and their colleagues, was accomplished using nuclei from adult cells that are permanently in the G0 stage, that is, they typically exist in a nondividing, quiescent state, making them more amenable to genetic reprogramming. They tried cloning with cumulus cells, nerve cells, and Sertoli cells from the testis. Only cloning with cumulus cells was successful. The success rate ranges from 1 surviving animal in 80 embryos transplanted to 1 in 41.

Yanagimachi attributes their success to using G0 cells as well as leaving the transplanted eggs for a few hours, preventing the egg from dividing in the meantime, before activation. He believes that the delay in activation gives the transplanted chromosomes time to reprogram.

▶ First Male Clone

The famous Dolly and other less heard of mouse, sheep, goat and cow clones were produced not only from female adult cells, but from cells harvested form various parts of reproductive systems such as mammary cells, ovarian cells, and thus all clones were female.

The female monopoly on mammalian reproduction and cloning was challenged when Teruhiko Wakayama and Ryuzo Yanagimachi cloned the first male animal, a mouse named Fibro. Fibro was cloned from flattened connective cells known as fibroblasts, and thus the name of the male clone. The fibroblast cells were taken from the tails of adult male mice of the brown-coated agouti strain; the ova were collected from females that had been hormonally stimulated to overproduce eggs, in other words, to superovulate just like in human *in vitro* fertilization procedure.[235]

---

[235] T. Wakayama, and R. Yanagimachi, "Cloning of mice from adult tail tip cells", *Nature Genetics*, 22, 1999, pp. 127-128.

Figure 7. Teruhiko Wakayama and three of the cloned mice. (Photo: Courtesy of University of Hawaii).

Just as in cloning using cells harvested from female reproductive systems, cloning using cells harvested from male body parts also suffer a very low success rate: about half the embryos divided sufficiently to be implanted into surrogate mothers. Of the 274 implanted embryos, only 3 reached full term, and of the three, only one, Fibro, survived. This high mortality rate is believed to have something to do with transferred nuclei failing to perform all the tasks necessary in their new environment – incomplete communication between the placenta and the embryo or fetus may be the major cause of the problems.[236]

•••••   •••••   •••••

In summary, in this chapter I have provided the four most popular cloning techniques used by referencing to the original path breaking experiments. As we saw in the preceding two chapters, cloning techniques have been perfected and cloning experts now have a better understanding of some of the species-specific reproductive physiology. These four techniques, particularly the Roslin technique, have been modified in various ways to clone different animals: the cat, the rabbit, the banteng, the mule, the horse and others, and now they are being applied to clone the human...

---

[236] Sara Abdulla, "First male clone", *Nature*, Science Update, June 3, 1999.

It is not as if human clones do not appear every day. In the United States alone, there are about 4 million births in a year. Of this, 1 in 400 is a pair of identical twins. A simple arithmetical calculation shows that about thirty twins are born each day. We can then say that thirty clones are born each day, in the United States alone![237]

---

[237] Richard Bronson, "A case of identical twins", In: Ethical Issues on Human Cloning, *Contexts*, 6(3), March 1998.

10 Reproductive Human Cloning

"Supermodels could one day have a whole new human cloning career, selling cells from their bodies to make hundreds of 'perfect' human clones for tomorrow's parents."

- Dr. Patrick Dixon, physician and futurist

► Super Clones?

When people hear of cloning mammals such as Dolly the cloned sheep, the first reaction is what about extending the technique to clone human beings? Humans are mammals and all the techniques learned in cows, sheep, mules, horses and other mammal cloning could in principle be used to clone human beings.

Unbeknown to most of these inquisitive people, a natural kind of human cloning has been going on for thousands of years. About 30 human clones are born in the United States every day! This is so because statistically, 1 out of ninety births produces twins, 1 out of 8,000 births produces triplets, and 1 out of a million births produces quadruplets…

Of the 1 out of 90 births of twins, about 1 out of 400 births are identical twins. Identical twins are natural clones and they have the same DNA, including mitochondrial DNA, in all of their cells. Even so, identical twins are not exactly identical: environment and freewill play major roles in making each of us – including identical twins – unique.

If someone were to take cells from an adult super person, say Abraham Lincoln, Albert Einstein, Mother Theresa, Princess Diana or Michael Jordan, and clone these cells to produce multiple clones of Lincoln, Einstein, Theresa, Diana or Jordan, by the time these cloned cells became children and then adult human beings, they might resemble the donor in looks, but they would not share all of the other traits, personalities and talents. The clones will be different from their respective donors because they would have grown up in an environment very different from their donor parent's environment and they would have made different choices in their lifetime. Lincoln II, III, IV… would not have to endure so many sicknesses; Einstein II, III, IV… might decide to be dancesport competitors instead of becoming physicists; Mother Theresa II, III, IV… might not end up in India; Lady Diana II, III, IV… might not be Princess Diana; and one of the Jordan II, III, IV… might finally be a professional golfer! And Lincoln II, III, IV… would not be exact duplicates of one another; similarly for Einstein II, III, IV…; and so forth. So forget about a super team of Jordan basketball players.

If resemblance is all that will be preserved between the progenitor and clones, a lurking commercial opportunity is that supermodels could one day have a whole new human cloning career – selling cells from their bodies to make "perfect" human clones for tomorrow's parents. Indeed we could soon clone a supermodel without her knowledge or consent – from a drop of her saliva or blood or cell…[238] Having said this, we should emphasize that the definition of "beauty" varies from culture to culture, and changes from time to time. At this very moment in time, some African tribes value obesity while the Western culture prefers a slender look. This will change with time – the sex symbol of the 1960s Marilyn Monroe certainly had quite a different facial look and body shape from the supermodel of the 1990s Cindy Crawford.

Lest we forget, we must also emphasize that beautification technologies, including facelifts, face remake-up, breast implants… are rather advanced. Not to mention all the lifestyle medications such as Propecia (for hair growth) that are available over the counter. In fact, many of the supermodels are the base clienteles of this lucrative industry.

▶ Tinkering with Conception

Experts in human fertility are already helping infertile married couples and couples of the same sex conceive through various outside-of-the-body fertilization, and implantation techniques. *In vitro* fertilization (IVF) is one such assisted conception technique. To improve the odds of a successful assisted conception, should these experts be encouraged to use a type of cloning? They could, for instance, take a 2-3 day old human embryo when it has only about eight cells, and is *in vitro* (outside of the body and in laboratory

---

[238] Patrick Dixon, *Futurewise: The six faces of global change*, (HarperCollins Publishers, 1998).

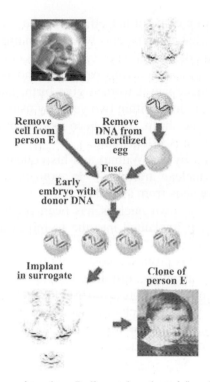

Figure 1. The technique used to clone Dolly can be adapted for cloning humans. First, the DNA of a donor cell is injected into an unfertilized egg whose nucleus has been previously removed. After the resulting embryo has grown to no more than 10 cells, the embryo is split 3-4 four times. In this way a number of viable embryos are available for transplantation. The available embryos are then transplanted into a surrogate mother until a successful conception takes place. The surrogate mother carries the embryo to term and the clone is born naturally or by Caesarean birth. The clone is an innocent-looking boy (oldest picture of Einstein in archive) to show that the development of a clone is a function of both nature (genetic) and nurture (environments and freewill). The clone of a genius may develop into a rowdy dancer. The egg donor and surrogate mother are faded sketches to avoid resemblance to anyone, living or dead. (Einstein photos: Courtesy of Albert Einstein Archive).

dish), separate (or split or twin) these eight cells and then culture or cryopreserve these cells. If one of the cells did not lead to a successful pregnancy, then they could try another cell, and another cell if needed... This was the original purpose of Jerry Hall and Robert Stillman of George Washington University when they performed artificial twinning experiments in 1993.

If the woman's ovaries fail to produce eggs, should she be allowed to use an egg from a donor whose DNA had been removed and replaced with her own DNA? Or replaced with the DNA from a selected male? In the first case she would give birth to a clone of herself, and in the second to a clone of the selected male. Then what about parents who lose an only child in an accident or to a disease? Should they be allowed to clone this child?

Indeed, Italian professor Severino Antinori, who is world renowned for helping several post-menopausal women give birth, announced in 2001 his plans to create a human clone within two years by using the same technology used to produce Dolly the sheep. It has been more than half a decade, but the first perceivable heartbeat is still pending.[239]

The technique used by Ian Wilmut and his colleagues to clone Dolly is called somatic-cell nuclear transfer. In somatic-cell nuclear transfer, researchers take the nucleus from a somatic cell of the donor and inject it into an egg, or ovum, whose own nucleus has been previously removed. The resulting embryo is then implanted into the womb of a surrogate mother to carry to term.

If Antinori delivers his promise, most likely the first cloned infant will be showered with media attention, just like Louise Brown, the first *in vitro* fertilization (IVF) baby born more a quarter of a century ago. And most likely within a few years the clone would just be one of hundreds or thousands of such cloned children. It is likely that cloning will be routine, but only in situations where there is such a need.

► Natural Birth Versus Cloning

Since the time as a child when we start asking our parents about "the birds and the bees," we are both curious about and interested in sex. When we grow from teenagers through mid-life or older, we are capable of sexually reproducing.

Though some cultures may not like to admit it, sex plays a major role in much of every culture, constantly manifesting itself in social gatherings and in fashion, literature, music, television and movies. When HAL[240] was walking down San Francisco one day, he saw an advertisement of a sex shop selling sex toys, sex aids and contraceptives that says "If you have the hardware, we have the software for you," in flashing neon lights. This may be overdoing a little, but there is nothing to be ashamed of in regards to sex. From a biological standpoint it serves a very special purpose: to merge two sets of genetic information – one from the father (sperm) and one from the mother (egg) – to make a baby that is genetically different from either parent.

---

[239] Michelle Nichols, "Angers over plans for world's first cloned human", *TheScotsman*, March 22, 2001.
[240] HAL are the initials of the author, Hwa A. Lim. HAL is also the name of the supercomputer HAL9000 in the flick Odyssey 2001. Please write to hal_lim@yahoo.com for comments.

Hardware and Software of Sexual Reproduction

The primary goal of sex is to merge the sperm and the egg via a process called fertilization to make a baby. In many organisms, fertilization occurs outside of the body. External fertilization is practiced in most fish and amphibians: females lay eggs on the riverbed, the male comes along and sprays the eggs with sperm and fertilization takes place. In reptiles and mammals, fertilization takes place inside the body of the female. Internal fertilization increases the chances of successful sexual reproduction. Because we humans use internal fertilization, our sexual organs are specialized for this purpose.

When a normal fetus is conceived in the womb, it has two sets of organs: one that can develop into the female sex organs (Mullerian ducts) and one that can develop into the male sex organs (Wolffian ducts). Which of these two sets of sex organs develop depends on the presence of the male hormone testosterone and in humans, the default sex is female.

This is the point where what we learn in elementary biology about X and Y chromosomes comes in: if the embryo is a male (XY chromosomes), then testosterone will be produced to stimulate the Wolffian duct to develop male sex organs while the Mullerian duct will degrade; if the embryo is female (XX chromosomes), then no testosterone is made. In such a condition, the Wolffian duct will degrade while the Mullerian duct will develop into female sex organs. The female clitoris is the remnants of the Wolffian duct.

Sex-organ development is determined by the third month of development but abnormalities can occur. If the embryo is a male (XY), but there is a defect such that no testosterone is produced, then the Wolffian duct will degrade, and the Mullerian duct will develop into a non-functional female sex organ. On the other hand, if the embryo is female (XX chromosomes) and testosterone is made, the Mullerian duct will degrade while the Wolffian duct will develop into a non-functional male sex organ. This is how we get people of the "wrong" sex organs, of different sex drive and of different sexual orientations.[241]

In normal development, from the time of puberty on, a male makes sex cells – in the form of sperm cells – continuously throughout his life. In contrast, by the time a female is born, she has made all of the eggs that she will ever have. As she reaches puberty, the eggs begin to develop and get released during menstrual cycles, and these processes continue until menopause, interrupted only by pregnancies.

In both males and females, the production of sex cells involves meiosis, a type of cell division whereby the two sets of genetic instructions are reduced to one set for the sex cell (haploid or having only 23 chromosomes). During

---

[241] Craig C. Freudenrich, "How human reproduction works", *Howstuffworks*, 1999-2003.

meiosis, the cell randomly sorts chromosomes from both sets in one cell division and then reduces them by half in another division. Therefore, each sperm or egg that the body produces is unique and different – it contains a different combination of the mother's genes and father's genes. Any wonder why two siblings can look and act totally different from one another even though they come from the same parents.

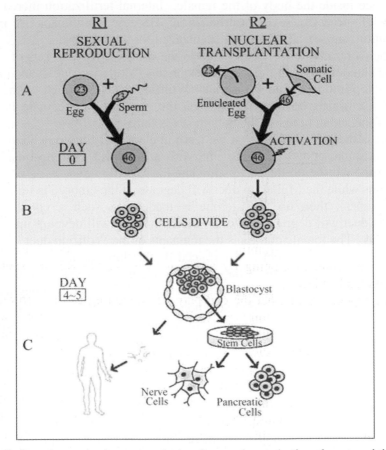

Figure 2. Sexual reproduction versus cloning: In sexual reproduction, the egg and the sperm are haploid or having 23 chromosomes each. When the sperm unites with the egg, the resulting zygote has a full set of 46 chromosomes – 23 from the egg and 23 from the sperm. In the form of cloning used in cloning Dolly the sheep, the nucleus of a somatic cell of diploid or having 46 chromosomes is implanted into an enucleated egg. The resulting "zygote" is also diploid, but the chromosomes are from only one source, the donor. Note that from fertilization (in sexual reproduction) and implantation (in cloning) onward, the development processes are exactly the same, whether in sexual reproduction or in cloning.

During fertilization, when a sperm meets an egg to form a zygote, the zygote now has a full set of chromosomes (diploid or having 46 chromosomes). Cell divisions begin immediately.

Mitosis is a process of cell division that results in the production of two daughter cells from a single parent cell. The daughter cells are identical to one another and to the original parent cell.

Hardware and Software of Cloning

To clone is to make an identical genetic replica. However, there are confusions over cloning. For example, a commonly asked question is "does cloning involve sperm-egg fertilization?" The answer is both "yes" and "no," as we presently explain.

Since the success of cloning Dolly the sheep, when people say "clone" they often mean "cloning by somatic cell nuclear transplantation," shown as R2 in the figure in which no sperm-egg fertilization takes place. However, it is conceivable in R1 and in R2 that at stage B when the resulting embryo has grown to no more than 10 cells, the embryo is split 3-4 four times. This process is called artificial twinning, like in spontaneous splitting of embryos in natural twins. If the split embryos are all implanted into surrogate mothers to be carried to term, the offspring will be twins or clones of each other. Thus in cloning by artificial twinning via R1, fertilization of a sperm and an egg takes place, while in cloning by artificial twinning via R2, no sperm-egg fertilization take place.

And depending on what the end purpose is, cloning can be reproductive cloning or therapeutic cloning: If the end result is a complete human being, it is reproductive cloning; if the end result is stem cells for therapeutic purposes, such as in regenerative medicine, it is therapeutic cloning.

In somatic nuclear transplantation approach to cloning (R2), a somatic cell from the donor is fused with an enucleated egg cell. Note that the somatic cell already has the full set of chromosomes (diploid or having 46 chromosomes). This should be contrasted with a sperm (haploid or having only 23 chromosomes) fertilizing an egg (haploid) to get a diploid embryo. If two sperm fertilize an egg (23+2×23) to get a triploid, the embryo aborts spontaneously, similarly for other fertilizations of an egg by multiple sperm. The key to a successful birth is a diploid embryo in the case of a singlet, or diploid embryos in the case multiplets.

The key to the success of cloning using nuclear transplantation of adult cells is a phenomenon called nuclear reprogramming or chromatin remodeling. Chromatin is the portion of a cell nucleus that consists of a mixture of proteins,

DNA and small molecules. Most nuclei do not suffer irreversible changes during differentiation, but research has found that chromatin inside the nucleus of an adult cell differs considerably from that inside an egg or that inside an early embryo. It has been observed that during cells differentiation, clusters of atoms, including methyl groups, latch onto DNA and some of these clusters deactivate specific genes on chromosomes. In addition, some of the proteins forming the core of chromatin have been observed to change over time.

In other words, for adult cells to return to a stage similar to an early embryo, a rejuvenation formula has to be found. Such a rejuvenation formula seems to exist inside the egg cell: proteins and other molecules floating within the egg can interact with a transplanted nucleus to restore its vitality. How this rejuvenation is achieved is still not completely understood. Experiments with frogs show that one egg protein induces the nucleus of an adult frog cell to shed certain chromatin protein and incorporate others, and in the process turn on several genes that have previously been silenced.[242]

In principle, to clone an animal from an adult cell, an approach is for scientists to find ways to help the egg with this chromatin remodeling or nuclear reprogramming. The trick, developed by Keith Campbell, a colleague of Ian Wilmut in the cloning of Dolly, involves starving the donor cells of almost all nutrients for five days. Depriving the donor cells induces the cells to abandon their normal growth cycle and to enter a quiescent stage, a stage known by the trade name of G0 stage. In such a stage, few, if any, genes are activated and Campbell suggests that the starvation prompts some initial nuclear reprogramming that makes it easier for the egg cell to complete the reprogramming task.

In the Honolulu technique for cloning, the donor cells are of special type – they are either those that remain naturally in G0 phase such as sertoli and brain cells, or those that remain either in G0 or G1 phase such as cumulus cells. In the chromatin transfer technique specific somatic cell proteins that may affect embryonic development, rates of pregnancy loss and survival of cloned offspring, are removed from the chromatin. This improves the success rate of cloning over somatic cell nuclear transplant technique.

Physiology of Human Cloning

A study by Keith Killian and colleagues at Duke University actually shows that a single genetic difference may make human cloning less complicated than in sheep and other mammals. The gene in question is insulin-like growth

---

[242] Alan P. Wolffe, *Chromatin*, (Academic Press, 1999).

factor 2 receptor (IGF2R), which is an imprinted gene. In contrast to the usual way in which genetic information is passed on to a child, an imprinted gene retains a chemical mark of its heritage (thus imprint) depending on whether it comes from the mother or the father. Imprinting ensures that only one copy of the gene, either that from the mother or that from the father, is turned on.

According to the study, humans and their primate kin possess two active copies of IGF2R that may prevent them from fetal overgrowth or large offspring syndromes (LOSS), a problem that has thwarted efforts to clone animals. LOSS can kill the mother, the fetus, or both. Since humans are not imprinted at IGF2R, fetal overgrowth would not occur if a human were cloned. Killian found that primates and their close relatives lost the imprinted gene some 70 million years ago, while other mammals, including mice, sheep, cows and pigs still carry it.[243]

Other scientists assert that the study is flawed, citing that other genes may contribute to the problem of LOSS. Indeed, one gene could explain up to 90% of failed clones. Hans Scholer and colleagues at the University of Pennsylvania found in most cloned mouse embryos, if a gene called *Oct4* switches on at the wrong time or at the wrong place, the clone would not be able to undergo early stages of implantation. In fact, incorrect *Oct4* levels turn cells into the wrong tissue type so the embryos cannot survive.[244]

In normal embryos, *Oct4* is turned on when the embryo is of only four cells and is gradually switched off in subsequent cell divisions. Cloned embryos usually have too few *Oct4* or have *Oct4* in wrong cells. By selecting only embryos with correct *Oct4* activity can increase the rate of success of cloning, but may not resolve the clone's health problems.

Cloning involves reprogramming during which the pattern of gene activity overwrites that of the adult pattern of gene activity. It is thus conceivable that besides *Oct4*, other essential genes may not be accurately reprogrammed, as the following experiment seems to indicate.[245]

In contrast to the Keith Killian study at Duke University, Richard Schultz and Marisa Bartolomei of the University of Pennsylvania have found that small differences in the amount of salt or amino acids used in the culture can cause certain imprinted genes in an embryo to behave aberrantly. Though Schultz's study focused primarily on *in vitro* fertilization, it is plausible to assume that similar aberrations can happen in the culture of cloned embryos.

---

[243] Kate Wong, "Genetic difference may make humans easier to clone than sheep", *Scientific America*, August 15, 2001.

[244] M. Bolani, S. Eckardt, H. Scholer, and K. McLaughlin, "*Oct4* distribution and level in mouse clones: consequences for pluripotency", *Genes and Development*, 16, (2002), pp. 1209-1219.

[245] Helen Pearson, "Gene marks clones for death", *Nature*, May 15, 2002.

The current practice also exacerbates the problem: embryos are not implanted as soon as possible into surrogate mothers; they are allowed to divide for 4-5 days in culture into blastocysts in an effort to select the most viable to increase the chances of a successful pregnancy.[246]

In a species-specific study, Gerald Schatten of University of Pittsburgh reported in April 2003 that monkey eggs are different from other animal eggs in which cloning has succeeded. In monkey eggs, when chromosomes are removed, crucial proteins are inevitably removed with them. These egg proteins are necessary for an embryo to develop normally. With the proteins missing from the egg, the inserted donor chromosomes cannot function properly.[247]

Knowing that cattle, pigs, mice and a cat have been cloned, a joke is that God in her infinite wisdom said "I will let you clone domestic species for agriculture, and I will let you clone mice and even a cat to catch the cloned mice. But I draw a sharp line between primates and all other mammals."[248] This joke, as can be guessed by now, was made before Idaho Gem (the first mule clone), Prometea (the first horse clone) and Snuppy (the first dog clone) came into beings.

There might be other fundamental barriers to clone humans using nuclear transfer. One is the differing speed at which developmental events occur among species. One difference is that a developing embryo may not immediately make use of its genes: mouse embryos activate their genes quickly while sheep embryos wait for much longer. The sheep embryo's delay in activation may have been crucial to the success of cloning Dolly because the delay allows ample time for nuclear reprogramming. Human embryos fall somewhere in between mouse embryos and sheep embryos – human embryos do not make use of their genes immediately (like in mouse embryos) but activate the genes sooner than sheep embryos during development.

There are thus still many ill-understood physical and biological problems with cloning, not only in animals that have yet to be cloned, but also in animals such as barnyard animals (cattle and pigs) that have been cloned with regularity.

▶ Key Human Cloning Mavericks

Looking at the menagerie of animals that have been cloned and the rate at which cloning techniques are being perfected, many people have predicted that within the first five years of the 21st century a group of scientists or a clone

---

[246] Rosie Mestel, "Some studies see ills for *in vitro* children", *Los Angeles Times*, January 24, 2003.

[247] *Science*, 300(5617), April 11, 2003.

[248] This joke is attributed to Gerald Schatten.

company somewhere in the world would likely announce the birth of the first cloned human baby. Sure enough, on December 26, 2002 Clonaid announced that the first human clone, nicknamed Eve, was born to U.S. parents. Eve, the seven-pound baby girl was cloned from the DNA of a skin cell of a 31-year-old American woman and the embryo was implanted in her own womb to grow to term.

In the midst of all controversies arising from a lack of verification by independent scientists, Clonaid announced on January 3, 2003 that they had a cloned girl born to a Dutch lesbian; and on January 22, 2003 that they had a cloned Japanese boy. Like the first claim, all these subsequent claims have yet to be verified.

The first five years of the 21st century have come and gone, with the exception of Clonaid claims, so far there is not a solitary known claims. In the stirs of all the human cloning news, who are the key players and mavericks, and who are the people who have set up cloning clinics or clonics for commercial purposes: –

Richard Seed

A few months after the announcement of Dolly the sheep, in January 1998, G. Richard Seed of Chicago announced he would be opening a cloning laboratory to aid infertile couples. Seed has a Ph.D. in nuclear physics from Harvard University and turned to reproductive technology more than twenty years ago when he founded a company to transfer embryos from prize cows to surrogate mothers. Later he was part of a team that conducted the first human embryo transfer. After perfecting the technique on animals, in the 1980s he incorporated a company, Fertility & Genetics, to apply to humans the embryo transfer technology. The technique involves moving fertilized eggs from healthy women, inseminated days before, to those women with fertility problems. The technique failed to compete with IVF and the company went defunct.

Figure 3. Richard Seed has plans to clone humans. He would also like to be one of those cloned.

In September 1998, Seed was 76 years old when he announced that he would clone himself using his wife's uterus. His stated motivation in producing a clone of himself was to neutralize criticism that he was using an unproven process to take advantage of desperate families. He also spoke about cloning his wife and his plans to set up a cloning laboratory in Japan by August 1999 to produce cloned pets and to clone extinct and endangered animals.[249]

As far as we know, he is still looking for financing, has no cloning facilities, or personnel trained to carry out cloning. There has been no report of any advances made to date.

Claude Vorilhon and Brigitte Boisselier

The Raelian Group also made an announcement to clone a human in 1997. The Group is said to be a sect based in Geneva, with 60,000 followers distributed in 90 countries. The leader, Rael, born on September 30, 1946, is a Frenchman whose original name was Claude Vorilhon. He is a racecar-driver-turn-prophet. Allegedly on December 13, 1973, in the crater of a volcano near Clermont-Ferrand in central France, he met a UFO (unidentified flying object) from which he received instructions from the Elohim, the scientifically advanced aliens aboard the UFO. According to Rael, the Elohim created all life on the earth – from microbes to man – by genetic engineering 25,000 years ago. Rael was asked by the UFO to set up an embassy on Earth to facilitate the return of the extra-terrestrials.

Figure 4. Brigitte Boisselier, chief executive of Clonaid, and Claude Vorilhon, leader of the Raelian Group meeting the press after the announcement of the birth of the first human clone.

---

[249] Nancy Ashe, "Peas in a pod: The cloning option, Part 3: The major players", *Adoption at About.com*, 2001.

The Group founded Clonaid in 1997 with the mission to create the first human clone. Clonaid is believed to be located in the Bahamas, may be in Canada or even in Nevada. Clonaid advertises that they would clone people for a fee of $200,000. In 1997, they had no facilities, no financing and no personnel.

By October 2000, they claimed to have a financing of $500,000 from a couple trying to clone their deceased son who had died at 10 months old from a botched operation to fix his faulty heart. They also claimed to have the scientific know-how to do cloning and to have lined up some 50 women ready to serve as surrogate mothers. In March 2001, Rael announced that Clonaid might soon be listed on the stock market with a potential worth of $100 million. Clonaid also claimed a list of some 1,000 potential customers willing to pay $200,000 for a clone. Clonaid has a web site to offer:

❑ Storing genetic materials for future cloning attempts;
❑ Cloning pets and race horses;
❑ Selling eggs internationally from a catalog showing photos of donors.

To date Clonaid has yet to go IPO, and has yet to be listed on any stock markets.

The Chief Scientific Officer, Brigitte Boisselier, who first heard of the Raelian creation story in 1992, was born in France in 1956. She has a master's degree in biochemistry and a doctorate in physical chemistry from the University of Dijon in France, and a doctorate in physical chemistry from the University of Houston.[250]

In 1997, Boisselier told the Paris daily *Le Monde* that it was okay to clone humans. She was laid off from her twelve-year job as a sales manager in the French chemical giant Air Liquide. She sued her ex-employer for religious discrimination and eventually won on appeal. Shortly after having been laid off, she lost custody of her youngest daughter, Iphigenie, to her ex-husband Panos Cocolios. She then retreated to Montreal with her son, Thomas, to be near some of her Raelian friends.

Her passion for teaching soon flared and she signed on for a one-year stint at State University of New York (SUNY) at Plattsburgh in 1999 before signing a three-year contract in June of 1999 to teach at SUNY Hamilton. Boisselier resigned from SUNY Hamilton in mid May 2001.

Boisselier's scientific team includes a geneticist, two biologists, and an *in vitro* fertilization specialist. Clonaid has no trouble lining up surrogate mothers, including Boisselier's eldest daughter Marina. The Raelians are forbidden to ingest drugs of any kind, including caffeine, but they are allowed

---

[250] Bryn Nelson, "Profile of Brigitte Boisselier", *Newsday*, April 22, 2001.

to have a glass of wine now and then. They believe drug-free, healthy women would make ideal surrogate mothers.

According to Boisselier, the birth of Eve was a result of ten implantations: five babies were spontaneously terminated during the first few weeks of pregnancies. The extreme high rate of success and the lack of any DNA verification create a lot of skepticism about the authenticity of the claim. Some believe that Clonaid was pressured into making the announcement because there are a number of scientists trying to make clones, and that Antinori had announced that he would make a clone by mid-January 2003. Clonaid had one last shot and they just took it. By making the announcement in December 2002, Clonaid could claim the distinction of being the first!

The Raelian Group detached itself from Clonaid in 2002 soon after the announcement of Eve and other clones, and the controversies that followed. Boisselier has since assumed the chief executive position of Clonaid.

Severino Antinori

In December 1998, Severino Antinori of Rome announced his interest in human cloning to help infertile couples.

Antinori graduated from the University of Rome in medicine. He was trained in the Queen Elena Maternal Institute as an obstetrics and gynecologist. He married biologist Caterina Versaci and in 1982 the two set up an infertility clinic in Rome.

In 1986, he pioneered the use of intracytoplasmic sperm injections (ICSI) – sperm injection techniques to combat male infertility and later used mice to incubate the sperm of men by placing infertile men's sperm cells in mouse testicles to enhance the cells' maturation so that they can be used for artificial insemination. In 1989, Antinori implanted his first menopausal subject with a donor egg. He became internationally famous for helping the 63-year-old menopausal Rosana Della Cortes in becoming pregnant via *in vitro*

Figure 5. Antinori Severino, Italian embryologist, and Panayiotis Zavos, his American partner in a plan to clone humans.

fertilization and delivering a boy on July 18, 1994. She is the oldest woman known to have given birth until the record was broken on April 9, 2003.[251] Thus to the Italian press he is "the father of impossible children," but to those who find his work grotesque, his unorthodox approach is irresponsible and accuse him of wanting to imitate Hitler.[252]

Today, Antinori is head of a string of Rome-based fertility clinics. He is also chairman of the privately funded Italian-U.S.-Israel human cloning consortium. His partners in the consortium are Panayiotis Zavos and Avi Ben-Abraham. He has proposed carrying out human cloning in an unnamed Mediterranean country or on a boat on international waters. He claims that it is immoral to clone humans just for the sake of it, but it is justifiable to clone as an effort to help infertile couples.

In November 2002, Antinori announced that he had successfully used cloning to induce pregnancy in three women, with the birth of the first child expected in January 2003. Some believe this announcement prompted Clonaid to jump ahead to announce the birth of Eve. To date, Antinori has not announced any clone births, at least not yet. If Antinori's announcement in November 2002 had really goaded Clonaid to jump ahead with the announcements of Eve (December 26, 2002), of a Dutch girl (January 3, 2003) and of a Japanese boy (January 22, 2003), then Clonaid had done themselves in when it was announced that Dolly was euthanized on February 15, 2003!

Panayiotis Zavos

Panayiotis Michael Zavos was born in Cyprus, Greece. He was formerly a reproductive physiologist at Kentucky Center for Reproductive Medicine, University of Kentucky, and an *in vitro* fertilization expert at the Andrology Institute of America in Lexington, Kentucky. He now runs a fertility clinic in Kentucky.

On January 25, 2001, Zavos and Antinori held a press conference in Lexington announcing they would clone humans in an undisclosed foreign country. The scientists created a furor and the heat forced Zavos to resign from a long-time position at the University of Kentucky. A forum organized by the team in March 2001 in Rome to publicize the project erupted in

---

[251] "India is 'world's oldest mother'", *BBC News*, April 9, 2003. If the claim is fully verified, then retired schoolteacher Satyabhama Mahapatra from Nayagarh in Orissa, India is the oldest mom in the world. She gave birth to a healthy boy weighing 6 pounds 8 ounces at 65 years old by using an egg from her 26-year-old niece Veenarani and the sperm of Veenarani's husband. The case is made all the more remarkable by the fact that the average life expectancy in India is about 63 years old.

[252] Nigel Cooper, "Life Stories: Severino Antinori", *Channel 4 News*, December 2001.

shouting matches, and shortly afterward, the Italian medical association threatened to yank Antinori's license to practice medicine.

During the U.S. Congressional hearing of March 2001 – later in the month of the shouting matches in Rome – to consider an outright ban on human cloning, Zavos and Boisselier met for the first time but they spoke little. Boisselier assumed the silence was due to Zavos' discomfort with her religious beliefs, or the threat of competition.

During the same Congressional hearing, Avi Ben-Abraham was seated in the first row, directly in line with the television cameras. According to subcommittee aides, Ben-Abraham appeared unexpectedly, in the company of the principal witness Zavos, politely offering to help answer the subcommittee's questions.

On January 18, 2004, Zavos caused a stir when he announced he had implanted a cloned embryo in a 35-year-old woman, and prominent reproductive biology experts said they were dismayed at repeatedly having to respond to such claims.[253]

Avi Ben-Abraham

Within months after Antinori and Zavos decided to team up, in March 2001 Antinori, Zavos and Avi Ben-Abraham announced plans to introduce cloning by 2003 as treatment for male infertility. It is reported that the three: Antinori, Zavos and Ben-Abraham are backed by R.H. Wicker, the CEO of the Human Cloning Foundation, who would like himself to be cloned.

Figure 6. Panayiotis Zavos, Ave Ben-Abraham and Brigitte Boisselier testifying before the U.S. Congress in 2001.

---

[253] "Scientists blast maverick cloning claims", *The Associated Press*, January 20, 2004.

Ben-Abraham, born November 18, 1957 in Kfar-Saba, Israel is the youngest to receive a Doctor of Medicine degree. He graduated with the MD *summa cum laude* on March 4, 1976 from the University of Perugia, Italy, at the age of 18 years 3 months. This record is now found to be shrouded in doubt, so is the medical degree.[254]

Ben-Abraham was a former Los Altos Hills, California resident and former president (1986-1993) of the American Cryonics Society – a non-profit firm based in California that freezes peoples' bodies and tissues with the hope they can be brought back to life later. The flamboyant young medical researcher hosted parties in Los Altos Hills with such celebrities as Muhammad Ali, Elliott Gould and Joe Kennedy, and was active in raising money for U.S. presidential candidates.

No one has heard much from Ben-Abraham when he left the area to become an investment banker and later headed up a multimillion-dollar biotech business, Ben-Abraham Technologies, based in Lincolnshire, Illinois, USA. Nanoparticles, which are the size of microorganisms, have interesting surface properties applicable to molecular transport and drug delivery. The technology led to the founding of Structured Biologicals in December 1996, which grew into Ben-Abraham Technologies Inc. In November 11, 1999, Ben-Abraham Technologies changed its name to BioSante Pharmaceuticals.[255]

The last time Ben-Abraham returned to his home country Israel was in 1999, he was already seen as something of a rising star in the Likud Party. He ran for national political office and came close to winning a seat in the Israeli parliament. He returned to Los Altos, California to regroup. Then he headed back to the Middle East the following year. This time, he made world headlines with the announcement that he planned to clone a human with his Italian and American partners.

But this was not the first time he stirred the media with cloning. Ben-Abraham got a lot of media attention in November 1993 when he made the suggestion to clone Jesus Christ from DNA found on the Shroud of Turin. According to the Israeli newspaper *Ha'aretz*, Ben Abraham told an interviewer that the cloning project is the most important development in the history of mankind and [it] represents the closing of a circle that began with Adam and Eve in the Garden of Eden.

---

[254] John Crewdson, "Academic record refutes lofty claims", *Chicago Tribune*, July 22, 2001.
[255] "Freeze frame", *Metro*, May 13-19, 1999.

▶ Human Embryo "Clones"

Mavericks of human cloning efforts do not always publicly announce their intentions. Some of these mavericks are working fervently in secrecy, while others are working in the private sector and therefore do not make claims because of corporate policy or because they do not want to jeopardize their intellectual property position.

Details of the first hybrid human embryo clone were released on June 18, 1999 even though the watershed achievement in biotechnology was reportedly to have happened in November 1999, months afterwards. Advanced Cell Technology (ACT) took enucleated cow egg cell and replaced the nucleus with the nucleus of a cell from Jose Cibelli, a research scientist. The new hybrid cell was then chemically activated to behave like a new embryo and start dividing in the normal way up to the 32-cell stage (for twelve days) at which time the hybrid embryo was destroyed. The technique they used was the Roslin technique. In a normal pregnancy, an embryo implants into the womb wall after 14 days. In principle, if the hybrid embryo had been allowed to continue beyond implantation and if everything went well, it would have developed into Cibelli's serial identical twin or clone![256]

On December 16, 1998, a South Korean research team at the infertility clinic of Kyunghee University Hospital in Seoul said they had grown an early human embryo using a cell donated by a woman in her 30s. Lee Bo-yon, a researcher of the team reported that the embryo divided into four cells before the experiment was aborted. The technique they used was the Honolulu technique.[257]

So the Koreans were the first to have cloned a pure human embryo. However the biggest piece of the news is not that ACT cloned the first hybrid embryo, but the fact that they kept the experiment secret for three years, and presumably they had been doing it at least a couple of years before that. Three to five years before 1998 is 1994 or 1995. If we recall that Dolly was born on July 5, 1996 but the news of her birth was not announced until seven months later, such late announcement of the embryo clone experiment was not too surprising. But the chronology would put these scientists ahead of Dolly in having successfully performed human nuclear transfer into an unfertilized egg, albeit a cow egg![258]

---

[256] "Details of hybrid clone reveled", *BBC News*, June 18, 1999.
[257] "First human embryo cloned in Korea or Britain?", *Globalchange*, 1999.
[258] "Cow-human clones: Human cloning from human cell and cows egg", *Globalchange*, June 1999.

The huge media rush to Dolly the sheep came only because Ian Wilmut and colleagues got DNA verification of their experiment sooner. The lesson from Dolly and the first hybrid embryo clone is: the headlines we read in newspapers or see on TV are recent histories – what cloning experiments were being conducted secretly in 2002, 2003... that we would not know until a few years later (we use three in this case) in 2005, 2006...? Had it not been because of the business competitive pressure from the disclosure of Dolly, we probably would still not know the successful clandestine hybrid embryo experiment!

On November 26, 2001, ACT announced that it had created a human embryo clone that had been hailed as an incredible scientific achievement by some, but decried as a dangerous step by others. Not only did the team describe how it made embryo clones using nuclear transfer methods pioneered in animals, it also revealed details of how human eggs were encouraged to start dividing on their own without fertilization from a sperm or the transfer of genetic material from another cell. This latter process known as parthenogenesis occurs in insects and microbes but not naturally in higher animals. Eggs usually dump half their genetic material to become haploid but if gathered early enough contain a full set of genes (diploid) and can in principle, start dividing if the conditions are right.[259]

Whatever it is: the hybrid clone, the Korean clone, and the ACT clone, in truth, the embryonic development went as far as six cells, far short of the 150 or so cells in blastocysts that represent the first essential step of therapeutic cloning. Thus whether these clones would develop into a fetus had they not been terminated are questionable. ACT even went as far as publishing their results in an electronic journal, *e-biomed. The Journal of Regenerative Medicine*. Some accused ACT of wishing to cloak its work in scientific respectability by so doing.[260]

Clone Abortion

There are a total of 192 countries in the world. Of these, over 170 countries are grappling with the issue of cloning and they still have yet to pass legislation on human cloning. But most have said they would support limiting the research and development budgets of state-funded researchers if they continue to perform cloning experiments.

---

[259] "Controversy over human embryo clone", BBC News, November 26, 2001.
[260] Robert A. Weinberg, "Of clones and clowns", *The Atlantic Monthly*, June 2002.

To date, the successes of cloning come with health problems: the placentas of cloned fetuses are routinely 2-3 times larger than normal; the offspring are usually larger than normal... These abnormalities, the reasons for which no one understands, is so common that it has its own name – large offspring syndromes (LOSS). Because of this, many clone fetuses must be delivered by Caesarean section due to their unusual sizes.

As far as we know, one of the most successful reproductive cloning attempts was reported by ACT. In 2001, ACT produced 496 cow embryos by injecting nuclei from adult cells into eggs that had been stripped of their own nuclei. Implantation led to 100 established pregnancies, of which 30 went to term. Five newborns died shortly after birth and a sixth died several months later. The 24 surviving calves developed into cows that were healthy by all criteria examined, except that most, if not all, had enlarged placentas and as newborns, they suffered from respiratory distress, typical of LOSS.

Working with barnyard animals is one matter, but working with human subjects is quite a different matter. In every human cloning experiment, researchers risk facing possible punishment from funding sources or facing the music of legal restrictions. In addition, the stigma of being attached to a cloning disaster can be tremendous: if a cloned baby has any defects, the researchers may not be able to return to the project for years and may be banned for life, not to mention other researchers in the same area may not be able to touch the area for 20 years!

A precedent case is gene therapy. Diseases arising from a single faulty gene, that is, monogenic diseases, are prime candidates for treatment using gene therapy. The idea is just simply to replace the faulty gene with the correct gene. A standard approach is to place the desired DNA in the shell of a neutralized virus and set the virus loose in the patient. Hopefully the gene will get inserted in the right place in the patient's genetic machinery. But in 30 years, gene therapy has partially cured only a few patients, and in 2002, two of gene therapy's poster children contracted leukemia as a result of their treatments. Subsequently, in January of 2003, the Food and Drug Administration (FDA) of the U.S. suspended gene therapy trials in humans, making the field a troubled area.

Thus researchers take extreme caution, exercise the highest precautionary measure and abort clones at the first sign of any potential problem. In this light, the unsubstantiated claims of Clonaid that it had cloned five human clones will likely set research on human cloning back by several years if the claims are later found to be groundless. In fact, most people believe the Clonaid claims are untrue. The euthanization of Dolly on February 15, 2003 because of her terminal lung problems will also throw caution to policy-makers who are deciding on the fate of human cloning. These facts may help explain why since 2003 there is a relative silence of announcements of attempts to clone humans, and no new cloning mavericks have come forward.

Why Hybrid?

There is one benefit of hybrid cloning. Most people are very uneasy about cloning human embryos and then dismembering them to obtain stem cells for therapeutic purposes. These same people might feel more comfortable with a hybrid solution, if it were shown that hybrid cow-human stem cells, for example, were viable for producing tissues and organs.

Maisam Mitalipova, a pioneer of "mix-and-match" hybrid cloning at the University of Wisconsin, claims that hybrid embryos do not give good quality stem cells but researchers are working hard to improve the quality. Nonetheless, there are other disturbing issues. For a beginner, how many human genes does a cow have to have before we give it human rights?

To understand this better, let us look at a human-cow hybrid clone as a working example. Most of the DNA of a human cell resides in the nucleus, but not all; a small part of 60 genes or so of the human cell's DNA is carried independently of the nucleus in structures called mitochondria. Mitochondria are power generators in the cytoplasm but reside outside the nucleus of the cell. They grow and divide inside cells and are passed on from one generation to another.[261] The human cell has a total of 50,000 genes. In a human-cow hybrid, a human cell nucleus is implanted into an enucleated cow egg cell. Technically, a small part of the resulting hybrid human clone genes, the mitochondria genes, belong to the cow. Judging from the success of the human-cow hybrid, it appears that cow mitochondria may well be compatible with human embryonic development.

But this is not the key issue: the disturbing issue is that what proportions of this hucow (human-cow) will turn this hucow into a cow? Similar arguments apply to, for example, a humonkey (human-monkey) which is within our capability now. Scientists have already succeeded in other hybrids: geep (goat-sheep),[262] and camas (camel-llamas).[263]

---

[261] Bill Stonebarger, "Cloning: how and why", Hawkhill Film #927, 1998.

[262] In 1981, Steen Willadsen was the first to clone an artificial chimera. He did this by mixing a sheep and a goat getting the result of a "geep" – it had the body shape and the head of a goat, and a dappled coat which had large patches of sheep's wool.

[263] Miral Fahmy, "Enter the camas", *Reuter*, March 26, 2002. Lulu Skidmore, chief scientific officer at the Camel Research Centre, the United Arab Emirates has been breeding camas for the past five years to create an animal with the sought-after fleece of the llama and the endurance of the larger camel.

Cows and humans are more dissimilar genetically. But humans and monkeys have 97% common genetic material. So if the right 1.5% is transferred, we could end up with a humonkey (human-monkey) or a monman (monkey-human). The issue gets even murkier if we take a species closer to the human, 99.8% or better similarity. Then that 0.1% mitochondria gene will be within the similarity!

▶ The "Clone" Stork – Real or Rael?

To date, all the five reported human "clones" are claims made by Clonaid, and as of date, not a single one of the clones has been genetically tested for authenticity.

The first clone, a baby girl 7 pounds at birth was born the day after Christmas at 11:55am on Thursday, December 26, 2002 to an American couple at an undisclosed location. The baby girl has been nicknamed Eve. The cloned embryo was produced using a skin cell from the 31 years old mother using the Roslin technique and was electrically stimulated to start dividing. The skin cell of the mother was inserted into her own egg. The mother, allegedly has a daughter from a previous marriage, currently has a husband who is sterile, leading the couple to resort to cloning. Boisselier said Eve was the result of one in 10 implantations performed by Clonaid. Five babies were spontaneously aborted during the first few weeks of pregnancy.

Presumably from one of the five viable pregnancies, the second clone, a baby girl, was born on the night of January 3, 2003 to a Dutch lesbian couple. The cloning, however, did not take place in the Netherlands where cloning is illegal.

The third clone was born to a Japanese couple, a boy this time, on January 22, 2003. The boy was cloned from the preserved cells and tissues of the deceased two-year-old baby of the Japanese couple, who are in the 40s. The two-year-old boy had died in an accident two years earlier.

After the "birth" of Eve, Clonaid claimed that five clones were expected to be born by February 5, 2003, presumably from the five viable pregnancies.[264] Sure enough, a fourth cloned baby was born from Saudi Arabian parents on January 27, 2003; and on February 4, 2003, a fifth clone was born, right on schedule and five out of five![265]

To date, none of the first five "clones" has been DNA tested to prove that they are genuinely clones. However, Clonaid announced that the five children

---

[264] "Clones: And then there were three", *The Associated Press*, January 22, 2003.
[265] From Clonaid website, www.clonaid.com.

of the "first generation" are in excellent health, and that twenty couples would be involved in the second generation.

Unfortunately, all the media stirs were soon drowned by the news of the euthanization of Dolly at age six on February 15, 2003 after being diagnosed with progressive lung disease, the U.S.-led war against Iraq,[266] and the rampage of severe acute respiratory syndrome (SARS) in Asia.

The fact still remains that all the Raelian claims are too good to be true: First, the Raelians were right on target to have five clones by February 5, 2003. Second, the success rate was way too high for the current cloning community to find plausible. Third, all the clone cases cover the likely scenarios of the need for cloning – a girl clone of a fertile mother and an infertile father implanted in her own womb; a girl clone for a lesbian couple; a boy clone of a deceased boy; a clone for a Saudi Arabia parents. The last case is likely because it is commonly believed that Muslims are more tolerant towards cloning...

Without any verifications, we can only wonder if all these clones are real, or they are just Rael? If they are just Rael, then this in itself is a clone gate of scandalous proportion.

●●●●●   ●●●●●   ●●●●●

While keeping reproductive cloning in mind, we now turn to another, more acceptable form of cloning – therapeutic cloning.

---

[266] The US-Iraqi War started on the 3rd day of the 3rd week of the 3rd month of the 3rd year of the 3rd millennium (March 19, 2003, Wednesday). Saddam Hussein's government fell after three weeks.

11 The Alchemy of Stem Cells

"The ability to turn blood into liver would be the envy of the alchemists of former times. Turning stem cells into 'therapeutic gold' will probably rest on our ability to identify the mechanisms by which tissue-derived stem cells respond to environmental cues and execute new developmental decisions."

- Stuart H. Orkin, *Nature Medicine*, November 2000

▶ Organ Transplant

On November 27, 2005, a startling feat of medicine – the world's first partial face transplant – was performed by Dr. Bernard Devauchelle, the head of maxillo-facial surgery at the university hospital in Amiens, France. The recipient was a thirty-eight-year-old Isabelle Dinoire whose face had been disfigured by a dog on May 27, 2005. The donor was Maryline St. Aubert, 46, who was declared brain dead when part of her face was taken.[267,268]

Figure 1. Dr. Bernard Devauchelle, head of maxillo-facial surgery, Amiens, France.

---

[267] Craig S. Smith, "As a face transplant heals, flurries of questions arise", *The New York Times*, December 14, 2005.
[268] "Face transplant recipient speaks to media", *UPI*, February 7, 2006.

Early human tissue transplantation is not a myth, according to at least one medical historian. More than 2,000 years ago, Indian surgeons used skin flaps from the forehead and the cheek primarily in the operation of rhinoplasty (nose rebuilding).[269] It was not until the nineteenth century that Western medicine significantly utilized skin transplantation: in 1804, early transplantation pioneers reported successful skin grafts on animals; and by the end of the century, numerous skin grafts were performed from human to human.

Organ transplants in humans came much later – the first attempt was not done until the first decade of the twentieth century. Early unsuccessful attempts include transplantation of kidneys from animals to humans. A major breakthrough occurred in the 1940s when Peter Brian Medawar (1915-1987) and Frank Macfarlane Burnet (1899-1985) explained how the human body's immune system recognizes and rejects foreign materials that enter it. This knowledge led to tissue typing in which the donor and recipient tissues are matched for compatibility.

David Milford Hume (1917-1973) performed the first human-to-human vital organ transplantation in 1951. He transplanted the kidney of a cadaver into a failing patient in an unsuccessful attempt to save the patient's life. Hume made a few more unsuccessful attempts until February 11, 1953 when he and Joseph Murray (1919-) achieved moderate success – one of the patients lived six months with a transplanted kidney. In 1954, Murray performed what is regarded as the world's first successful kidney transplant: Richard Herrick of Boston, Massachusetts received a kidney from his identical twin brother Ronald and survived for eight years before dying from a heart attack. By 1975, over 25,000 kidney transplants had been carried out worldwide. The patient who survived longest lived twenty years beyond the transplantation. Today about 20,000 transplants are performed each year. David Hume's brilliant career came to a tragic and premature end when, on the evening of May 19, 1973, the small plane piloted by Hume struck a mountain near Van Nuys, California en route to Richmond, Virginia.

Ten years after the successful kidney transplant, William Waddell and Thomas E. Starzi (1926-) at the Veteran's Administration Hospital in Denver performed the world's first liver transplant. The recipient, William Grigsby, at forty-seven, died three weeks later. Liver transplants are not as common as kidney transplants. By the early 1980s, only a few hundred liver transplants had been performed. The longest survival was seven years.

The least successful transplanted organ is the lung. James Hardy (1918-) at the Mississippi Medical Center performed the first lung transplant in June

---

[269] Russell Scott, *The Body as Property*, (Viking Press, New York, 1981).

of 1963. The transplantation had a legal aspect to it because the patient, John R. Russell, was a convicted murderer serving a life sentence. Russell died eighteen days later. By the 1980s, only about forty lung transplants had been attempted, and the longest survival was ten months.[270]

The first successful heart transplant was performed in South Africa. On December 3, 1967, South African surgeon Christiaan Barnard (1922-2001) successfully transplanted the heart of an auto accident victim, a young woman, into a fifty-five-year-old Louis Washkansky at Groote Schuur Hospital in Cape Town. Washkansky lived for eighteen days with the new heart. Barnard's second patient, fifty-eight-year-old dentist, Philip Blaiberg, lived for eighty-four weeks. On September 2, 2001, an acute asthma attack killed the pioneering South African heart surgeon while he was vacationing in the Cyprus resort town of Paphos.[271]

Organ transplants, in which the organ of interest is removed form one human to be transplanted into another human, finally came of age in the 1980s, thanks to better surgery techniques, better understanding of the human immune system, and development of effective drugs to combat rejection.

But tissues and organs need not be harvested from a human. They can be grown!

▶ Growing Human Organs?

During the earliest stages of human development, when the embryo consists of fewer than a dozen cells, the genes inside every cell nucleus have their full potential. These embryonic cells, in the jargon of biologists, are totipotent for they each have the ability to develop into a complete human being.

As the embryo develops further, the embryonic cells lose this totipotent ability as they begin to specialize by turning on or off genes: a group of the embryonic cells may commit themselves to develop into the nervous system, while another group commits to become the muscles, and so forth. This gradual specialization of cells, also known as differentiation, poses a provocative question: does a nucleus from a differentiated cell retain the know-how to develop into an entire human, or the know-how to develop into an organ or a tissue? Scientists have generally dismissed this idea of nuclear equivalency, suggesting that as cells specialize, they irreversibly alter their DNA or even discard genes, making their nucleus nonequivalent.

---

[270] Andrew Kimbrell, *The Human Body Shop: The engineering and marketing of life*, (HarperCollins Publishers, New York, 1993).
[271] "Christiaan Barnard died of asthma, not heart attack", *ABC News*, September 5, 2001.

Now we have learnt from experiments with barnyard animals and model animals such as mice about development. As our knowledge of how cells differentiate accumulates, other more radical, even to the extent of bizarre, choices may be at our disposal. For example, if we separate cells at just the right moment during early embryonic development, we might be able to grow human organs like the heart, the kidneys, the liver, and human tissues like skin and muscles, all *in vitro* (in the laboratory). Should scientists be allowed to tinker with human embryonic tissue in hopes of producing such human organ factories?

The most important fruit of research in cloning will be basic knowledge about life and particularly, human development. Knowing more about how cells differentiate, and how genes get turned on or off during development is important. If we learn enough about how cells do this, we might be able to devise gene-regulating drugs that would stimulate the growth of nerve cells in the living body (*in vivo*). This kind of drugs would do wonders for people who are now paralyzed from a spinal cord injury, for example. Christopher Reeve, the actor in the movie "Superman," ironically was paralyzed due to a spinal cord injury arising from a fall off the back of a horse. This kind of therapy could have reversed his paralysis. Christopher Reeve finally succumbed in October 2004, after many years of struggling with quadriplegia.

We might be able to awaken genes that create bone marrow to help cancer victims fight leukemia. Or stimulate the development of new brain cells in people such as Michael J. Fox, the actor in the movie "Back To the Future" who is suffering from Parkinson's disease, [272] or else he might just not be able to move steadily back and forth in time. Or Ronald Reagan, the former U.S. president who was suffering from Alzheimer's disease, [273] and who seemed not to be able to remember anything except remembering to seek the Fifth Amendment [274] in the

---

[272] Parkinson's disease is a disorder of the brain, characterized by shaking (tremor) and difficulty with walking, movement, and coordination. The disease is associated with damage to a part of the brain that is involved with movement. It comes about when nerve cells in the brain that produce a chemical called dopamine die, causing movement and walking difficult. It affects up to 5 million people worldwide.

[273] Alzheimer's disease (AD) is a slowly progressive form of dementia, which is a progressive, acquired impairment of intellectual functions. Memory impairment is a necessary feature for the diagnosis.

[274] Fifth Amendment to the U.S. Constitution: No person shall be held to answer for a capital, or otherwise infamous crime, unless on a presentment or indictment of a Grand Jury, except in cases arising in the land or naval forces, or in the Militia, when in actual service in time of War or public danger; nor shall any person be subject for the same offence to be twice put in jeopardy of life or limb, nor shall be compelled in any criminal case to be a witness against himself, nor be deprived of life, liberty, or property, without due process of law; nor shall private property be taken for public use without just compensation.

Iran-Contra Affair investigation during his presidency.[275] Ronald Reagan died in 2004.

This form of therapy may no longer be a fantasy in the near future. Companies have already patented more than 30 molecules that can activate genes responsible for, among other things, the embryonic development of the brain, the sperm and bone cells.

▶ Stem Cell

As far back as 1938, Hans Spemann discussed nuclear transplantation as a form of fantastical experiment to verify nuclear equivalency.[276] In recent years, this form of nuclear transplantation has intensified as a form of cloning. Unfortunately, the public, when they hear the word cloning, they immediately associate it with reproductive cloning. In fact, there are at least three basic types of cloning:

- ☐ Hybrid cloning: genetically engineer a human gene into barnyard animals to produce milk containing human therapeutic proteins. The barnyard animals function as a bioreactor in this case;
- ☐ Therapeutic cloning: cloning of tissues and organs, most often from stem cells. The controversy in this area revolves around the creation of human embryos for the sole purpose of harvesting stem cells;
- ☐ Reproductive cloning: this is a form of cloning for producing a whole human.

In our discussion, we shall be very specific about which type of cloning so that there will be no ambiguity. We will also concentrate mainly on therapeutic cloning in this discussion; the other two types of cloning have been previously discussed.

Indeed, development does not cease after birth. Some cells in the body, including the nerve cells, the muscle cells of the heart, and the lens cells of the eye, cannot reproduce. Once these cells are established, they function until they die or the organism of which they are a part dies.

---

[275] Iran-Contra Affair or Iran-Contragate is a secret arrangement in the U.S. history in the 1980s to provide funds to the Nicaraguan contra rebels from profits gained by selling arms to Iran. The Iran-contra affair consisted of two separate initiatives during the administration of President Ronald Reagan: The first was a commitment to aid the contras who were conducting a guerrilla war against the Sandinista government of Nicaragua; The second was to placate "moderates" within the Iranian government in order to secure the release of American hostages held by pro-Iranian groups in Lebanon and to influence Iranian foreign policy in a pro-Western direction.

[276] John Travis, "A fantastical experiment", *Science News*, April 5, 1997.

Other cells, including those in the liver, blood vessels, pancreas and body tissues, however, multiply by simply replicating, as is evident in liver cells: when liver cells die, other liver cells divide to make up for the loss.

There is yet another category of cells which originate from undifferentiated cells known as stem cells. When a stem cell divides, each of its two daughter cells has two choices: it can remain a stem cell, or it can differentiate into another cell type. Blood cells – white and red blood cells – originate from a single type of stem cell. In other words, stem cells are the body's primordial master cells that have the capacity to grow into other specialized cells. Stem cells remain uncommitted until they receive signals from the body to develop into specialized cells.

In this sense, stem cells are sometimes called magic seeds since they possess the ability to replicate indefinitely and morph into any kind of tissue found in a human body. They are nature's blank slates, capable of developing into any of nearly 230 cell types that make up the human body. Scientists believe they will lead to cures for debilitating diseases once thought untreatable.

While a post-doctoral fellow at University College London in 1974, Gail Martin succeeded in keeping fragile stem cells alive in a petri dish. In 1981, at University of California at San Francisco, she succeeded in isolating for the first time stem cells from mouse embryos. But the stem cell controversy did not get pushed into public arena until almost a decade ago. In 1998 James Thomson of the University of Wisconsin succeeded in isolating stem cells from human embryos, setting off the hope that such stem cells may be used to cure a host of devastating human ailments. The difference between Martin and Thomson is the difference between mice and humans!

Derivation of Stem Cells

Once isolated, stem cells can be grown in the laboratory and stored for future use. Each reservoir of cells, derived from a single embryo, is known as a cell line. Current technology permits derivation of pluripotent human cell lines from a number of sources. Pluripotent stem cells are isolated:[277]

☐ Directly from the inner cell mass of human embryos at the blastocyst stage. The sources of the human embryos are *in vitro* fertilization clinics. James Thomson of University of Wisconsin pioneers this approach.

---

[277] Michael Shamblott, et al, "Derivation of pluripotent stem cells from cultured human primordial germ cells." *Proceedings of National Academy of Sciences*, 95 Nov. 1998, pp. 13726-13731.

□ From fetal tissue obtained from terminated pregnancies. The stem cells are taken from the region of the fetus that is destined to develop into sexual organs such as the testes or the ovaries. John D. Gearhart of John Hopkins University is a leader in this approach.

□ Directly from inner cell mass of cloned embryos at the blastocyst stage. Somatic cell nuclear transfer (SCNT) is another way pluripotent cells can be isolated. In this approach, a somatic cell – any cell other than an egg or a sperm cell – is fused with an enucleated (nucleus removed) egg cell. The resulting fused cell and its immediate descendants are believed to be totipotent.

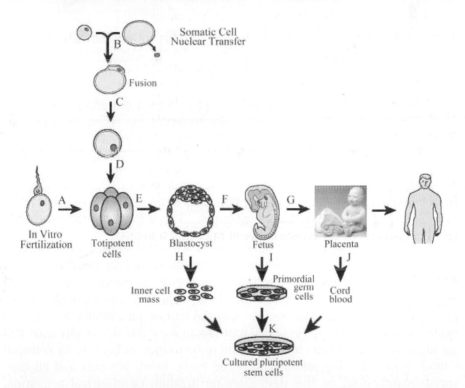

Figure 2. *In vitro* fertilization (A) and somatic cell nuclear transfer (B) both lead to totipotent cells. 4-5 days into development, totipotent cells become a blastocyst, containing an inner cell mass of about 150 cells. Pluripotent stem cells can be harvested in various sways: 1) the inner cell mass of the blastocyst can be harvested (H); 2) cells from an aborted fetus committed into becoming testes or ovaries can also be extracted (I); 3) the cord blood during birthing can be banked for future use (J). AEFG is *in vitro* fertilization reproduction; BCDEFG is reproductive cloning; BCDEHK is therapeutic cloning.

□ From placenta cord blood that has been cryogenically preserved in liquid nitrogen. During birthing process, the cord blood is drawn by inserting a needle into the vein of the umbilical cord during the short window when the stem cell concentration is at the highest. The cord blood is then cryogenically preserved for future use.

□ From fat such as those from liposuction. Marc Hedrick of University of California at Los Angeles was the first to discover stem cells in fat.

Banking Embryonic and Fetal Parts

Embryonic and fetal tissues have several attributes that make them ideal candidates for medical purposes:

□ They are unquestionably human. This offers an advantage in elevating rejection.

□ They exhibit tremendous developmental potential. Embryos and fetuses are, after all, growing organisms.

□ Fetal tissues are in abundant supply. Over 1.6 million fetuses are aborted in the U.S. alone each year. Embryos can be grown or cloned.

□ Substantial evidence indicates that fetal tissues are effective only when they are alive. Abortions normally occur at scheduled times, it is possible for researchers to obtain tissues in the shortest interval possible, thus increasing the likelihood that the tissues will be alive. Placenta cord blood is collected right after birth during the narrow window of time when the stem cells are most abundant.

A cursory review of human development will shed some lights as to why stem cell research causes so much controversy. In sexual reproduction when a sperm fertilizes an egg, the fertilized egg begins to divide. Similarly, in nuclear transplant cloning, when the donor DNA is transplanted into an enucleated egg, the egg and nuclear material can be electrically activated to divide. The fertilized egg (in sexual reproduction) or the electrically activated egg (in nuclear transplant cloning) is said to be totipotent because its potential is total: it can develop into a fetus – the body, head, placenta and all that. Approximately four or five days after fertilization or electrical activation, several cycles of cell divisions will have taken place and the cells develop into a hollow sphere with an inner mass of about one hundred and fifty cells called the blastocyst.

In further development, the outer layer of cells will develop into the placenta and other supporting tissues needed for the fetal development, while the inner cell mass will develop into other parts of the human body. The inner cell mass is said to be pluripotent rather than totipotent and is a ball of primordial master cells called stem cells. In further development, the ball of stem cells, alone without the layer of cells, cannot develop into an organism because none of them will develop into the placenta and supporting tissues necessary for the fetal development in the uterus. The pluripotent stem cells can however undergo further specialization into stem cells that are capable of developing into cells of particular function. For example, blood stem cells, skin stem cells, and other human body's 230 different cell types.[278]

Embryonic and fetal parts are more easily accessible from fertility clinics than we care to admit. To increase the odds of success, during infertility treatment, more than enough – usually more than ten eggs are collected. Eggs are very fragile and it is very difficult to freeze and preserve them. All the eggs collected are normally fertilized and fertilized eggs not transferred into the mother's womb end up frozen for later use. Many thousands of these human embryos are frozen at the some 300 infertility clinics in the United States and many more are frozen in infertility clinics in other parts of the world. These unused frozen fertilized eggs or embryos can either remain frozen indefinitely, be donated for research, be donated for other infertile couples, or be thawed and discarded. Given these options, many couples see the value of donating them for research.[279]

Lifeline – Cord Blood

At the blastocyst stage, the inner cell mass may be extracted for medical use, what about the outer layer of cells of the blastocyst, destined to develop into the placenta and other supporting tissues for fetal development? The placenta is the organ that develops in the uterus during pregnancy, and through the umbilical cord, it links the blood supply of the mother and the baby. It thus functions as the baby's lifeline to its mother. The placenta and the umbilical cord are rich sources of cord blood. Besides giving life inside the womb, they perform other useful services: doctors use cord blood to assess the health of the unborn.

About ten years ago, scientists found another good use for cord blood – blood in a newborn's umbilical cord and placenta. The cord blood is a rich source of stem cells! In 1998, French doctors demonstrated the pluripotentiality of stem cells when they first successfully gave cord blood

---

[278] *Stem Cells: A primer*, National Institutes of Health, May 2000.
[279] Jeffrey P. Kahn, "Making a market for human embryos?" September 4, 2000, Center for Bioethics, University of Minnesota.

Figure 3. (Left to right) Henry Low (Exec. Director, CryoCord), Dr. Azman F. Shafii (Founding Dir., NanoBiotech Sdn. Bhd.), Wenddi-Anne Chong (Exec. Director, CryoCord), HAL (the author), and Ravindran B.P.M. (General Manager, CryoCord) at the CryoCord booth during the Ground Breaking Ceremony of BioValley, Malaysia, May 20, 2003.

from a newborn to his five-year-old sibling suffering from severe anemia syndrome and skeletal defects. In 2001, another five-year-old boy was treated for thalassemia minor with cord blood from a non-relative. Thalassemia minor is a genetic blood disorder that predisposes the victim's red blood cells to becoming deformed and thus these blood cells are weaker than normal.[280]

Until about five years ago, after birth, the placenta is expelled with other redundant tissues collectively known as the "afterbirth." After nine months of faithfully performing its function, the placenta is finally discarded unless the mother decides to preserve the placenta for memory sake. Of late the childbirth garbage is preserved for storing umbilical cord blood from which stem cells are extracted as potential cures of debilitating diseases. Countries including the U.S., U.K., Sweden, Israel, Japan, China, Taiwan, Vietnam, South Korea, Singapore and Malaysia have adequate facilities for banking cord blood. Private companies offering such services include CryoCord Sdn. Bhd. in Malaysia, Cryolife Cord Blood Banking in Hong Kong, CordLife Pte. Ltd. in Singapore, and Sino Cell in Taipei. They each offer very similar services, but may follow different guidelines or use different storage methods. For example, CordLife follows the American Association for Blood Banking (AABB) while CryoCord follows the European Directive guidelines; also CryoCord prides in adopting a multi-vial stem cell storage technique which will offer multiple opportunities to utilize the stem cells in the future.

Cord blood banking is considered a form of bio-insurance because by storing the baby's cord blood, this baby's biological resource is being saved

---

[280] Gina Abuyuan-Llanes, "Lifeline", *Health Today*, April 2002, pp. 30-32.

and may someday save the baby's life or someone else's life. To bank cord blood, expecting parents are advised to contact the cord blood banking company no later than two weeks before the expected delivery date. During the delivery, the doctor will collect the blood from the umbilical cord within the window of time when stem cells can be found in their highest concentration. The procedure is safe, simple and non-invasive.

For a normal birth, the blood is collected while the placenta is still *in utero* by taking advantage of the uterine contractions. For a Caesarean delivery, the blood can be collected prior to the removal of the placenta by hanging the umbilical cord over the side of the mother and collecting from the umbilical vein. The collection can also be done within ten minutes after the removal of the placenta by accessing the veins from the fetal side of the placenta.

The entire collection process generally takes only a few minutes. It should be possible to collect between 40cc to 150cc of blood. The amount of blood obtained is affected by factors including the size of the cord, where the cord is cut, the health of the mother and the health of the child. Cord blood has been obtained from premature babies as early as 30 weeks into gestation.

After delivery, the collection kit with the baby's cord blood is sent to the banking company to perform standard screening tests including bacterial contamination screening. It is important to ensure the cord blood is received within 24 hours after collection as cell viability decreases after this time.

In the banking company, red blood cells and plasma are removed and the white cells including the stem cells are harvested. These cells are viably frozen and then cryogenically stored in liquid nitrogen at −196°C. Under these conditions, the cord blood theoretically has an infinite life. Cryo-preserved cord blood stem cells have remained viable and has been successfully used in transplant after 7-8 years of preservation; cord blood cells preserved for more than 15 years have been shown to be viable and appear to be identical to cord blood cells preserved for merely a week.

The cost of banking cord blood is rather affordable. Currently, for example CryoCord, charges an initial cost of RM2,500 (or $600) which includes the collection kit, special courier service, processing, testing and first year storage of the baby's stems cells in 5 cryovials. The annual storage fee is RM250 (or $60).

Cord blood's application is potentially much wider, but for now the widest application of cord blood is to repair bone marrow after it degenerates from aggressive chemotherapy or radiation. According to a study by Mary M. Horowitz, a member of the International Bone Marrow Transplant Registry team, transplants using cord blood are just as effective in saving the lives of children with leukemia as transplants using bone marrow cells. Cord blood may, however, offer certain advantages over bone marrow transplants:

- ☐ Increased immune tolerance – Cord blood stem cells are "naïve" and have a muted immune system, the stem cells can be transplanted into the donor, a relative, or a non-relative with a lower risk of rejection. In bone marrow transplant (BMT), the immune system of the recipient may attack the transplanted tissue, a disease called graft versus host.
- ☐ Higher availability – Cord blood stem cells are easier to obtain. Getting stem cells from bone marrow requires a major surgical procedure.
- ☐ More time-saving – A matching procedure is necessary for BMT and may take more than six months if relatives are not available. A cord blood sample can be collected, cryo-preserved and made available for treatment within two to four weeks.
- ☐ Greater ability – Cord blood stem cells regenerate new blood cells faster than bone marrow stem cells. Ounce for ounce, there are ten times as many blood-producing cells in cord blood.

As pluripotent as embryonic stem cells are, they also have disadvantages: First, transplanted embryonic stem cells sometimes grow into tumors; Second, the human embryonic stem cells that are available for research may be rejected by a patient's immune system. An improvement on tissue-matching could be achieved by either creating a bank of stem cells from more human embryos, or by cloning a patient's DNA into existing stem cells to customize them.

Multipotent Adult Stem Cells

Though pluripotent stem cells are relatively abundant in early human development, in children and adults, multipotent stem cells are also found. Late in embryonic or fetal development, subsets of lineage-restricted, but not fully differentiated, cells are set aside as reserve multipotent adult stem cells (ASCs). These reserve cells may reside within a tissue or circulate in the blood, and some are used for juvenile growth after birth while others for regeneration throughout life. An excellent example of multipotent stem cells is blood stem cells found in the bone marrow or in smaller quantities circulating in the blood stream. These multipotent stem cells perform the life-critical role of continually replenishing red blood cells, white blood cells and platelets into the blood stream. Indeed, stem cells are needed to constantly replenish worn out cells in our body, in a process called regeneration. Besides blood ASCs, multipotent ASCs can be found in other types of adult tissue as well. Regeneration via reserve ASCs is the most common avenue of tissue regeneration in multicellular organisms, including humans.[281,282]

Ectodermal and endodermal derivatives regenerate from epithelial ASCs that reside in the basal layer of the epithelium. Examples are skin epidermis,[283] hair follicles, the epithelium of the respiratory and digestive tracts,[284,285] certain parts of the central nervous system,[286] and acoustic sensory hair cells.[287] Even though the liver can regenerate to a certain degree, it also contains a population of stem cells (hepatic stem cells or HeSCs) that are activated when the liver is damaged beyond the capacity for regeneration by differentiated liver cells.[288]

Mesodermal derivatives, such as skeletal muscle, bone and blood regenerate from non-epithelial hematopoietic stem cells (HSCs), mesenchymal stem cells (MSCs) and muscle stem cells (satellite cells) associated with the tissue or residing in the bone marrow.[289,290]

For reserve ASCs to serve throughout life, they have to be self-renewing. They typically divide asymmetrically to give rise to a cell with a more restricted lineage and another stem cell. The developmental potential varies, depending on the tissue each ASC serves.

They normally differentiate into a range of phenotypes within their lineage, called their prospective significance, or fate. For example, hematopoietic stem cells, which reside in the bone marrow, are multipotential and give rise to erythrocytes, myeloid cells and the several cell types of the immune system. Others, such as liver stem cells, are bipotent and give rise only to hepatocytes and bile duct cells. Still others, such as epidermal stem cells, appear to be unipotent, giving rise only to one cell type, in this case keratinocytes.[291]

However, some adult stem cells previously thought to commit to developing into one line of specialized cells have been shown to be able to

[281] E. Fuchs, and J.A. Segre, "Stem cells: a new lease on life", *Cell*, 100, 2000, pp. 143-156.

[282] I.L. Weissman, "Stem cells: units of development, units of regeneration, and units in evolution", *Cell*, 100, 2000, pp. 157-168.

[283] U.B. Jensen, S. Lowell, and F. Watt, "The spatial relationship between stem cells and their progeny in the basal layer of human epidermis: A new view based on whole mount labeling and lineage analysis", *Development*, 126, 1999, pp. 2409-2418.

[284] C.S. Potten, C. Booth, and D.M. Pritchard, "The intestinal epithelial stem cell: The mucosal governor", *Intl. J. Path.*, 78, 1997, pp. 219-243.

[285] J.M.W. Slack, "Stem cells in epithelial tissues", *Science*, 287, 2000, pp. 1431-1433.

[286] F.H. Gage, "Mammalian neural stem cells", *Science*, 287, 2000, pp. 1433-1438.

[287] D.W. Roberson, and E.W. Rubel, "Hair cell regeneration", *Curr. Opin. Otolaryngol. Head Neck Surg.*, 3, 1995, pp. 302-307.

[288] G.K. Michalopoulos, and M.C. DeFrances, "Liver regeneration", *Science*, 276, 1997, pp. 60-66.

[289] F.M. Hansen-Smith, and B.M. Carlson, "Cellular responses to free grafting of the extensor *digitorum longus* muscle of the rat", *J. Neurol Sci.*, 41, 1979, pp. 149-173.

[290] A.W. Ham, and D.H. Cormack, *Histology*, 8th ed., (J.B. Lippincott, Philadelphia, 1978), pp. 377-462.

[291] David L. Stocum, "Regenerative biology and medicine", *Cellscience Reviews*, 1(1), July 2004.

develop into other types of specialized cells. Recent experiments in mice have indicated that when neural stem cells are placed in the bone marrow, they appear to produce a variety of blood cell types; studies in rats have shown that when stem cells found in the bone marrow are placed in the liver, they are able to produce liver cells. These findings suggest that even after a stem cell has begun to specialize, the stem cell may, under proper conditions, be more flexible and more potent.

Such being the case, why has it not been more aggressive pursuit of adult stem cells? While adult stem cells can be potent, there are some significant limitations to what we may be able to accomplish with them. Adult stem cells

- ☐ Have not been isolated for all tissue types.
- ☐ Are present in minute quantities, and are thus difficult to isolate and purify.
- ☐ May decrease in quantities when an individual ages.
- ☐ May not proliferate as effectively as younger cells.
- ☐ May contain too many genetic abnormalities from environmental exposure for them to be useful.
- ☐ May not be suitable for early stage cell specialization research for they have progressed further along the specialization pathways.
- ☐ Are multipotent and are not pluripotent.

Table 1. Differences between embryonic stem cells and adult stem cells. Examples of adult stem cells are: hematopoietic stem cells (HSCs), hepatic stem cells (HeSCs), neural stem cells, muscle stem cells, spermatogonial stem cells (SSCs), epidermal stem cells. (Source: Lecture notes by Choon Kit-Too, et al, May 2005).

Human Embryonic Stem Cells (hESCs)	Human Adult Stem Cells (hASCs)
They are abundant since they can undergo prolific cell growth.	SCs are difficult to obtain from organs, are limited in number, difficult to access and purify.
Cell lines are available, convenient to scale-up employing bulk culture.	Absence of cell lines.
Exhibit wide plasticity.	Exhibit narrow plasticity.
Pluripotent.	Multipotent.
Possible to have differentiated tissues through somatic cell nuclear transfer.	Not possible to have customized SCs or differentiated tissues.
Ethical issues in some nations.	No ethical issues associated with them.
Find applications in: (1) transplantation therapy, (2) pharmaceutical screening, (3) production of gametes and embryos, (4) research to study diseases such as infant cancer.	Find application in transplantation therapy.

Attempts to use stem cells from a patient's own body for treatment may overcome immunological rejection, but it may have other complications. Such efforts would require isolation of the stem cell from the patient and then grow the cells in culture to sufficient quantities for treatment. For some acute disorders, there may not be enough time to isolate and grow enough cells for treatment, while in other genetic disorders, the genetic disorder is likely present in the patient's stem cells.

Researches on adult stem cells are progressing along several directions. As noted above, there are many different types of adult stem cells that were thought to generate cells from different tissues, for example blood-forming stem cells have been thought to make only blood cells. However, studies over the past four or five years have observed a contribution of bone marrow cells to unrelated tissues, like heart, causing some scientists to believe that adult stem cells can mature into specialized cells of unrelated tissue types, in a process known as "transdifferentiation." Thus, for example, hematopoietic or blood stem cells could give rise to mature neurons, or vice versa if they were placed in the appropriate environment.

The ability of bone marrow cells to contribute at very low levels to other tissues was interpreted as indicating that these cells had the plasticity to make new cells in other tissues. As a result, clinical trials have been initiated in which bone marrow cells have been injected into heart muscles in an effort to stimulate the formation of new heart cells after a heart attack.

An exemplary of research on adult stem cells is that by Catherine Verfaillie. Verfaillie and colleagues at the University of Minnesota made a breakthrough discovery that a rare cell, apparently an immortal stem cell, which they termed multipotent adult progenitor cell (MAPC), can morph into heart, brain, liver and other types of cells. The team isolated the rare cell in bone marrows from mice, rats and humans.[292] This discovery is significant because it shows that stem cells taken from adults can be versatile enough or have the plasticity to produce a wide variety of specialized cells, which in turn might be used to treat diseases.[293]

Groups arguing against human cloning immediately capitalize on the discovery to cite Verfaillie's results as evidence that there is no need to clone humans to make embryonic stem cells. But scientists suspect that there are

---

[292] Yuehua Jiang, Balkrishna Jahagirdar, Lee Reinhardt, Robert Schwartz, Dirk Keene, Xilma Ortiz-Gonzalez, Morayma Reyes, Todd Lenvik, Troy Lund, Mark Blackstad, Jinbo Du, Sara Aldrich, Aaron Lisberg, Walter Low, David Largaespada, and Catherine Verfaillie, "Pluripotency of mesenchymal stem cells derived from adult marrow", *Nature*, 418, July 2002, pp. 41-49.

[293] Yuehua Jiang, Dori Henderson, Mark Blackstad, Angel Chen, Robert F. Miller, and Catherine M. Verfaillie, "Neuroectodermal differentiation from mouse multipotent adult progenitor cells", *PNAS*, 100, 2003, pp. 11854-11860.

some subtle differences between the newly discovered adult cells and embryonic stem cells, differences of which might manifest therapeutically.

Verfaillie, who was already anointed by *The U.S. News and World Report* as one of the top 10 scientific innovators in 2001 before this discovery of MAPCs made international headlines, cautioned that despite the discovery, scientists should not abandon embryonic stem cell research. Others including Austin Smith, join in, "…it is very exciting that we can now compare the two: [embryonic and adult stem cells]..."

In October 2003, a study by Sean Morrison of the University of Michigan, Manuel Alvarez-Dolado and Ricardo Pardal in the laboratory of Arturo Alvarez-Buylla of the University of California, San Francisco, cast doubts on the plasticity of adult stem cells. The new finding indicates that bone marrow contributes to other tissues by fusing with pre-existing cells rather than by forming new cells. The phenomenon of fusion would give the appearance that bone marrow stem cells are altering themselves to become mature cells in other tissues, when in fact they are not.[294]

The study is also significant in another way: investigators have long used cell fusion as an experimental tool to explore the relative influence of one cell's cytoplasm over another's nucleus, but they never suspected that fusion occurs naturally. It is possible that nature uses fusion to enable cells such as neurons with damaged nuclei to obtain donor nuclei from blood cells that fuse with them. This may be a rescue mechanism for cells that are long-lived and that have no other alternative to avoid death. If this is the case, cell fusion may provide therapeutic benefits![295]

This study further corroborates the voices of Sean Morrison, Catherine Verfaillie, Austin Smith and others who emphasize the importance of using a wide range of studies, including using embryonic and adult stem cells, to determine the properties of stem cells.

► Stem Cells from Fat

Fat stem cells? Fat chance. Not that these stem cells are fat; they are stem cells derived from fat. If so, getting rid of fat may help one's health in more ways than one might expect. Cytori Therapeutics, a San Diego biotechnology company, has developed a machine called the Celution System. This platform

---

[294] M. Alvarez-Dolado, R. Pardal, J.M. Garcia-Verdugo, J.R. Fike, H.O. Lee, K. Pfeffer, C. Lois, S.J. Morrison, and A. Alvarez-Buylla, "Fusion of bone-marrow-derived cells with Purkinje neurons, cardiomyocytes and hepatocytes", *Nature*, October 12, 2003.
[295] "Doubt cast on adult stem-cell plasticity studies", *Howard Hughes Medical Institute Research News*, October 12, 2003.

technology can extract stem cells and other regenerative cells out of fat so they can be re-injected into the body to repair tissue damaged by heart attack or disease.

The company's goal is to be one of the first to commercialize the use of stem cells from fat and bring new therapies to patients. "There are broad markets for our platform technology," Chief Executive Christopher Calhoun said. "Stem cells work in a variety of ways and are going to work for treating a variety of conditions, by turning into muscle or bone or cartilage. And that on its own creates a lot of different opportunities."[296]

Cytori did not start out obsessed with fat and stem cells. The company was founded in 1996 as MacroPore Biosurgery and focused on orthopedic implants, which is now one of the product lines. In 2000, the company went public and started trading on the Frankfurt stock exchange. Two years later, the founder Calhoun met University of California at Los Angeles (UCLA) professor Marc Hedrick, who first discovered stem cells in fat. Hedrick had founded StemSource. The companies merged in the same year and Hedrick became president. In 2005, MacroPore changed its name to Cytori and began trading on the Nasdaq.

In January 2006, the company gained European regulatory approval to begin using its machine in clinical trials on humans. Its current plans include using the system in clinical trials outside the U.S. in applications on the heart, and beginning studies in Japan using fat-derived stem cells in reconstructive surgery. Meanwhile, the company is pursuing U.S. regulatory approval for its system as a medical device. If the company can clear that approval hurdle, it will also begin clinical trials in the U.S., using the fat-derived stem cells in cardiac applications, spinal disc regeneration, gastrointestinal disorders and peripheral vascular disease.

The company hopes regulators will consider it a medical device company since its platform technology does not manipulate the stem cells in any way; the machine would be used only to extract fat, separate stem cells and inject patients' own stem cells back into them. Device approval process is a much faster and less complicated than the approval process for drugs or therapies.

The stem cells Cytori deals with are "adult" stem cells (ASCs), which are programmed to become a specific cell type: bone, blood, nerve or tissue. Studies have shown that ASCs sit within reservoirs in fat, bone and blood until they are called upon to repair a broken bone or damaged tissue.

---

[296] Terri Somers, "San Diego's Cytori banks on a new process to help damaged heart", *The Union Tribune*, February 3, 2006.

Each type of stem cell has its own unique qualities. They each require different methods for retrieval. For example, only a small amount of stem cells can be pulled from bone marrow. To create a sufficient amount for therapy, the marrow must be sent to a lab where it is used to grow more cells, in a process called culturing. This can take two to four weeks. It is expensive – estimated to cost about $30,000 – and requires the patient to make a second visit to the doctor to receive the therapy.

Stem cells from fat are more plentiful, easier to retrieve and therefore less expensive. Most people have fat they would not mind getting rid of around their waist or their love handles, but even thin people have enough fat to provide stem cells.

In the retrieval process, pinkish-orange fat tissue is first transferred into a small centrifuge containing enzymes that help digest it. The centrifuge separates the yellow-tinged fat cells from the pink regenerative cells that include stem cells, collagen and other tissue scaffolding. The desired regenerative cells are then skimmed off and placed in a larger centrifuge, which goes through a relatively quiet spin cycle to rinse the cells with a saline solution and further separate the stem cells from other cellular material. The final product of stem cells is the color of pink lemonade.

Weekly, Californians are making nonmonetary donations to Cytori. The company receives liters of raw material from people who agree to allow the fat removed from their bodies during liposuction be used in research, just like infertility clinics are donating unused embryos. The stem cells in about one cup of fat can generate what Cytori currently considers to be a therapeutic dose. The average liposuction procedure for cosmetic purposes provides about one liter.

A competitive advantage of the system is that it would allow a doctor to remove a small quantity of fat from a patient and process out the stem cells and regenerative cells within an hour, so that they could be injected into the patient in real time. Giving the patient his or her own stem cells as a therapy should avoid the issue of rejection by the autoimmune system.

In November 2005, Cytori received $11 million by entering into collaboration with Olympus. The company envisions that its final system, which the Japanese company Olympus is helping Cytori refine, would be placed in a hospital surgery suite or catheter lab. As a doctor is snaking a catheter into a patient's heart, the machine could be separating the stem cells from fat for injection when the doctor is ready.

Making the system affordable is the key to grabbing investor interest. Stem cell therapy is considered a form of personalized medicine, meaning therapies are tailored to individual patient's needs and genetics. Investors have objections to putting their money behind personalized medicine because typically therapies are thought to cost about $30,000.

"To be commercially viable, and reach the point where insurance might cover it, the cost to the hospital or doctor needs to be about $5,000 or less," said Stephen Dunn, an analyst with Dawson James Securities in Boca Raton, Florida. Adult stem cell therapies, such as the one Cytori is pursuing, seem to be in that range.[297]

Several studies have shown that stem cells and other regenerative material from fat can help build blood supply and restore blood flow to cardiac muscle that has been damaged by a heart attack. However, there could be other questions. As a cardiologist, stem cell researcher Dr. Judy Swain would like to see supporting data showing that the restorative effects of the therapy for heart ailments is a result of the stem cells and not the other matter, like growth factor, in the space between the cells.

"I'd like to see them label the cells that are injected and show whether the cells stay where they are injected," said Swain, who is dean of translational medicine, University of California at San Diego (UCSD). "Do they divide? Do you get more? Are they in sync with the other electrical impulses in the heart muscle?"

"Most other tissue and organs, you can put extra cells in and if they are not doing any good, at least they will not do anything bad. With the heart, those extra cells have the potential to do something bad electrically," Swain said. Cytori executives say they do not know if the fat-derived stem cells alone are responsible for the restorative results shown in animal studies.

The company, however, is already making plans for the future. Once the company passes its regulatory hurdles, studies on the efficacy of the therapy and characterizing the cells functions are priorities they will tackle with collaborators. If the stem cell business takes off, Cytori already has a secondary business up and running: a stem cell bank.

"We're really not marketing it right now," Calhoun said. "But younger fat appears to be much richer in stem cells." People could have their stem cells harvested early, he said, bank them and rest easier knowing they are there should they be needed. Other companies are also looking at stem cells from bone marrow and umbilical cord blood, but Cytori thinks the money is in the fat.

---

[297] Terri Somers, "San Diego's Cytori banks on a new process to help damaged heart", *The Union Tribune*, February 3, 2006.

▶ Medical Uses of Stem Cells

There are at least three broad categories of applications of stem cells:

First, stem cells can be used in delineating the complex events during human development. Stem cells develop into specialized cells and during the specialization process genes are turned on and off. Identifying the factors involved in the cellular decision-making process that results in cell specialization has become a primary goal of studying stem cell development. Currently we know genes are turned on and off but we know little about the decision-making genes or what turns the genes on and off. In some of the most serious medical conditions, including cancer and birth defects, aberrant cell specialization and cell division take place. An understanding of normal cell processes will unlock the key to the mystery of the aberrations that cause these deadly diseases.

Second, stem cells can be used to streamline drug safety tests. Conventionally, new drugs go through three phases of clinical trials before they are released onto the market: phase I being tests conducted on animals, phase II on patients, and phase III on a larger human population including healthy individuals. The three phases of clinical trials are very time-consuming and can take as many as fifteen years. Few of the new drugs will pass the clinical trials while most will fail at different phases of the trials. To be more cost-effective and to expedite the clinical trial process, new medications can be initially tested on human cell lines. Only those that are not toxic but appear to have curative effects in cell lines would pass for further testing using laboratory animals and human subjects. MAPCs may also be used to compare the toxicity and efficacy of new drugs on embryonic and adult cells.

The process of testing on cell lines will not replace testing in whole animals and testing in humans, rather it would eliminate unlikely candidates early and thus streamline the process.

Third, stem cells can be used in cell and tissue therapies, and organ transplants. This is perhaps the most far-reaching application of pluripotent stem cells. Many diseases and disorders originate from disruptions in cellular function, destructions of tissues or organs of the body. Current remedies use donated tissues and organs to replace ailing or destroyed tissues and organs. Unfortunately, the number of people suffering from these disorders far outstrips the number of tissues and organs available for transplantation. Stem cells, stimulated to develop into specialized cells, offer the possibility of a renewable source of replacement cells, tissues and organs to treat a myriad of diseases, conditions, and disabilities.

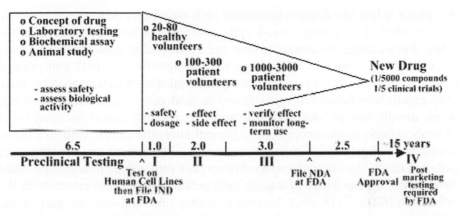

Figure 4. The long and expensive procedure for gaining FDA approval for marketing pharmaceutical products can be streamlined by first testing drug targets on human cell lines. Only those drug targets that pass this test need proceed further.

Stem Cell Gold Mines

To further underscore the important role stem cells play in therapy and medicine, an example is in order here. One common misconception about cancer is that cancer cells always beget cancer cells. This has an unfortunate impact on the development of therapies because the misconception implies that unless cancer cells are eradicated from the body, the host will die. Treatments have therefore focused on eliminating cancer cells entirely through surgery, chemotherapy, or radiation therapy. Oncologists first poison the patient and then rescue the individual, then another dose of poison followed by another rescue, and so on, with the hope that through repeated cycles the patient will recover and the tumor will suffer incremental damage, thus resulting in a cure. Though these treatments have met with successes, they exact a high toll on the patients because chemotherapy and radiation do not only destroy cancer cells, but also normal cells.

There has been a breakthrough in cancer therapy research, pioneered by Barry Pierce. Pierce sees an uncanny resemblance between the unchecked proliferation of cancer cells and the process of cell renewal. In cancer, a stem cell or some other kind of cell capable of division is somehow transformed into a malignant stem cell that produces cancerous offspring. Differentiation into other cell types may be blocked with the malignant stem cells producing more stem cells in an uncontrolled avalanche of proliferation.[298]

---

[298] Steve Olson, *Shaping the Future: Biology and human values*, (National Academy Press, Washington, DC, 1989).

This is where the dogma that cancer cells only beget cancer cells is wrong. In a number of situations cancer cells differentiate into apparently normal cells. For example, teratocarcinomas are cancers that arise from malignant sex cells. This form of cancer shares the characteristics of both cancers and embryos, that is, the tumors contain both undifferentiated cancerous stem cells and differentiated cell types found in embryos. The differentiated cell types, usually benign, are derived from differentiating cancerous stem cells! In other words, in the process of differentiating, the cancer cells somehow shed their malignancy. Similarly, leukemia, colon cancer and breast cancer cells can be chemically induced to differentiate into nonmalignant cells. And remarkably, certain kinds of cancer cells when inserted into embryos at the right time in the right place lose their malignancy and become part of the embryos.

All the above instances point to an entirely new approach to cancer treatment using differentiation therapy – if cancer cells could be induced to differentiate, it might be possible to eliminate cancer cells without harming other cells in the body. For this therapy to be of any practical applications, two conditions must be met: First, the inducing agent must be very specific for the particular cancer cell since inducing agents that cause all normal stem cells to differentiate would be disastrous; Second, the differentiated cancer cells must remain benign. To date, induced differentiated cells do not always remain benign.

Cell and tissue therapies, and organ transplants have stirred a lot of controversy. Besides ethical issues, part of the controversies arises from current knowledge of these therapies. For tissue therapy and organ transplants to be effective, we have yet to understand cell specialization and immune rejection. We have to understand the cellular events that lead to cell specialization in humans so that we can direct stem cells to become the types of tissue or organ needed for transplantation.

Cells in the human body display a protein called human leukocyte antigens (HLA) that help the immune system tell friend from foe. HLA are proteins that are located on the surface of the white blood cells and other tissues in the body. Like organ transplants, embryonic stem cells infused in a patient can trigger a potentially fatal reaction unless the immune system is suppressed. But suppression of the immune system triggers a host of side effects that can doom the stem cell treatment to failure. We therefore have to know how to overcome immune rejection because human stem cells derived from different sources are genetically different from the recipient. Immune rejection can be partially overcome if we know how to modify stem cells to

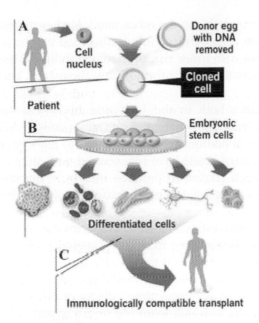

Figure 5. In therapeutic cloning, the nucleus of any somatic cell of the patient may be used to create the required cell type: (A) The nucleus of a somatic cell of the patient is fused with an enucleated egg. Four to 5 days into cell division, cells from the inner cell mass of the blastocyst are extracted. (B) The cells are cultured for later use to stimulate into different cell types. (C) The differentiated cells, immunologically compatible, can be transplanted into the patient with almost no risk of rejection.

minimize tissue incompatibility. One way to do so is to knock out or shut off the HLA genes. Such modified cells could be universal donors like type-O blood accepted by the immune systems of all patients. Alternatively, we can create tissue banks with the most common tissue-type profiles to increase the chances of a match.

Somatic cell nuclear transfer (SCNT) and cord blood banking may have an advantage in addressing the issue of tissue incompatibility. Suppose a patient has a degenerative disease. Using SCNT (otherwise known as therapeutic cloning), as the name stipulates, the nucleus of virtually any somatic cell from the patient could be used for therapeutic cloning to create a culture of pluripotent cells. These cells could then be stimulated to develop into the required cell type. Since these cells would essentially be genetically identical to the patient's (and donor's in this case) cells, the chances of rejection could be drastically reduced.

Cord blood stem cells, on the other hand, have a muted immune system, that is to say, stem cells derived from cord blood can be transplanted into a non-relative patient with lower risk of incompatibility. Thus banking of cord blood is a form of bio-insurance for not only the owner, but also for others in need of transplants.

Limited research both in the U.S. and elsewhere has already shown glimmers of the enormous potential of stem cells. Mice paralyzed by degenerative nerve conditions have been cured to walk again using stem cell therapy; damaged mice hearts have regenerated with stem cell injections, and stem cells have been coaxed to produce insulin in diabetic mice. In other words, stem cells can be used in regenerative medicines to cure degenerative disorders.

It is still difficult to estimate in absolute terms the commercial potential of stem cell research. But judging from the potential applications of stem cells, the impact will be enormous and the windfall to owners of cell lines will be tremendous. Almost all realms of medicine will be affected by this new innovation, all the way from drug discovery to organ transplant to regenerative medicine in which worn out or ailing parts of a human body are rejuvenated. Stem cells can be cajoled into virtually every kind of tissue, offering a potentially renewable source of replacement cells, tissues and body parts to treat a myriad of diseases, conditions, and disabilities including Parkinson's and Alzheimer's diseases, spinal cord injury, stroke, burns, heart disease, diabetes, osteoarthritis and rheumatoid arthritis.

To date, there are about 30 viable cell lines. There is still a need to generate more cell lines for obvious reasons:

- ❑ Current existing cell lines may have problems. Of the six cell lines owned by Geron, Inc., only three are stable enough to be deemed useful.
- ❑ New ways of deriving cell lines are being developed. Newer ways may improve the quality of existing stock.
- ❑ Of the order of about a hundred cell lines are needed to represent the genetic diversity in humans.

To generate more cell lines, more embryos will be needed. Though we can tap into the potential supply of many thousand (~100,000) spare embryos from the some 300 infertility clinics in the U.S., and many more spare embryos from infertility clinics in other parts of the world, there are reasons for making new embryos. The supply of stem cells is too limited is one reason, but a more important reason is the need for unique characteristics to match recipient patients to avoid rejection.

To keep cells undifferentiated, stem cells are currently grown on mouse tissue, which acts as soil or feeder, and nourished with a liquid derived from cows and pigs. The medium generates an as-yet unknown cocktail of proteins that signal cells to remain immature. The procedure causes concerns over the exposure of human stem cells to mouse cells and possibly unknown pathogens from mouse, cows and pigs. The risk concern also pretty much eliminates the stem cells from clinical trials.

Researchers in Singapore and Israel are making new grounds in making animal-free human stem cells. One of the new animal-free stem cell systems uses human tissue as feeders, and nourished with liquid derived from blood and plasma. Such an animal-free system not only quells concerns over the risk of contamination, but also provide more consistent crop of cells. In contrast, growing stem cells on mouse tissue often has erratic results. Another added advantage is that stem cells grown on human tissues stay stem cells for nine days while those grown on mouse cells grow into other types of cells after only seven days.

Growing Breast from a Stem Cell

A complete, functioning breast has been grown from a single stem cell, by researchers in Australia. Basically, the newly grown tissues do everything that a normal breast does. They respond to pregnancy, producing milk and milk proteins. This is the first time it has been shown that one stem cell can give rise to all the cell types needed for normal breast development.

Now before your imagination runs wild, it was done in a mouse, but experts believe it will not be long before it happens in humans. This idea of trying to produce a complete functioning breast from one adult stem cell originated with Geoffrey Lindeman and Jane Visvader at the Walter and Eliza Hall Institute of Medical Research in Melbourne. They enlisted Kaylene Simpson, now at Harvard University, to work with them because she had the technical expertise needed to extract individual cells from breast tissue. Simpson acquired the expertise during her Ph.D. research at the Victorian Institute of Animal Science, also in Melbourne.[299]

Breast tissue, mouse or human, is like "a bunch of grapes in a big blob of fat." Separating stem cells from the blob turned out to be a tough problem. These cells do not like to be alone; naturally, they are in aggregates of 50 to 100 individuals.

---

[299] William J. Cromie, "Complete breast is grown from single stem cell", *Harvard University Gazette*, February 9, 2006.

The separation process requires finding marker proteins that would bind to the surface of breast stem cells, something that had never been done before the Australian experiment. This was necessary to separate them from the many other types of cells that make up a breast. In their experiment, the marked stem cells were tagged with fluorescent dyes, and then sorted by machines that recognize the fluorescence. Finally, the cells were injected into young mice where they grew into new breasts.[300]

The Australian researchers have also separated stem cells from tumorous mouse breasts, and they intend to identify the different genes involved in malignant versus normal growth.

The obvious next step is to find the same markers in human breasts, which could be used to identify stem cells in either live tissue or frozen samples. Once human breast stem cells can be isolated, scientists will attempt to grow a human breast in a laboratory. There are numerous potential applications. Comparing the differences in development between normal and tumorous breasts could lead to new treatments that attack the earliest stages of breast cancer, when therapy is most effective. Perhaps breast surgery could be avoided completely. There is also the possibility of growing new breasts to replace those lost to surgery. As such medical applications become more likely, commercial interest in using stem cells to enhance breast size, or customizing breast, could also grow larger.

Errant breast cells can become what researchers call "a tumor factory" where they perpetually divide and produce daughter cells. Such factories may be responsible for recurrence of human cancers after surgery and drug treatment or chemotherapy. Chemotherapy works by killing cells that grow rapidly, typical of cancer cell behavior. But tumor factory cells grow more slowly and may survive for months or years after treatment. The hope is that a close comparison between genes in normal and malignant breasts will identify the genes that make the difference. It then may be possible to design a drug that targets the faulty genes of rogue cells.

It is also conceivable that using similar techniques, it could be possible to isolate stem cells that control development of other internal organs. "I'm hoping that the isolation of stem cells that produce such massively complex organs as breasts will become a model for growing organs, such as the lungs and pancreas," both sites of lethal cancers, Simpson says.

---

[300] Mark Shackleton, François Vaillant, Kaylene J. Simpson, John Stingl, Gordon K. Smyth, Marie-Liesse Asselin-Labat, Li Wu, Geoffrey J. Lindeman, and Jane E. Visvader, "Generation of a functional mammary gland from a single stem cell", *Nature,* 439, January 5, 2006, pp. 84-88.

► Regenerative Medicine

Currently, the only way to compensate for diseased or injured tissues or organs is through bionic implants and organ transplants. Regenerative medicine is a new innovative technique to replace or repair defective or disease tissues or organs by *in vivo* (in the living body) means or *in vitro* (in the laboratory) design for *in situ* (in the natural environment) usage to restore the structure and function of those damaged tissues or organs. This new technique – for therapeutic use in the treatment of incurable, debilitating, and chronic disease – encompasses many novel approaches to treatment of disease and restoration of biological function through the following methods:

□ Using therapies that prompt the body to autonomously regenerate damaged tissues or organs (*in vivo* means),

□ Using tissue engineered implants to prompt regeneration (*in vitro* design and *in situ* usage),

□ Direct transplantation of healthy tissues or organs into damaged environments (*in vitro* design and *in situ* usage).

Collectively, these new treatments of regenerative medicine allow for at least two substantial advances over the current state of medicine. The first advance is the potential to *in vivo* (in the living body) regenerate currently irreparably damaged tissues or organs so that they return to full functionality. The second advance is to be able to produce tissues or organs *in vitro* (in the laboratory) to be used for transplantation purposes when regeneration is not possible. Creating tissues *in vitro* and then bringing these tissues *in situ* (in the natural environment) will also shed light on how to integrate laboratory-grown cells into actual living bodies. This technology has the potential to cure (yes, to cure, NOT to treat) diseases ranging from diabetes (through regeneration of islets) to the repair of cancerous tissues (by replacing the removed cancerous tissue with externally grown healthy tissue).[301]

Attempts to establish a regenerative medicine have been ongoing for over two decades. To date, the vast majority of work has been done with experimental animals, although human clinical trials have been made as well, such as those to restore several types of tissues, including islet cells of the pancreas to cure diabetes and dopaminergic neurons of the substantia nigra to cure Parkinson's disease.

---

[301] *2020: A New Vision – A Future for Regenerative Medicine*, (U.S. Department of Health and Human Services, Washington, D.C., January 2005).

Why Regenerative Medicine?

What regeneration does is to maintain or restore the original architecture of a tissue by recapitulating part of its original embryonic development. However, because of the energy cost and other reasons associated with regeneration, nature has provided us with a much more common mechanism for injury-induced repair, called fibrosis, which is the result of an inflammatory response. Fibrosis produces a fibroblastic granulation tissue that is then remodeled into a collagenous scar. In other words, it is a quick fix that patches a wound with scar tissue, lowering its functional capacity. Obviously, tissues that do not regenerate spontaneously are repaired by fibrosis. There are also situations in which tissues capable of regeneration may repair by fibrosis because they suffer wounds that exceed their regenerative capacity. Chronic degenerative diseases can also diminish the ability to regenerate, again resulting in fibrotic repair. Excellent examples of tissues that undergo fibrotic repair after damage are articular cartilage, the dermis of the skin, the pancreas, the spinal cord and most regions of the brain, the neural retina and lens of the eye, cardiac muscle, lung, and kidney glomerulus.

The goal of regenerative medicine is to be able to replace tissues, organs and appendages with bioartificial constructs or to guide the repair process along a regenerative pathway, rather than a pathway leading to scar tissue formation. Virtually any disease that results from malfunctioning, damaged, or failing tissues may be potentially cured through regenerative medicine therapies. What has been overlooked in the midst of all the excitement, however, is that regenerative medicine will not become a reality without a fundamental understanding of the mechanisms of regeneration, which are driven by advances in molecular, cell and developmental biology, and information science and technology. This understanding is far from complete. Thus for now this new field of tissue or organ restoration should be called "regenerative biology" to highlight that understanding the biology of regeneration is a prerequisite to establishing a regenerative medicine, just like the full understanding of the basic science comes before any new technology.

The objective of regenerative biology is to define the factors and understand the mechanisms that lead to a regenerative response and how these factors and mechanisms differ from a fibrotic response to injury. Regenerative medicine then seeks to apply this new knowledge to devise therapies that will stimulate the functional regeneration of damaged human tissues that do not regenerate spontaneously, or whose regenerative capacity has been compromised.[302]

---

[302] David L. Stocum, "Regenerative biology and medicine", *Cellscience Reviews*, 1(1), July 2004.

From an economic point of view, there are at least three reasons to develop regenerative medicine: First, it could help curb healthcare spending; second, it would alleviate loss of productivity or quality of life; and third, it would help stay competitive in a new market sector.

What truly distinguishes regenerative medicine from many current therapies is that regenerative medicine has the potential to provide a cure to failing or impaired tissues or organs. Many of today's skyrocketing healthcare costs are results of recurring treatments for chronic diseases and their subsequent complications. One such example is insulin therapy for type 1 diabetes, and glucose therapy for type 2 diabetes. While insulin and glucose can help patients manage diabetes, these therapies do not cure diabetes, nor do they prevent long-term complications, such as kidney failure.

In diabetics, pancreatic islets do not produce the proper insulin levels. Through regenerative medicine, instead of treating the disease, insulin-producing pancreatic islets could be regenerated *in vivo* or grown *in vitro* and implanted, creating the potential for curing the patient and completely eliminating the need for future treatments. Because regenerative medicine focuses on functional restoration of damaged tissues, not abatement or moderation of symptoms, it cuts healthcare costs.[303]

The cost of tissue damage due to regenerative incompetence is enormous in terms of healthcare. It is estimated to exceed $400 billion in the U.S. alone. This includes lost economic productivity, diminished quality of life, and premature death. The healthcare costs of spinal cord injuries, which are some of the most devastating known, exceed $8 billion per year and $1.5 million per patient over a lifetime in the U.S. Diabetes, heart, liver, and renal failure, emphysema, macular degeneration and other retinal diseases, diseases of the central nervous system such as Parkinson's, Huntington's, and Alzheimer's, arthritis, burns, and various sports injuries that damage ligaments, tendons, and joints are other major contributors to these fiscal and human costs. Thus medical science seeks ways not only to prevent and abate underlying disease, but also to restore the structure and function of damaged tissues and organs.[304]

Additionally, there is the opportunity to create a tremendous new global industry. The current world market for replacement organ therapies is in excess of $350 billion, and the projected U.S. market for regenerative medicine is estimated at $100 billion. The worldwide market for regenerative medicine is conservatively estimated to be $500 billion by 2010. Furthering

---

[303] *2020: A New Vision – A Future for Regenerative Medicine*, (U.S. Department of Health and Human Services, Washington, D.C., January 2005).
[304] David L. Stocum, "Regenerative biology and medicine", *Cellscience Reviews*, 1(1), July 2004.

this field would create a new sector of the healthcare industry while creating a new generation of life-saving products.[305]

Already, Japan, the European Union (EU), China and Australia have begun national initiatives and efforts to spur the advancement of their regenerative medicine programs. These commitments range from policy directives in the EU to extensive financial investment by the Japanese government focused on the city of Kobe and surrounding Kansai region targeted to develop a region of expertise in tissue engineering and regenerative medicine.[306,307,308]

Regenerative Medicine and Stem Cell Research

The failure of a tissue to regenerate could be due to various factors. The two most obvious are a lack of regeneration-competent cells, the lack of an environment favorable to regeneration, or a combination of both. Regeneration-competent cells can exist as both differentiated or undifferentiated reserve cells. This suggests that virtually all adult cells (with the exception of B and T cells) have the potential to engage in regeneration. This idea is compatible with the fact that present in all cells is a complete genome. This genome can, in principle, be reprogrammed to regenerate tissues and appendages,[309] or even complete organisms.[310,311]

Thus, potential strategies of regenerative medicine include:

- ☐ Stem cell transplantation – this is an *in situ* approach, that is, stem cells, which are regeneration-competent, are allowed to grow in their natural environments,
- ☐ Implantation of bioartificial tissues synthesized in the laboratory – this is an *in vitro* approach, that is, design tissues in the laboratory for implantation into the body, and

---

[305] Ratner, Buddy. "What are the opportunities in the field of tissue engineering/regenerative medicine?" Workshop on Tissue Engineering and Regenerative Medicine, U.S. Department of Health and Human Services, 28 March 2003.

[306] Larry V. McIntire, "What is the Current State of Regenerative Medicine?" Workshop on Tissue Engineering and Regenerative Medicine, U.S. Department of Health and Human Services, March 28, 2003.

[307] The Kobe Medical Industry Development Project Outline, City of Kobe, Japan, October 8, 2003.

[308] Mission Statement, Enterprise Directorate-General Unit F3, October 8, 2003.

[309] D.L. Stocum, "Amphibian regeneration and stem cells", In: Regeneration: Beyond the Stem Cells, *Curr. Topics in Microbiol. and Immunol.*, 280, 2004, pp. 1-70.

[310] J.A. Byrne, S. Simonsson, and J.B. Gurdon, "From intestine to muscle: nuclear reprogramming through defective cloned embryos", *Proc. Natl. Acad. Sci., USA*, 99, 2002, pp. 6059-6063.

[311] I. Wilmut, N. Beaujean, A. de Sousa, A. Dinnyes, T. King, L.A. Paterson, D.N. Wells, and L.E. Young, "Somatic cell nuclear transfer", *Nature*, 419, 2002, pp. 583-586.

□ The induction of regeneration from the body's own cells by rendering the injury environment or responding cells regeneration-competent – this is an *in vivo* approach, that is, regeneration is induced in the body.

Activation of stem cells to divide may occur differently in maintenance regeneration and in injury-induced regeneration. In maintenance regeneration, the stem cells divide in response to environmental signals, undergoing continual but slow divisions, feeding a constant excess of progeny into a new environment. In tissues that do not normally undergo rapid turnover, stem cells remain dormant until injury-induced signals mobilize them.

A current debate is whether some adult stem cells (ASCs) can be reprogrammed to transdifferentiate by extracellular signals from other cells?[312,313,314] This is just a whimsical way of asking if an ASC can be converted into a different cell under favorable ambient conditions. If so, it would be possible to use just one type of autogenic stem cell, say bone marrow cells which are easily accessible, for transplantation to a lesion site to generate the cell types that make up the local tissue.

In fact, the bone marrow and connective tissue compartments of multiple organs harbor multipotent and/or pluripotent stem cells, which have been isolated after extensive culturing of bone marrow and connective tissue cells. These stem cells possess a number of qualities of embryonic stem cells (ESCs), though they are different from ESCs.[315,316] They have been shown to differentiate into many, if not all, cell types, both *in vivo* and *in vitro*. Whether these cells actually exist as such in the bone marrow or connective tissue or are the result of dedifferentiation in culture is not yet clear. Recent evidence, however, suggests that experimental results that have been interpreted as transdifferentiation are actually due to fusion with host cells.

Either way, the use of these ASCs in regenerative medicine would obviate two of the major problems of ESCs – the need to combat immunorejection and the bioethical issues associated with the production of ESCs. The use of ESCs for therapeutic purpose, better known as therapeutic cloning, is one form of regenerative medicine.

---

[312] E. Fuchs, and J.A. Segre, "Stem cells: a new lease on life", *Cell*, 100, 2000, pp. 143-156.

[313] I.L. Weissman, (2000) "Stem cells: units of development, units of regeneration, and units in evolution", *Cell*, 100, 2000, pp. 157-168.

[314] D.L. Stocum, "Tissue restoration through regenerative biology and medicine", *Adv. in Anat. Embryol. Cell. Biol.*, 176, 2004, pp. 1-101.

[315] Y. Jiang, B.N. Jahagirdar, R. Reinhardt, R.E. Schwarts, C.D. Keene, X.R. Ortiz-Gonzalez, M. Reyes, T. Lenvik, T. Lund, M. Blackstad, J. Du, S. Aldrich, A. Lisberg, W.C. Low, D.A. Largaespada, and C.M. Verfaille, "Pluripotency of mesenchymal stem cells derived from adult marrow", *Nature*, 418, 2002, pp. 41-49.

[316] H. Young, and A.C. Black Jr., "Adult stem cells", *Anat. Rec.*, 276A, 2004, pp. 75-102.

▶ Regeneration, Stem Cells and Aging

Here is a heartening truth: Whatever a person's age, the body is actually many years younger. This is the regeneration that we hear about so much, which means that most of the body's tissues are under constant renewal. But here comes the not so good news: regeneration is not for all cells. A few cells endure from birth to death without renewal. This explains why people behave their birth age, not the physical age of their cells. Examples of such cells include some or all of the cells of the cerebral cortex.

Though regeneration is known to occur, it is only recently that a novel method has been developed to estimate the age of human body cells. The inventor, Jonas Frisen, a stem cell biologist at the Karolinska Institute in Stockholm, has successfully used the method to estimate the average age of various cells in an adult's body.[317]

Dr. Frisen was attempting to resolve the dispute over whether the cortex ever makes any new cells. Existing techniques for determining how old human cells are depend on tagging DNA with chemicals, but they are far from reliable. This led Dr. Frisen to look for an alternative and better way. In fact, a natural tag – carbon 14 – is already in place in the atmosphere from above ground nuclear weapon tests until 1963, and from cosmic rays entering the earth's atmosphere. Carbon 14 is unstable and has a half-life of 5730 years.

Photosynthesized by plants worldwide, which are then consumed by animals and people, the carbon 14 gets incorporated into the DNA of cells each time the cell divides and the DNA is duplicated. While most molecules in a cell are constantly being replaced, the DNA is not. This offers a great advantage of working with a cell's DNA. Since all the carbon 14 in a cell's DNA is acquired on the day its parent cell divided, the extent of carbon 14 enrichment could be used to figure out the cell's age. Because individual cells do not have enough carbon 14 to produce a detectable signal, the method has to be performed on an aggregate of many cells, that is, on tissues. By calibrating against carbon 14 incorporated into individual tree rings in Swedish pine trees, Dr. Frisen worked out a scale for converting carbon 14 enrichment into calendar dates.

After validating the method with various tests, Dr. Frisen and colleagues proceeded to their first tests with a few body tissues. Cells from the muscles of the ribs of people in the late 30s, have an average age of 15.1 years. The epithelial cells that line the surface of the gut, as determined by other

---

[317] Nicholas Wade, "Your body is younger than you think", *The New York Times*, August 2, 2005.

methods, last only five days, but the average age of those in the main body of the gut, or jejunum, is 15.9 years.

A contentious issue is whether the heart generates new muscle cells after birth. The conventional view is that it does not, but Dr. Piero Anversa of the New York Medical College in Valhalla has challenged this view. Dr. Frisen has found the heart as a whole is generating new cells, but he has not yet measured the turnover rate of the heart's muscle cells.

Another contentious issue is whether brain cells renew. The prevailing belief is that, with the exceptions of two specific regions – the olfactory bulb that mediates the sense of smell, and the hippocampus, where initial memories of faces and places are laid down – the brain does not generate new neurons after its structure is complete. A few years ago, Elizabeth Gould of Princeton University reported finding new neurons in the cerebral cortex, and proposed the elegant idea that each day's memories might be recorded in the new neurons generated that day.

In dating cells from the visual cortex, Dr. Frisen finds these are exactly the same age as the individual, leading him to conclude that new neurons are not generated after birth in this region of the cerebral cortex, or at least not in significant numbers discernable by his method. Cells of the cerebellum, however, are slightly younger than those of the cortex. This is in agreement with the idea that the cerebellum continues developing after birth.[318]

The human body may look like a fairly permanent structure. Most of it is actually in a state of constant flux as old cells are discarded and new ones generated in their place. Depending in part on the workload endured by its cells, each kind of tissue has its own turnover time. The epidermis (surface layer of the skin) is recycled every two weeks or so. The reason for the relatively quick replacement, as explained by Elaine Fuchs, an expert on the skin's stem cells at the Rockefeller University, is that "the skin surface is the body's saran wrap, and it can be easily damaged by scratching, solvents, wear and tear." The red blood cells, bruised and battered after traversing nearly 1,000 miles through the maze of the body's circulatory system, last only 120 days or so on average. They are then dispatched to their graveyard in the spleen.

The liver is the detoxifier of all the natural plant poisons and drugs that pass a person's lips. Its life on the chemical-warfare front is quite short. According to Markus Grompe, an expert on the liver's stem cells at the

---

[318] Kirsty L. Spalding, Ratan D. Dhardwaj, Bruce A. Buchholz, Henrik Druid, and Jonas Frisen, "Retroactive birth dating of cells in humans", *Cell*, 122, July 15, 2005, pp. 133-143.

Oregon Health & Science University, an adult human liver probably has a turnover time of 300 to 500 days,

Other tissues have lifetimes measured in years, but they are still far from permanent. For a structure that looks as permanent as the skeleton, the body's twin construction crews of bone-dissolving and bone-rebuilding cells combine to remodel it constantly. The entire human skeleton is replaced every 10 years or so in adults.

The same may not be said of the neurons of the cerebral cortex, the inner lens cells of the eye and perhaps the muscle cells of the heart. On present evidence, they seem to be the only pieces of the body that last a lifetime. The inner lens cells, for example, form during embryonic development and then lapse into inertness for the rest of their owner's lifetime. In fact, they are so inert that they dispense altogether with their nucleus and other cellular organelles.

Now comes the burning question. If the body remains so perpetually youthful and vigorous, and so eminently capable of renewing its tissues, why does not the regeneration continue forever? A theory is that the DNA accumulates mutations and its information is gradually degraded. A second theory blames the DNA of the mitochondria, which lack the repair mechanisms available for the chromosomes. A third theory is that the stem cells that are the source of new cells in each tissue eventually grow feeble with age.[319]

Table 2. Our body may look like a permanent structure, but most of it is actually in a state of constant flux as old cells are discarded and new ones generated in their place. (Data collected from various sources.)

Cell	Lifetime	Remarks
Gut surface	5 days	The epithelial cells that line the surface of the gut last only for a few days.
Skin	14 days	The entire skin surface is replaced every two weeks or so.
Red blood	120 days	Red blood cells last only 120 days or so before being sent to the spleen for destruction.
Liver	1 year	Liver cells have a turnover time of 300 to 500 days.
Bone	10 years	The entire skeleton system is replaced every 10 years or so.
Muscle	15.1 years	Rib muscle cells have an average life span of 15 years.
Gut	15.9 years	Cells of the main body of the gut, or jejunum, regenerate every 16 years or so.
Heart	Permanent	Muscle cells of the heart may last a lifetime.
Brain	Permanent	Neuron cells of the visual cortex do not regenerate.
Eye	Permanent	Lens cells of the eye last a lifetime.

[319] Nicholas Wade, "Your body is younger than you think", *The New York Times*, August 2, 2005.

"The notion that stem cells themselves age and become less capable of generating progeny is gaining increasing support," Dr. Frisen said. He hopes to see if the rate of a tissue's regeneration slows as a person ages, which might point to the stem cells as being the single impediment to immortality.

▶ Master Cells and Master Genes

Parthenogenesis – the development of an egg into an embryo without sperm fertilization – is common among reptiles, but in mammals, the eggs fail to develop.

Jose Cibelli of Advanced Cell Technology created a lot of controversy after creating human embryos in November 1999. He has done it again. In early 2003 Cibelli's team reported results from their experiments with macaque monkeys, close genetic relatives of humans. The researchers created a line of stem cells through parthenogenesis. This was achieved by tricking the egg chemically to begin the cell division process.[320]

Making embryos through parthenogenesis could bypass concerns about therapeutic cloning in which a clone is created with human cells solely to harvest the stem cells, a process that destroys the embryo. Cells from the monkey's egg, the one success out of 77 eggs, developed into a stem cell line. The stem cells were prodded to grow into brain cells and heart cells. Cibelli acknowledges that stem cells derived through parthenogenesis could be used only to make new tissue for women of childbearing age because only they produce ova. The immune systems of other recipients of these stem cells would likely trigger rejection of the foreign tissue.[321]

In February 2003, Thomas Zwaka and James Thomson of the University of Wisconsin reported the first success in developing a technique for genetically manipulating human embryonic stem cells (hESCs). The technique is similar to the one scientists use in mice to study the role of particular genes in disease and in normal development. The so-called "knock-out" mice, which are developed from embryos with one or more genes removed, have helped reveal the origins of a wide variety of conditions such as obesity and cancer. Though such procedures are routine in working with mouse embryonic stem cells (ESCs), the knock-out technique, also known as homologous recombination, has proven much more difficult in a human

---

[320] J.B. Cibelli, K.A. Grant, K.B. Chapman, K. Cunniff, T. Worst, H.L. Green, S.J. Walker, P.H. Gutin, L. Vilner, V. Tabar, T. Dominko, J. Kane, P.J. Wettstein, R.P. Lanza, L. Studer, K.E. Vrana, M.D. West, "Parthenogenetic stem cells in nonhuman primates", *Science*, 295 February 1, 2002, pp. 819.
[321] Dan Vergano, "Technique might quell stem cell research concerns", *USA Today*, January 31, 2002.

setting. One problem with hESCs is they tend to remember a little too much, and they can form a random mass known as a teratoma instead of the desired tissue.

Zwaka and Thomson succeeded with a gene that causes Lesch-Nyhan disease, a rare kind of mental retardation involving an enzyme deficiency that manifests in self-mutilating behavior such as head banging. Earlier researchers had wanted to study this gene in mice, but when mice were engineered to lack the same gene that causes the disease in humans, the mice showed no symptoms. Now with this new way to manipulate hES cells, the basic molecular mechanisms behind the disease can be studied in human cells.

The method could also be used to remove the genes that cause the body's immune system to reject foreign tissue to create universal donor batches of cells. The universal cells can be transplanted into any patient with reduced risk of immunological rejection.[322]

Though not using stem cells, but still at the gene level, Betty Pace of the University of Texas at Dallas, in an effort to overcome the current U.S. FDA's suspension of gene therapy, has come up with a novel way to cure sickle cell anemia using fetal genes. Instead of shoehorning new DNA into cells like in gene therapy, she comes up with a way for the body to heal itself. To counter sickle cell anemia, she uses a method to provoke the body to activate the fetal hemoglobin gene. This gene, which makes a protein that helps growing fetuses to siphon the oxygen they need from their mothers' bloodstream, goes dormant at birth. What makes the protein interesting is that it never carries the sickling mutation, and if enough of the hemoglobin is present in adult blood, sticky polymers will not form, even in the presence of mutated proteins made by the adult sickle cell gene. For this work, Pace was voted one of the Brilliant 10 by *Popular Science* in 2003.

Instead of just activating a fetal gene, it is conceivable that a master gene or master genes of some sort may be responsible for the differences between an adult cell and an embryonic cell. To see this, let us recall a striking attribute of stem cells is they can multiply for years in laboratory dishes, suspended in timeless youthfulness, and still retain their potential to turn into any organ or tissue they might be called upon to become. In sharp contrast, ordinary cells grow visibly older with time and they cannot help but turn into one kind of organ or tissue with time.

In May 2003, scientists claimed they had discovered the long-sought "master gene" in ES cells that is largely responsible for giving those cells

---

[322] T.P. Zwaka, and J.A. Thomson, "Homologous recombination in human embryonic stem cells." *Nature Biotechnology*, 21(3), February 10, 2003, pp. 319-321.

their unique regenerative and therapeutic potential. This discovery brings scientists closer to the holy grail of biology – the ability to turn ordinary cells into those that possess all the biomedical potency of hES cells.

Working independently, Shinya Yamanaka of Nara Institute of Science and Technology, Japan, and Austin Smith at the Institute for Stem Cell Research of the University of Edinburgh, Scotland have been conducting a series of experiments on a gene that is active only in ES cells. They later agree to call the gene *nanog*, a reference to the mythological Celtic land of Tir Nan Og,[323] whose fairy-like residents stay forever young.[324,325]

Unlike all genes which are stretches of DNA code that direct cells to make proteins, *nanog* belongs to a special class of genes whose proteins attach themselves to specific regions of a cell's DNA strand. In so doing, the gene can turn on or off other genes in that stretch of DNA, affecting the production of other proteins that affect the activity of other genes. As a result, *nanog* functions as a master gene or a regulator that can single-handedly control the activity of a whole collection of genes.

In one crucial experiment, Smith's team inserted copies of the human *nanog* gene into mouse embryonic stem cells and subjected the cells to laboratory conditions that normally make the cells mature and become one kind of tissue. The human *nanog* gene prevented the maturation process. This indicates that if scientists could reawake the dormant *nanog* gene in adult cells, they might be able to reprogram the gene activity patterns in the cells and turn them into cells that, for all practical purposes, are embryonic stem cells.

The work, conducted on mouse ES cells and on human equivalents, has already revealed more about the mysterious capacity of ES cells to retain indefinitely their youthful potential – pluripotency – to become any kind of cell. This work, however, will not bring a quick end to the political controversy over hES cell research. Rather research involving human embryos will become even more important than ever for at least a while as

---

[323] Tir Nan Og is the land to which the Irish faeries known as Tuatha de Danann fled when their lands were taken by the Milesians. In Tir Nan Og they spend their days feasting, gaming, lovemaking and partaking of beautiful music. The faeries can even enjoy the thrill of battle, for anyone slain is resurrected the following day. It is the paradise that mortals can only dream of.

[324] Kaoru Mitsui, Yoshimi Tokuzawa, Hiroaki Itoh, Kohichi Segawa, Mirei Murakami, Kazutoshi Takahashi, Masayoshi Maruyama, Mitsuyo Maeda, and Shinya Yamanaka, "The homeoprotein nanog is required for maintenance of pluripotency in mouse epiblast and ES cells", *Cell*, 113, May 30, 2003, pp. 631-642

[325] Ian Chambers, Douglas Colby, Morag Robertson, Jennifer Nichols, Sonia Lee, Susan Tweedie, and Austin Smith, "Functional expression cloning of nanog, a pluripotency sustaining factor in embryonic stem cells", *Cell*, 113, May 30, 2003, pp. 643-655.

scientists turn their attention to the master gene to learn how it works in its natural embryonic state. Scientists have yet to identify the signal that tells *nanog* to turn on early in an embryonic existence. They still do not know how to make a drug or chemical cocktail that would switch on *nanog* in an older cell. In current experiments, *nanog* activity can be turned up or blocked with genetic technique that so altered the cells as to make them impractical for use in humans.[326]

When the recipe for the cocktail is found, the understanding will eventually eliminate any further need to destroy embryos to extract stem cells because embryonic-stem-cell-like cells could then be derived from any cell, just by turning on its *nanog*!

More discoveries are coming out of the Institute for Stem Cell Research at the University of Edinburgh. When studying mouse cells, Brian Hendrich and team members found that a protein called Mbd3 plays a role in the process by which embryonic stem cells become specialized cells. It appears that cells require Mbd3 to stop self-renewing and become specialized cells.

The protein has been known to be necessary for an embryo to develop, but it is in this study that the researchers found the precise stage the protein is needed. This discovery brings further advances in the understanding of how embryonic stem cells differentiate to become all the different types of cell in the body. It could then lead to replacement therapies for specific kinds of diseases and injuries.[327]

●●●●●   ●●●●●   ●●●●●

On many an occasion, our instinctive reactions and emotions are at odds with modern age. When it comes to cloning, stem cell research and regenerative medicine, science and religion clash head on, and politicians get caught in between on a tightrope. But the lure of the huge potential benefits, both in economic terms and in healthcare, makes stem cell research and regenerative medicine priorities of many nations. And in at least one instance, a shenanigans of scandalous proportion.

---

[326] Rick Weiss, "Stem cell master gene found", *The Washington Post*, May 30, 2003, pp. A01
[327] "Protein plays crucial role in embryonic stem cells become specialized", *Stem Cell Monthly*, February 8, 2006.

12 Stem Cell Entities

"I am a Buddhist, and I have no philosophical problem with cloning. And as you know, the basis of Buddhism is that life is recycled through reincarnation. In some ways, I think, therapeutic cloning restarts the circle of life."

- Hwang Woo-suk, in an interview after his announcement of having cloned a human embryo, Seattle, February 2004

▶ Stem Cell Nations

Derivations of stem cells were made possible by advances in *in vitro* fertilization (IVF) techniques. Though scientists have vast experiences in deriving mouse embryonic stem cell lines, they cannot simply follow the recipe. In the case of mouse, blastocysts develop about a week after fertilization. The blastocysts are flushed from pregnant mice to get the inner cell mass.

Such a flushing method would never be morally acceptable in pregnant women. Researchers knew they had to find new ways to derive embryonic stem cells from test tube embryos. The big issue then was how long would a human embryo last outside the body. Until the mid 1990s, a normal routine in IVF clinics was to transplant embryos into a patient about 3 days after fertilization. But human blastocysts take four to five days to develop. The hurdle then was to keep embryos alive longer in test tubes, long enough for blastocysts to form. Researchers in the U.S. and Australia finally overcame the hurdle, setting the race to isolate human stem cells.

It is thus no coincidence that current leading nations in stem cell research are nations that have been very active in IVF, and those that have benefited from returning scholars from these nations.

On July 16, 2001, Former First Lady Nancy Reagan wrote to President George Bush to ask the president not to ban stem cell research. But on August 9, 2001, President Bush of the U.S. adopted a brand of political and regulatory tightrope of allowing stem cell research, but with restriction to existing cell lines. At 9:00pm EDT, the president announced his decision to allow federal funding to be used for research on existing human embryo stem (hES) cell lines as long as prior to his announcement (1) the derivation process had already been initiated, (2) the embryo from which the stem cell line was derived no longer had the possibility of development as a human being. In addition, the following criteria must be met to qualify for funding:

- The stem cell must have been derived from an embryo that was created for reproductive purposes;
- The embryo was no longer needed for these purposes;
- Informed consent must have been obtained for the donation of the embryo;
- No financial inducements were provided for donation of the embryo.

The Human Embryonic Stem Cell Registry (HESCR) lists the following as satisfying the president's criteria:

Goteborg University, Goteborg, Sweden	19
Cy Thera, Inc., San Diego, California	9
Geron Corporation, Menlo Park, California	7
Reliance Life Sciences, Mumbai, India	7
ES Cell International, Melbourne, Australia	6
Karolinska Institute, Stockholm, Sweden	6
Wisconsin Alumni Research Foundation, Madison, Wisconsin	5
BresaGen, Inc. Athens, Georgia	4
Technion University, Haifa, Israel	4
Maria Biotech Co. Ltd., Seoul, Korea	3
National Centre for Biological Sciences, Bangalore, India	3
Pochon CHA University, Seoul, Korea	2
University of California, San Francisco, California	2
MizMedi Hospital, Seoul, Korea	1

This type of government policy of placating real and imagined ethical concerns, as expected, invited criticisms from opponents and supporters. The reactions can be broadly divided into four groups:

- Religious group – This group of religious conservatives, often linked to anti-abortion/pro-life groups, opposes the destruction of human embryos to harvest stem cells. They thus believe stem cell research should be banned completely. It views the decision compromises the sanctity of life.

❑ Patient – This group, including Hollywood stars like Michael J. Fox and Christopher Reeve, supports stem cell research and believes the restriction will stifle research into finding cures for debilitating and degenerative ailments including Parkinson's disease, Alzheimer's disease, diabetes, spinal cord injury, and others.

❑ Scientific community – This group has mixed feelings on the decision to limit federal funding to research on existing colonies of cells. First there are those who feel relieved that Bush had actually given in from his campaign stance of banning completely human cloning and stem cell research. Second, there are those who believe restricting research to existing cell lines will impede research. The restriction will block scientific progress if further research unearths problems in the existing cell lines or points to ways to improve the stock of existing and newly derived cell lines.

❑ Business community – This group wants the work to proceed because of potential medical benefits. They believe restrictions will slow down the development of commercially viable repair kits to treat debilitating human ailments. The restriction to existing cell lines will provide windfall to the few companies who currently have access to these cell lines.

Bush's placating policy allows government funding of research on stem cells harvested prior to his August 9, 2001 decision, but removes any incentive to destroy more embryos going forward. In so doing, he "ensures" researchers' needs could be partially met in the future by cells grown from existing stock. Because of this limited funding, the research may go offshore to countries where the restriction is less stringent. The HESCR list also, in many cases, thrusts those who are listed to prominence in the field of stem cell research.

Sweden

Sweden is considered to be at the forefront of stem cell science and technology, an area many other "big science" nations including Singapore, Israel and India are pursuing actively. Sweden has all the right ingredients for future expansion and success of stem cell science and technology: strong public support, a favorable bioethical climate, a tradition of science and research, and strong government funding. The country is ahead of other countries with the establishment and funding of a national stem cell bank. The stem cell bank received $1 million for three years in September 2002.[328]

---

[328] "Sweden's stem cell success", *Cell News*, October 19, 2002.

As of last count, in this country of 8.9 million, the country has more than 33 research groups and about 300 researchers at nine institutions. In Sweden, the researchers themselves hold the patents on their discoveries, not their institutions. There is none of the placating policy, and none of the debate roiling, and none of the interruptions by wars like that in the U.S. The framework and guidelines for such research have been worked out quietly and reasonably fast: the guidelines allow stem cells to be taken from embryos that will no longer be used for *in vitro* fertilization. The creation of embryos for therapeutic cloning is also ethically defensible.

Added to the favorable climate is the long tradition of biomedical science and research. Among Swedish institutions, the Goteborg University's Sahlgrenska Medical Center has 19 lines, and Stockholm's Karolinska Institute has 6 lines on the U.S. National Institutes of Health's The Human Embryonic Stem Cell Registry (HESCR).

The stem cell pioneer, Patrik Brundin of the Wallenberg Neuroscience Centre at Lund University has been transplanting neural tissues from aborted embryos of six to 8 weeks old into patients with Parkinson's disease for more than 15 years. Not surprisingly, in March of 2002, the Michael J. Fox Foundation for Parkinson's Research awarded $4.4 million for the production of a cell line designed to advance the study and treatment of the disease.

United Kingdom

The United Kingdom has since 1990 allowed stem cell research using leftover embryos from IVF. Laws in the U.K. also allow for the creation of embryos for research. Most of this research falls under the control of the Human Fertilisation and Embryology Authority (HFEA). The 1990 law covers the use of such embryos for research in reproductive biology.

In 1999, Geron, which controls the rights to seven stem cell lines in the U.S., bought Roslin Biomed in Scotland, a spin-off of the Roslin Institute, which cloned Dolly. There are now some 20 Roslin scientists working on Geron-funded stem cell research. David Greenwood, Chief Financial Officer of Geron, cited the company invested in response to the very receptive environment in Britain and the fact that public money might be available.

Based in Guildford, ReNeuron uses embryonic stem cells from the tissue of aborted fetuses to repair damaged brain cells. After raising $7 million in venture capital in 1999, ReNeuron Holdings PLC became the first European stem cell company to go public. The company raised about $30 million in the initial public offering.

A new interpretation of the 1990 law in 2001 expanded the law to include many types of basic research, including stem cell research. In January 2001, the British Parliament voted to allow therapeutic cloning but concluded no research permits would be granted until a House of Lords committee, headed by the Bishop of Oxford, Richard Harries, completed a review of the scientific and ethical issues. The House of Lords' endorsement of therapeutic potential of stem cells set a precedent for the rest of the world. In 2002, the British House of Lords committee voted to reject the argument that advances in research on adult stem cells meant research on embryonic stem cells is unnecessary. The committee also added that embryos left over from IVF should be used in preference to embryos cloned for the purpose, but that cloning should be allowed where there is a demonstrable and exceptional need which cannot be met in other ways.[329]

The HFEA, the agency that regulates the human embryonic stem cell (hESC) research granted the first two licenses to scientists at Edinburgh University and King's College, London, allowing them to culture stem cell lines from embryos left over from IVF. The two main academic funding bodies, the Medical Research Council (MRC) and the charity Wellcome Trust announced grants worth several million pounds for embryonic stem cell research. The MRC also immediately called for tenders from laboratories to set up and run an independent U.K. national stem cell bank, which was established in 2002, in London. The stem cell bank makes a variety of characterized and newly derived stem cell cultures available to researchers. The HFEA would oversee the selection of cell lines to be established and included in the bank, and the MRC would run the bank.

All these pave the way for the U.K. to take the lead in this field as the only country with a precise regulatory framework governing such research. U.K. has other advantages. As the home of both the world's first test-tube baby and Dolly the sheep, U.K. has an established reputation in embryology and fertility. British researchers pioneered much of the early work on mouse embryonic stem cells in the past two decades. It is not just the British Parliament is supportive, Britain's Wellcome Trust, the world's largest medical charity, has an annual budget of $1.3 billion.

Australia

Alan Trounson at Monash University, Melbourne is a well-known embryologist and *in vitro* fertilization (IVF) expert. As late as January 1998,

---

[329] Nuala Moran, "UK government OKs research involving embryonic stem cells", *BioWorld International*, March 6, 2002.

he had been able to get hES cells to form, but was unable to maintained them for long. Because the State of Victoria outlawed research using human embryos, Trounson collaborated with Ariff Bongso at the National University of Singapore, where no such prohibition existed. In August 1998, Trounson dispatched Reubinoff to Singapore to create a stable line of hES cells. According to a report, "within weeks, they succeeded in deriving the cell line that Reubinoff then kept warm in his shirt pocket on the flight to Australia."[330]

When President George Bush of the U.S. announced in August 2001 his list of approved hES suppliers, Australia was suddenly catapulted into the limelight of international stem cell research. Bush named 10 companies and research groups worldwide as legitimate sources of hESC sources because their hESCs had been made before his announcement. Of the 10, only two sources had published evidence that their cells were bona fide hESCs. These two sources are the University of Wisconsin and Monash University.[331]

Not surprising, the Australian government has seized the opportunity to position itself as the world center of stem cell research. In the past decade, biotechnology has flourished in the country, with the number of biotechnology firms doubling to 300, of which 35 are publicly listed. In 2002, recognizing that the country has competitive advantages and a critical mass in stem cell research to compete on the international stage, the government picked stem cell research as its biotechnology flagship. To pave the way for an environment conducive to stem cell research, the country passed clear, progressive legislation allowing hESC researchers controlled access to frozen embryos left over from IVF. The government provided A$43.5 (or $30 million) to establish the National Stem Cell Center (NSCC). The center harnesses the country's talents in adult and embryonic stem cells, protect the intellectual property and work with its commercial members. The Executive Vice Chairman of the NSCC is no one else other than the famed Alan Trounson, who also functions as the Global Strategic Science Advisor of the center.

Australia is currently the second largest supplier of stem cells to researchers in the U.S. In fact, the University of Adelaide has calculated that Australia presently contributes about 25% of the world's research on stem cells, very significant compared with Australia's contribution of about 2.5% of the world's research generally. Of the four grants (a total of $3.5 million) that the U.S. National Institutes of Health awarded in April 2002 to help finesse the supply of hESCs, the largest went to the U.S. affiliate of BresaGen of Australia, Cellsaurus, Inc., while the second largest went to ES Cell

---

[330] Gretchen Vogel, "Stem Cells: In the Mideast, pushing back the stem cell frontier", *Science*, 295 (5561), March 8 2002, pp. 1818-1820.

[331] "Australia leading the way for stem cell research", *ForeignDirectInvestment*, April 2, 2003.

International, an equal partnership of Monash University and the National University of Singapore. The other two awardees are the University of California at San Francisco (UCSF) and the Wisconsin Alumni Research Foundation (WARF) of Madison, Wisconsin.[332]

Singapore

Britain is the first country to legalize cloning to allow scientists to create cloned embryos for stem cell research. The guidelines are among the most progressive in the world.

The city-state of Singapore has adopted guidelines that are closely modeled on Britain's guidelines. The guidelines allow scientists to extract stem cells from adult human tissues, aborted fetuses and surplus embryos from fertility clinics as long as the embryos are less than 14 days old. The guidelines also allow scientists, under strict regulations and on a case-by-case basis, to obtain stem cells by cloning technology.[333]

There are fewer than ten stem cell research groups in the city-state. Ariff Bongso of the National University of Singapore (NUS) is probably the most well known. Bongso was part of the team that produced the first test-tube baby in Asia in 1983. In 1994, he became the first to successfully isolate stem cells from a 5-day old embryo. In 1998 James Thomson of the University of Wisconsin did the same, but he obtained a patent for the technique.[334]

National University of Singapore owns the rights to Bongso's discoveries. In 2000, NUS, in equal partnership with Monash University, Melbourne, Australia launched a company called ES Cell International to lay the groundwork for future commercialization of Bongso and Alan Trounson's work. ES Cell International has six of the world's 78 stem cell lines listed on the registry of the U.S. National Institutes of Health. The company has provided cell lines to more than 40 laboratories in the world.[335]

On August 5, 2002, Bongso and his team announced another world first for successfully growing a human embryonic stem cell (hESC) line entirely without mouse cells. The new human stem cell line has been grown on human feeder cells and cell nutrients. This is a critical step in getting stem cell

---

[332] Constance Holden, and Gretchen Vogel, "Stem Cell Lines: Show us the cells", *Science* 297, July 18, 2002, pp. 923-925.

[333] Trish Saywell, "Stem cell secrets unlocked", *Far East Economic Review*, October 24, 2002.

[334] A. Bongso, C.Y. Fong, S.C. Ng, and S.S. Ratman, "Isolation and culture of inner cell mass from human blastocysts", *Hum Reprod.*, 9, 1994, pp. 21107.

[335] Benjamin E. Reubinoff, Martin F. Pera, Chiu-Yee Fong, Alan Trounson, and Ariff Bongso, "Embryonic stem cell lines from human blastocysts: Somatic differentiation *in vitro*", *Nature Biotechnology*, 18(4), (2000), pp. 399-404.

research to deliver its promise of cures for diseases. The ability to use stem cell lines grown entirely without exposure to mouse cells eliminates one of the potential risks of pathogens jumping from the animal feeder cells to hESCs. Currently all the 78 existing stem cell lines listed on the HESCR registry of the U.S. National Institutes of Health are supported or grown on animal, typically mouse-feeder cell layers. The potential risk of animal pathogens pretty much rules these cell lines out from clinical trials. Bongso's method is being developed and used as a research tool with the U.S. Food and Drug Administration (FDA) to create additional cell lines for clinical applications. The method may eventually become a totally animal-free system gold standard.[336]

Bongso is also working on developing a safer storage system for stem cell lines. Currently, cells are stored by putting them in open plastic straws and freezing them in liquid nitrogen. This method is not perfect because the liquid nitrogen, if contaminated with organisms or viruses, could infect the cells. Bongso's approach is a smart electronic system that promises to cut out contamination risk.

Israel

There is no law regulating stem cell research, and embryo destruction for stem cell research is legal in Israel. According to the Jewish tradition, to procreate is very important. There is a lot of support for infertility treatments and there are a very large number of IVF clinics in Israel.[337]

Because stem cells are culled from human embryos, the research is at the center of heated debates. Due to this, scientists in many countries, including the U.S., have spent more time lobbying politicians and courting public opinion at the expense of their valuable time. And because of the uncertain future of the research, the scientists spend less time in their laboratories learning about stem cells.

While researchers in these countries are engaged in political battles over human stem cells, scientists in Israel have moved to the vanguard of the scorching field. Thanks to the liberal regulations and broad public support, scientists in Israel enjoy scientific freedom to explore this scientific research area that to some can be repugnant. Because of their head start, Israeli scientists have helped set the pace for the rest of the world. But there is a far-flung connection with the U.S.

---

[336] "Human embryonic stem cells grown without animal tissue", *MSNBC News*, August 5, 2003.
[337] Gretchen Vogel, "Stem Cells: In the Mideast, pushing back the stem cell frontier", *Science*, 295 (5561), March 8 2002, pp. 1818-1820.

James Thomson of the University of Wisconsin and his colleagues announced the isolation of stem cells from human embryos in November 1998. Although the success happened as far east (in Melbourne, Australia) and as far west (in Wisconsin, U.S.) from Israel as one can get and, both teams had Israeli collaborators. Benjamin Reubinoff of the Hadassah Medical Center and Joseph Itskovitz-Eldor of the Rambam Medical Center at the Technion University in Haifa are two key players in the landmark isolation of stem cells from human embryos in 1998. Thomson was working with Itskovitz-Eldor, who in 1997 had sent Thomson more than a dozen frozen embryos donated by Israeli couples in IVF clinics. Four of the five cell lines Thomson's team succeeded in isolating came from Israeli embryos.[338]

Itskovitz-Eldor's team continues to create new cell lines. The team is also embarking on a new method to allow embryos to develop several days beyond the blastocyst stage. There are two potential advantages of the new method: First, the new cell line might be able to grow in culture more easily or develop into target tissues more readily; Second, the new technique might also fall outside the broad patents owned by the University of Wisconsin that cover cell line derivation using Thomson's method.

Nissim Benvenisty of Hebrew University in Jerusalem has also contributed greatly to stem cell research. Benvenisty is best known for knowing how to transfer what he knows in mouse cells to human cells. After Thomson's announcement, his first project was to test whether human embryonic stem cell (hESC), like those from mice, can form clusters of differentiating cells called embryoid bodies. Thomson's group had suggested hESCs could not form clusters. By suspending the human cells in liquid rather than flat on a dish, Benvenisty showed human cells can form clusters.

In October 2000, Benvenisty's team, in collaboration with Doug Melton of Harvard University, showed how growth factors prompt hESCs to mature into different cell types. In the spring of 2001, his team succeeded in inserting into stem cells a gene for green fluorescent protein, which glows in immature cells and shuts off as the cells begin to differentiate. This cell line should be a big help to research since it enables researchers to sort immature from mature cells.

Karl Skorecki of the Rappaport Institute at the Technion University is working on lines of embryonic stem cell that have been tweaked genetically to churn out loads of telomerase, a protein that adds "caps" to the ends of chromosomes to protect them from degradation after multiple divisions. The

---

[338] J.A. Thomson, J. Itskovitz-Eldor, S.S. Shapiro, M.A. Waknitz, J.J. Swiergiel, V.S. Marshall, and J.M. Jones, "Embryonic stem cell lines derived from human blastocysts", *Science* 282, November 6, 1998, pp. 1145-1147.

team has shown that telomerase, which is active in undifferentiated ESCs, normally shuts off when the cells begin to differentiate. The hope is that differentiated cells in which telomerase stays active can undergo more division, enabling researchers to grow larger batches of tissue.[339]

Israeli scientists are currently collaborating with scientists in six other countries, but have shared their hESC lines with only the U.S. The four: Benvenisty, Itskovitz-Eldor, Reubinoff and Skorecki have teamed up in a common project to study how human ES cells develop into four key tissues – blood, pancreas, neurons, and liver.

Japan

Like in many other countries around the world, the progress of medical research is often a controversial issue in Japan. In Japan, organ transplants from brain dead patients have only recently been permitted, and it was only in May of 2001 that Japan saw its first birth from a surrogate mother. The Japanese, however, have been taking a leading role in animal cloning studies, with its researchers claiming several world firsts. There is no doubt this country will emerge a key player in stem cell research.[340]

The government has established guidelines for stem cell research which stipulate that embryonic cells used in research should come from infertility treatment that would otherwise be discarded. In the spring of 2002, two major initiatives were taken to establish centers to advance stem cell research, one at Kyoto University and the other at the RIKEN Institute.

The Kyoto initiative is headed by Norio Nakatsuji, a professor at the Institute for Frontier Medical Science at Kyoto University. Nakatsuji is well known for his work on monkey stem cells. The initiative focuses on creating hESCs from fertilized eggs left unused in infertility treatment. The fertilized eggs will come from Kyoto University Hospital, Toyohashi Municipal Hospital in Aichi Prefecture, and Keio University in Tokyo, all of which are known for infertility treatment. Researchers at this initiative will enter the regenerative medicine race using domestically produced stem cell lines. On June 3, 2003, Nakatsuji announced that his group had isolated the first domestic hES cell.

The RIKEN initiative led to the establishment of the new Center for Development Biology (CDB) in Kobe. The CDB head is none other than the

---

[339] Hwa A. Lim, *Genetically Yours: Bioinforming, biopharming, and biofarming*, (World Scientific Publishing Co., New Jersey, 2002).
[340] "Japan sets to embrace stem cell research", *BBC News*, August 1, 2001.

cloning pioneer Teruhiko Wakayama, renowned internationally for cloning the first mouse, Cumulina, when he was still at the University of Hawaii in 1997. The emphasis is primarily basic science.

India

Two beneficiaries of President George Bush's August 2001 list of approved hES suppliers are in India: Reliance Life Sciences (RLS), Mumbai, and the National Centre for Biological Sciences (NCBS), Bangalore. The cell lines were derived from donated embryos, both frozen-thawed and fresh, from fertility clinics.[341]

The head of RLS is Firuza Parikh, the creator of India's first intra cytoplasmic sperm injection (ICSI) child. In August 2001, RLS was barely eight months old, backed by Reliance Industries Ltd with $5 million. It had succeeded in creating seven cell lines, making them third in rank among the ten institutions on the registry of the U.S. National Institutes of Health's list of legitimate hESC suppliers. According to RLS, diabetes, chronic heart diseases and stroke are major ailments where hESCs have significant therapeutic potential. RLS estimates by 2025, India will have 57 million cases of diabetes, an increase of 148% over the figure of 23 million in 2001.

At the U.S. announcement of 2001, NCBS had a much longer history. It had been working on stem cells since 1999 and had three documented stem cell lines. Isolating the stem cell lines has primarily been a joint effort of Mitradas M. Panicker of NCBS and a team of medical practitioners: Mehroo D. Hansotia, Sadhana K. Desai and Vijay Mangoli from Mumbai's Fertility Clinic. Panicker's team works on the aspects of basic research; the Mumbai team, which provided the frozen embryos, works on the clinical aspects and possible long-term medical applications.[342]

Another notable Indian player is L.V. Prasad Eye Institute of Hyderabad. The institute was founded in 1986 to realize its mission to achieve excellence, equity and efficiency in eye care; to conduct research into eye diseases and vision-threatening conditions, train eye care workers, product development and rehabilitate those with incurable visual disability. The focus has been on providing eye-care services to underprivileged populations in the developing world. The Institute got into limelight in 2001 when its doctors succeeded in transplanting a stem-cell-derived cornea to a patient who had lost his cornea.

---

[341] Anupama Katakam, "At the forefront in India", *Frontline*, 18(19), September 15-28, 2001.
[342] Ravi Sharma, "Crucial support from Bangalore", *Frontline*, 18(19), September 15-18, 2001.

One of the pioneers in stem cell research, Balkrishna Ganpatrao Matapurkar is leading stem cell studies at the prestigious Maulana Azad Medical College, New Delhi. Matapurkar is well known for holding a U.S. patent on a technique to regenerate organs and tissues. He was able to regenerate organs including uterus, urethra and ureter in animals and is now working to regenerate organs like kidneys and parts of intestines. He was also able to regenerate certain large tendons damaged in the abdominal region, which lead to the condition of hernia.[343]

Ruby Hall Medical Research Centre, a subsidiary of Pune-based Ruby Hall Clinic, and Denmark-based biotechnology company Mesibo formed a 49:51 joint venture to establish India's largest cord blood storage facility at Pune. The center will sell stem cells for autologous[344] use and for use of family members, where there is a good chance of matching the tissue types. The cost to the parent will be in the range of Rs20,000-25,000 ($450-$550) as a one-time fee and a Rs2,000 ($45) annual fee for preservation of the cord blood. The Danish company will provide technical assistance to the Indian subsidiary by way of transferring the know-how and a list of standard operating procedures to maintain high quality standards. Besides, Mesibo will also make available its latest research on stem cells to India for the purpose of clinical applications.[345]

Reliance Life Sciences, the aforementioned company, is another company in India to get into housing a cord blood repository for commercial purposes. The firm provides patients with ReliCord, an enriched blood.[346]

Cord blood can be used to treat disorders such as thalassemia, a fatal blood disease. According to WHO, the prevalence of thalassemia in India is around 7-8%, and 10,000 new cases are reported every year.

China

China bans research on embryonic tissue, although it allows research on stem cells obtained through umbilical cords. But enforcement of the laws is another matter.

A nation not known to have published too many scientific discoveries – even very important ones – in peer-reviewed English-speaking journals, much

---

[343] "Indian doctor gets US patent on human organ regeneration", *Indian Express*, June 7, 2000.
[344] Autologous: a procedure in which a patient's own tissue or organ is removed, treated, and then returned to the patient.
[345] Namita Shibad, "Ruby Hall to set up India's largest cord blood storage facility", *Express Healthcare Management*, 1, October 15, 2001.
[346] Sitaraman Shankar, "India Reliance's biotech plans spring to life", *Reuter*, May 23, 2001.

is still in the dark. However, many view the country as one of the potential havens for stem cell research.

In March of 2002, a team of scientists led by Lu Guangxiu of the Xiangya Medical College, South Central University of Changsha, Hunan Province, claimed to have cloned dozens of, of the order of 30, human embryos. Lu is reported to have cloned a human embryo two years before Advanced Cell Technology of the U.S. Lu, a fertility specialist, published her paper in China in 2000 describing her cloning efforts. The team has been able to grow embryos to 200-cell stage, large enough to easily harvest the ES cells. Those who know her work say her claims are credible. Xiangzhong (Jerry) Yang, head of the Transgenic Animal Facility at the University of Connecticut acknowledged that there is a blossoming human embryonic cloning research across China.[347,348]

At least five laboratories in China are engaged in hESC research. Chen Xigu, a leading cloning scientist at the Zhongshan Medical University, Guangzhou, has cloned 109 embryos, but has refrained from attempts to produce babies. Reproductive cloning is strictly prohibited under the Chinese law. The Shanghai No. 2 Medical University has also made a lot of progress in the field.

In a series of moves aimed at positioning China as a world leader in modern biological sciences, in December of 2002, the Chinese government approved the setting up of the country's first state-run stem cell bank in Tianjin. The bank has already gathered 6,000 samples. Compared to the 4.5 million samples in a typical bank in the U.S., the number is still very small. But the bank has plans to make it the largest in Asia in eight years.[349]

The Chinese can tap into the successes of the Chinese Beijing Genomics Institute (BGI), which is one of the sixteen organizations involved in the International Consortium of the Human Genome Sequencing, and which sequenced the rice genome in April 2002. Like India, China can tap into its huge population size; but unlike India, China is ethnically diverse, with the largest ethnic group the Han and 55 other minority ethnic groups.

Korea

On December 16, 1998, utilizing the Honolulu technique, a South Korean research team at the infertility clinic of Kyunghee University Hospital in Seoul

---

[347] "Stem cell advances likely within a year in China", *The Straits Times*, August 6, 2002.
[348] "China stem cell research leaps ahead", *The Wall Street Journal*, March 6, 2002.
[349] "China approves stem cell bank", *BBC News*, December 11, 2002.

claimed to have grown an early human embryo using a cell donated by a woman in her 30s. The embryo divided into four cells before the experiment was aborted.[350]

As in the U.S., human cloning is a focus of heated debate in Korea. When scientists at Kyunghee University announced in 2001 that they had begun research on human cloning, Korean religious and civic groups mobilized to stop such research. In 2003, the government adopted a law barring cloning for reproduction, while experiments that use embryonic stem cells have to receive reviews and authorizations from a government ethics panel.

Then on March 12, 2004, Dr. Hwang Woo-suk of the Seoul National University published a paper in *Science* describing the cloning of a stem cell line. He immediately won world acclaim for this "groundbreaking" work. In June 2005, Dr. Hwang reported in *Science* he and his team had cloned cells from 11 patients using the technique and whose promise to make paralyzed people walk had been engraved on a Korean postage stamp.

In August 2005, he published, this time in *Nature*, his clone of a dog by the name Snuppy (Seoul National University Puppy). With all these achievements, the Korean government invested copiously in his laboratory and in making him a national hero. Dr Hwang's work received $26.5 million in annual government funds.

By October 2005, his research center, the World Stem Cell Hub, opened with $132 million from the South Korean government. Plans were announced to open satellite cloning centers in San Francisco and London. The hub was being conceived not only as a stem cell bank, but also a collaborative research center, a place where people can go for research retreats and sabbaticals, something like a stem cell version of The Jackson Laboratory in the U.S. The Jackson Laboratory, based in Bar Harbor, Maine, is a nonprofit institute that supplies researchers around the world with some 2 million mice annually, but also provides training and conducts research.[351]

In early November 2005, things were beginning to fall apart. On November 24, 2005, as part of what he called "repentance," Dr. Hwang resigned as head of the new center. In short order it would be found that Dr. Hwang two *Science* papers were fraudulent, but his claim to have cloned the dog he named Snuppy is legitimate.

This not withstanding, when President George Bush made his August 2001 list of approved hESC suppliers, Korea has two entries: one from the private sector, Maria Biotech Co. Ltd., Seoul, with 3 cell lines; the other from

---

[350] "First human embryo cloned in Korea or Britain?" *Globalchange*, 1999.
[351] Stephen Pincock, "Hwang plans world stem cell hub", *The Scientist*, August 15, 2005.

the MizMedi Hospital, Seoul, with 1 cell line. MizMedi Hospital is the same hospital that supplied ova for Dr. Hwang's stem cell research. This cell line is not a subject of the controversy over Hwang's stem cell claims.

United States

In historical hindsight, in the 1970s, the U.S. did not allow cloning to be done on pathogens. As a result, the Europeans overtook the U.S. as the leader in studying viral infections. A very similar situation is beginning to unfold in stem cell research. Boston-based ViaCell, Inc., a premier cellular therapy company, announced establishing a new research and development center in Singapore on February 5, 2002. The Singapore Research Centre (SRC) will work with the U.S. research team to develop stem cell therapies for the treatment of cancer, immune deficiencies, genetic and neurological diseases, and diabetes.[352]

U.S. researchers have continued to look outside the country for stem cell research opportunities.[353] To complicate the matter, on September 6, 2001, within a month after Bush's televised announcement of federal support for stem cell research with restrictions, the U.S. government conceded that most of the 78 hESC lines they had claimed the world community had on and prior to August 9, 2001 might not be ready.[354]

The current stem cell technology puts cell lines in three phases: the proliferation phase in which the cells begin to grow; the characterization phase in which scientists identify the cells' biological characteristics; and the fully developed stage. For instance, highest on the HESCR registry list is the University of Goteborg in Sweden which has 19 hESC lines. All the cell lines meet the U.S. government's criteria, but only 3 are fully developed, 12 are proliferating, and 4 are in the characterization stage. In other words, only 24 or 25 of the 78 hESC lines disclosed on August 9, 2001 by the U.S. government are ready for experiments!

Mitradas M. Panicker of the National Centre for Biological Sciences (NCBS), India corroborated the fact by admitting that there is still a long way to go before anything concrete could be said about the three hESC lines that the NCBS developed in May of 2001, "From this stage (embryonic) to achieve human stem cells will take time, and how long it will take in culture, no one knows at the moment... Much of what has been made of the discovery in [my]

---

[352] ViaCell Press Releases, February 5, 2002.

[353] Eugene Russo, "Stem cell research climate", *The Scientist*, June 27, 2003.

[354] Sheryl Gay Stolberg, "US concedes some cell lines are not ready", *The New York Times*, September 6, 2001.

laboratory is mere media hype. What has caught the media attention is the fact that the NCBS and RLS have been chosen by the Bush Administration for the supply of stem cells... Yes, it is exciting but it is certainly not as spectacular as is being made out. We are still in the embryonic stage as far as this research is concerned, and further information will only be available once we are ready to publish our findings. The fact that Bush had to make a decision puts the NIH's decision in the limelight. And we were one of the 10 institutes chosen. But it is not as if these are the only 10 in the world. A number of laboratories in the U.K. have set up a bank of human stem cells. A fertility clinic in Boston has reportedly [it has] got 12,000 frozen embryos for Harvard University, which will generate cell lines..."[355]

The U.S. National Institutes of Health (NIH) was aware that the NCBS was working on stem cells and asked the center if it would like to register the isolated stem cell lines with it. As of September 2001, the NCBS has not taken any decision in this regard, even though the center is listed as one of the ten institutes in the HESCR registry compiled by the U.S. NIH.

Stability of the cell lines aside, there are also concerns in the standard procedure to grow the lines – scientists nourish them with mouse cells that have been killed with radiation. Nonetheless, the technique still raises concerns about safety because the cells could harbor viruses that would infect people. While the Food and Drug Administration (FDA) does not prohibit the use of mouse cells in human therapies, it imposes strict regulations on them, and officials do not know if the existing stem cell lines would meet the FDA's criteria.

The U.S. policy of paying for research on hESC but ban paying for collecting more of them is a policy rigged with potential problems. From a basic commerce standpoint, the U.S. government is effectively the market maker. As a public buyer, the U.S. is creating a demand to be filled by other suppliers from universities or the private sector. These suppliers will need embryos to extract stem cells, leading to a potential market emerging in embryo trade. The potential benefits of stem cell research are undeniable, but they should not come at the cost of commodifying human embryos.

In the U.S., currently, there are only a few stem cell research organizations in the public sector (Harvard University, Stanford University, University of California at San Francisco, John Hopkins University, University of Wisconsin, Tulane University) and the private sector (Geron Corp., Advanced Cell Technology, ViaCell, Cy Thera, Inc.). With U.S. federal funding for stem cell research, though restricted only to the existing 78 cell lines, it is likely

---

[355] Ravi Sharma, "Crucial support from Bangalore", *Frontline*, 18(19), September 15-28, 2001.

there will be an increase in the number of researchers in the field. Internationally, those countries with progressive guidelines on hESC research, such as the U.K., Australia, Singapore or any other countries later diving into the stem cell race will be beneficiaries of the strict U.S. policy. If the current U.S. position remains unchanged, these beneficiaries will soon pull ahead of the birthplace of stem cell research to lead the field.

A bill sponsored by Rep. Dave Weldon of Florida to ban all cloning research had passed in the House but stalled in the Senate. No federal legislation was passed in 2003. An encouraging development is that more than 60 senators and 200 representatives have expressed support for stem cell research that would go beyond what the president has authorized. One of the strongest proponents of stem cell research is none other than Senator Edward M. Kennedy, Democrat of Massachusetts, Ranking Member of the Committee on Health, Education, Labor and Pensions. The State of California has endorsed therapeutic cloning, with New Jersey, New York, and Massachusetts likely to follow suit.

In June 2003, the NIH's National Center for Research Resources (NCRR) announced a 5-year, $4.3 million grant to Tulane University in New Orleans to establish a center for the preparation, quality testing, and distribution of marrow stromal cells (MSCs) derived from human adult bone marrow and rat bone marrow using standardized protocols. These stem cells will be available to researchers for non-clinical use to explore ways to repair damaged tissues in the body and to improve gene therapy.

Case Study: Harvard Stem Cell Institute

Douglas Melton's lab extracts embryonic stem cells from fertilized human eggs as part of a search for a cure to Type 1 diabetes. There is an extra incentive for Dr. Melton to pursue this line of research. In early 2006, his son, Samuel, was about 6 months old when he came down with Type 1 diabetes, which means that his pancreas could not make insulin. About four years earlier, his daughter came down with this same disease.[356]

Because his lab is looking for new stem cell lines, and not necessarily working with the cell lines listed on the U.S. Human Embryonic Stem Cell Registry (HESCR), the lab is not entitled to federal funding. The Bush administration policy on stem cell research has made it more difficult, to say the least. Long before Bush's announcement of August 9, 2001, the lab had planned stem cell experiments. Afterward, they were able to go forward only

---

[356] Claudia Dreifus, "At Harvard's stem cell center, the barriers run deep and wide", *The New York Times*, January 24, 2006.

because the Howard Hughes Medical Institute, the Juvenile Diabetes Association and Harvard alumni provided private funding. However, because of federal policy, they had to set up this whole new laboratory that was separate from everything else here at Harvard University.

Because Melton's research is privately financed, he must certify that no federal money supports any of his efforts. Hence he installs key cards to enter and exit his lab. They provide a kind of barrier against people and things that have a federal tie. They had to separate the money in a really scrupulous way. They have an accountant who makes sure that not a penny of federal funds goes to their embryonic stem cell research.

Nonetheless, the lab has been very productive, having created 30 stem cell lines and offered them free to any scientist who want to do further research on them. In his speech of August 2001, President George Bush said that he was going to permit federal funding for scientists to work with some 60 pre-existing stem cell lines and it turned out that there were probably only about 10 or so usable lines. The rest were problematic: Some had abnormal chromosomes, like a cancer cell might have. If the goal is using stem cells to create pancreatic beta or muscle or nerve cells, the research must begin with high-quality normal cells.

Moreover, all the pre-existing lines were grown with mouse cells – which probably means they can never be used in human patients to treat diseases. Some of the lines, when researchers ask for them, come from a biotech company or some investigator who want to impose onerous restrictions on their use.

So Melton's lab makes cell lines because they believe there are not enough human stem cell lines around for researchers to really explore their properties. They are offering them free of charge because this is a long scientific tradition of making the fruits of one's research available to others.

Case Study: California Institute of Regenerative Medicine

After August 9, 2001, Congress and President Bush restricted federal funds – U.S. federal funds can only be used to support research on adult stem cells, or stem cells listed on the HESCR. However, a state can circumvent this restriction.

This is precisely what California did. The stem cell initiative, Proposition 71, also known as the Stem Cell Research and Cures Act, was an effort to circumvent federal restrictions on research involving human embryonic stem cells. In November 2004, 59% of Californian voters adopted Proposition 71, creating a state agency, the California Institute for Regenerative Medicine

(CIRM), to fill the gap in federal financing, with a $3-billion bond measure. In so doing, California has made itself the dominant player in stem cell research. To be sure the task at hand, often likened to setting up a state version of the National Institutes of Health, is formidable.[357]

Dr. Zach W. Hall was appointed permanent president of the new institute in September 2005 and is responsible for its scientific direction, subject to a 29-member governing board – the Independent Citizen's Oversight Committee. Members, appointed by state officials, are largely patient advocates and the deans or presidents of California universities and research institutes. Even so, the proposition was designed to insulate the governing board from politics, so it forbade legislative intervention for three years. It also made the committee responsible for the details of governance, including such relatively minor issues as compensation of institute staff and broader ones of grant disbursal for research and construction of laboratories.

Dr. Hall, 68 then, is a neuroscientist with managerial experience in government, industry and academia. He has been director of a National Institute of Health, chief executive of EnVivo Pharmaceuticals, and vice chancellor of the University of California, San Francisco.

Robert Klein, the real estate lawyer who spearheaded the campaign for Proposition 71 and the chief author of the Proposition, is chairman of the institute board. Mr. Klein is a real estate developer whose son has diabetes.

The effort, however, has been hobbled by litigations. It has been second-guessed by public interest groups and legislators. And it has been consumed by the bureaucratic minutiae required to set up rules for administering grants. But much progress has been made. Joan Samuelson, a member of the institute's board and an advocate for people with Parkinson's disease – a condition that she herself has and which is seen as a promising target for stem cell therapy – puts it succinctly, "Anything so important and so public is going to have a Greek chorus: the naysayers, whiners and complainers. We just have to stay the course."

And many of the rules and committees needed are now in place. For example, revenues of more than $500,000 from patented inventions by universities and nonprofit research institutes will trigger a 25% return to the state, in contrast to federal policies, which do not require payback. And patented inventions must be shared among researchers in California. Among the ethical standards that exceed those nationwide are those that ensure women

---

[357] Nicholas Wade, "California maps strategy for its $3 billion stem cell project", *The New York Times*, October 11, 2005.

are not coerced to donate eggs, compensated for out-of-pocket costs and are covered for any medical complications from egg donation.

What happens in California matters to the nation because the $3 billion to be spent on mainly embryonic stem cell research – $300 million annually for 10 years – is expected to dwarf funding from the federal government and from any other state. Whether stem cells fulfill their potential for medical breakthroughs, and whether the U.S. can maintain its preeminence in this new branch of biotechnology, could depend on California.

Already, Japan, the European Union (EU) (Great Britain, Germany, and Sweden), China and Australia have begun national initiatives and efforts to spur the advancement of their regenerative medicine programs. These commitments range from policy directives in the EU to extensive financial investment by the Japanese government focused on the city of Kobe and surrounding Kansai region targeted to develop a region of expertise in tissue engineering and regenerative medicine. Despite this strong foreign commitment to regenerative medicine, the U.S. presence in regenerative medicine is in danger of being eclipsed.[358]

California's bold and pioneering initiative has caused a backlash, as some states have enacted bans on publicly funded embryonic stem cell research. Yet others are using California's experience as an example of what to do – or what not to do – when trying to set up stem cell programs. Those include Connecticut, Illinois, New Jersey and Texas, which have recently approved small amounts of state funding for research. And in Florida, an effort is under way to promote a ballot initiative.[359]

A trial, heard by an Alameda County Superior Court judge, began on February 27, 2006. The dispute has frozen this California's bold experiment: to create an embryonic stem cell research industry that has been deemed off-limits for federal funding by the Bush administration. The crux of the two consolidated lawsuits – one by People's Advocate and the National Tax Limitation Foundation, the second by the California Family Bioethics Council – is that lack of direct state management and control of the taxpayer funds violates the state constitution. The state attorney general's office and the institute's lawyer counter that the measure amended the state constitution to allow for such independence. The Napa-based Life Legal Defense Foundation, which opposes abortion and euthanasia, was the counsel for the

---

[358] *2020: A New Vision – A Future for Regenerative Medicine*, (U.S. Department of Health and Human Services, Washington, D.C., January 2005).

[359] Andrew Pollack, "California's stem cell program is hobbled but staying the course", *The New York Times*, December 10, 2005.

first lawsuit; the second plaintiff was created by the Christian conservative California Family Council specifically to sue the institute.

Until the litigation is resolved, the state will not issue any of the $3 billion in bonds to pay for research. So far the institute has been subsisting on a $3 million loan from the state and a $5 million gift from the audio entrepreneur Ray Dolby. But that money would not last long and the institute, which has 19 employees, has frozen hiring. To maintain momentum, the high-powered Robert Klein, is raising $120 million in bridge loans for an initial round of research grants. And he is hoping to arrange for so-called anticipation bonds that are not backed by a payment guarantee by the state.[360]

Despite the hobble, Proposition 71 has nevertheless triggered a tsunami of interest, as universities eager for pieces of the research pie lured new talent to California and launched stem cell programs in anticipation of cash infusions. Philanthropists Eli and Edythe Broad donated $25 million in February 2006 to build a stem cell research center at University of California at Santa Cruz. The Broad Institute for Integrative Biology and Stem Cell Research is expected to be the largest such center in the state when it opens in 2008.

While all this legal mud wrestling is going on in the California courts, other states, including Illinois, are quietly and effectively distributing grants to researchers at some of their top institutions.

East Versus West

According to Simon West, Chair of Biotechnology Industry Organization's (BIO's) bioethics committee, problems on the issue of stem cell research remain in countries other than the U.S.[361]

There is strong opposition to therapeutic cloning within the European Union (EU). Peter Liese, a physician specializing in human genetics and pediatrics, and chairman of a European Parliament bioethics working group, stated on June 25, 2003 that the majority of the EU member states are against all kinds of cloning. The European Parliament is also strongly against cloning and asks for a European and worldwide ban.[362]

Issues of ethical disagreement typically fall to the discretion of member states. In this case, the EU law is unlikely to supersede the more liberal therapeutic cloning laws in member states like Sweden, Belgium and the United Kingdom. EU member states including Germany, Austria, Portugal,

---

[360] Lee Romney, "Stem cell institute's legality goes to trial", *The Los Angeles Times*, February 27, 2006.
[361] Simon West, speaking at session of "Global Partners: Stem cells", BIO 2003 Conference, June 22-25, 2003.
[362] Eugene Russo, "Stem cell research climate", *The Scientist*, June 27, 2003.

Ireland, Norway and Poland prohibit any kind of research that destroys embryos.

Elsewhere in the world, China, India, Israel, and Singapore are among potential havens with pro-therapeutic cloning laws. According to a report, unlike the public opinion in the West, which is against researches in stem cells, the public opinion in many Eastern countries including India is far more supportive. This may be partly due to the scientific temper inculcated by the epics and innumerable religious texts.[363]

On another front, the tricky business of intellectual property may work in favor of the U.S. Researchers prefer to bring their stem cell lines and groundbreaking work to the U.S. where the FDA regulations are very specific and researchers can be clear on the criteria they must follow in order to get a patent. The patent of Balkrishna Matapurkar (No. 6,227,202) is a case in point. He, at the Maulana Azad Medical College in New Delhi, India, had to apply for a U.S. patent since an Indian patent for the process was not possible owing to the country's patent laws.

The U.S. is not completely out of the running, of course. The restricted funding approved by President Bush could produce interesting results, although it is yet to see if existing stem cell lines are adequate for the experiments that researchers want to do. Private research in the U.S. is largely unregulated and will continue. On August 24, 2001, for example, the Howard Hughes Medical Institute decided to fund the procurement of human embryos left over from IVF clinics for Harvard University scientists to use to make new stem cell lines. Harvard scientists promise to distribute the lines free to all comers. Private companies such as Geron Corp., the leading biotech company involved in stem cell research, are also free to do as they please as long as they do not receive federal funding.[364]

Saudi Arabia announced in July 2002 to launch a stem cell research program that would include therapeutic cloning, currently prohibited in the U.S. With vast financial resources and a growing biotechnology infrastructure, the Saudis could become world leaders in developing medical treatments from stem cells. Saudi Arabia's decision to enter the race also spotlights stark philosophical differences between the world's two major religions, which could affect who will benefit from future medical therapies for diseases that affect everyone.

U.S. stem cell research policy, based on the predominantly Catholic view that life begins at conception, forbids federally funded scientists from

---

[363] Vinod Scaria, "Rediscovering a lost science?" *The Internet World of Vinod Scaria*, July 2003.
[364] "At risk: A golden opportunity in biotech", *Business Week*, September 10, 2001.

conducting therapeutic cloning. Islamic law, which governs about 1 billion Muslims worldwide, views life as beginning at 120 days after conception, so researchers working at the new center in Saudi Arabia are not faced with the same moral conundrum as those faced by U.S. scientists.

Sultan Bahabri, head of King Faisal Specialist Hospital and Research Center in Jeddah, and Hamad Al-Omar, a scientist and leader in developing Saudi Arabia's new biotechnology industry got the stem cell research center operating before 2003 ended. The center has an international staff of scientists.

Bahabri and Al-Omar and other Saudis are co-founders of Jeddah's BioCity, a private venture that includes numerous state of the art biotechnology laboratories and companies currently under construction. The BioCity Project has substantial financing and is intended to make Saudi Arabia a world leader in biotechnology.

Saudi Arabia already has a strong infrastructure for conducting embryonic stem cell research through its expertise with IVF. At about 4% of all births per year, Saudi Arabia has one of the highest IVF birth rates in the world. Comparatively the U.S. has 1%.[365]

▶ A Shenanigans of Scandalous Proportion

Almost exactly a year after the Raelian hoax of having cloned five human beings, another extraordinary tale of delusion and deception was in the making. The first clone-gate involved cloning of human beings; the second, cloning of human cells to make patient-tailored stem cells.[366]

To many people, Dr. Hwang Woo-suk does not seem nutty, squirrelly, or deceptive; in fact, he is very charismatic and handsome. His professional demise is a severe embarrassment for the Korean government, which invested copiously in his laboratory; his fall from grace is a blow to South Korea, where he had become the modern, high-tech face that the nation yearned to project to the world. He appeared in national promotional campaigns. Korean Air declared him a "national treasure," giving him and his wife first-class tickets for a decade.

---

[365] Tim Friend, "Saudis take lead on stem cell cloning", *USA Today*, July 8, 2002.

[366] The author has benefited greatly from the e-mails he was receiving over the course of the rise and fall Hwang. Hal was also fortunate to have met in person the chairman of Korean Foundation for International Cooperation of Science and Technology, under the aegis of the Korean Ministry of Science and Technology, at an advisory meeting in the U.S in early November 2005 to talk briefly, among other things, Hwang's achievements.

The blow to Korea's scientific reputation abroad may be cushioned by the fact that other Korean institutions, notably the television program "PD Notebook" and a group of skeptical young Korean scientists, took the lead in discovering the problems with Dr. Hwang's work that led eventually to an investigation by the Seoul National University.[367]

The Rise to International Stardom

Dr. Hwang Woo-suk was born on January 29, 1953 in the central Korean province of South Chungcheong.[368] When he was an infant, there was the Korean War; when he was five, the father passed away. One of six siblings, he worked at a farm to finance his studies when his widowed mother could not earn enough to provide for him and fellow siblings. Hwang matriculated at the prestigious Seoul National University after graduating from Daejeon High School, becoming a veterinarian.

Hwang first got into media splash in South Korea when he announced he successfully created a cloned milk cow, *Yeongrong-i*, in February 1999. This success was touted as the fifth instance in the world in cow cloning. His next claim came only a couple of months later in April 1999, when he announced the cloning of a Korean cow, *Jin-i*. In 2002, he succeeded in cloning miniature sterile pigs whose organs could be used for transplantation to humans. A notable caveat is that he failed to provide any paper for the research in either case. Despite the lack of any scientific data to prove the validity, Hwang's claims were well received by the South Korean media and public, which made him a national hero of sort.

Until 2004, Hwang focused his research in creating genetically-modified livestock that included cows and pigs. During that period, Hwang claimed to have created a BSE-resistant cow (2003), which has yet to be verified; he also promised to clone a Siberian tiger, a promise that has yet to be fulfilled.

---

[367] Nicholas Wade, and Choe Sang-Hun, "Researcher faked evidence of human cloning, Koreans report", *The New York Times*, January 10, 2006.
[368] There are confusions over the birthdate arising from different calendar systems. Hwang was born on January 29, 1953 in the Gregorian calendar. However, older Koreans often list their birthdate in the lunisolar Korean calendar, which in this case is December 15, 1952. This date is sometimes mistakenly quoted in English language publications without specifying that it is in the Korean calendar. Further confusion arises when the Gregorian year and Korean calendar month and day are used together to produce an incorrect birthdate of December 15, 1953.

Photo 1. South Korean president Roh Moo-hyun (center) and first lady Kwon Yang-sook (right) visited Hwang Woo-suk's (left) stem cell research lab at the Seoul National University, December 2003 (Photo: received via e-mail).

Hwang's meteoric rise from a peasant's son to international acclaim in the scientific firmament rests on three striking "achievements."

In the first, published on March 12, 2004 in one of the world's most respected journals, *Science*, Dr. Hwang and his team said they had, with great difficulty, cloned a line of human embryonic stem cells with eggs and somatic cells from a single female donor:

> Woo Suk Hwang, Young June Ryu, Jong Hyuk Park, Eul Soon Park, Eu Gene Lee, Ja Min Koo, Hyun Yong Jeon, Byeong Chun Lee, Sung Keun Kang, Sun Jong Kim, Curie Ahn, Jung Hye Hwang, Ky Young Park, Jose B. Cibelli, and Shin Yong Moon, "Evidence of a Pluripotent Human Embryonic Stem Cell Line Derived from a Cloned Blastocyst", *Science*, 303(5664), March 12, 2004, pp. 1669-1674. [Originally published in *Science* Express on February 12, 2004].

They started with 242 human eggs donated by 16 Korean women for the cloning, got about 20 embryos from which they tried to extract stem cells and managed to produce only one stem cell line. The scientists managed to produce embryos only when the person to be cloned was also the donor of the egg used. They, however, could not clone men or women who were not egg donors. If that remained the case, it would mean that therapeutic cloning would not be of benefit to men, or to women past menopause.[369,370]

Other biologists have tried these human cloning experiments, but to no avail. What Hwang's team did differently was they used a squeezing method when extracting the nuclear material from the ovum, which caused minimum damage to the egg. This is not an easy process because human eggs are very

---

[369] *Science*, 303 (5664), March 12, 2004.
[370] Andrew Pollack, Cornelia Dean, and Claudia Dreifus, "Cloning and stem cells: their research, medical and ethical issues cloud cloning for therapy", *The New York Times*, February 13, 2004.

Photo 2. (Left photo) Upon the invitation of Hwang, Scottish embryologist Ian Wilmut visited South Korea on April 6, 2005. During the visit, Wilmut gave a lecture to Seoul National University students, and also proposed a joint research to develop stem cell treatment for Lou Gehrig's disease. (Right photo) Hwang and the world's first male dog clone, Snuppy. To clone Snuppy, Hwang used the same technique Wilmut had used to clone Dolly the sheep nine years earlier. (Photos: received via e-mail).

sticky. Next, they used a different activation time to mimic the fertilization of the egg, and a special culture medium for growing the reconstructed egg.[371]

This technique is an essential first step in therapeutic cloning, in which one of a patient's adult cells is converted into an embryonic cell, and then the cell is cajoled into new adult cells to replace any damaged tissue. In other words, the great hope of researchers and patients – to obtain stem cells that were an exact match of a patient's, or patient-tailored stem cells – seemed easily within sight. This stunning research paper instantly changed the tenor of the debate over cloning human embryos and extracting their stem cells.

In the second report, which *Science* published on June 17, 2005, Hwang announced a considerable improvement in the efficiency of his human nuclear transfer procedure, in which he went from requiring 242 human eggs to create a single human cell to just 17 (he created a total of 11 stem cells using 185 eggs), an amazing improvement of almost fifteen-fold:

> Woo Suk Hwang, Sung Il Roh, Byeong Chun Lee, Sung Keun Kang, Dae Kee Kwon, Sue Kim, Sun Jong Kim, Sun Woo Park, Hee Sun Kwon, Chang Kyu Lee, Jung Bok Lee, Jin Mee Kim, Curie Ahn, Sun Ha Paek, Sang Sik Chang, Jung Jin Koo, Hyun Soo Yoon, Jung Hye Hwang, Youn Young Hwang, Ye Soo Park, Sun Kyung Oh, Hee Sun Kim, Jong Hyuk Park, Shin Yong Moon, and Gerald Schatten, "Patient-Specific Embryonic Stem Cells Derived from Human SCNT Blastocysts", *Science*, 308(5729), June 17, 2005, pp. 1777-1783. [Originally published in *Science* Express on May 19, 2005].

This was the result of a handful of technical improvements that mostly involved methods for growing cells and breaking open embryos. With the

---

[371] Claudia Dreifus, "A conversation with Woo Suk Hwang and Shin Yong Moon, 2 friends, 242 eggs, and a breakthrough", *The New York Times*, February 17, 2004.

improved technique, he had been able to generate embryonic stem cell lines from nine patients – male and female, young and old with ages ranged from 2 to 56 – afflicted with spinal cord injuries, diabetes and other diseases. His team produced stem cells that were exact matches for 9 of 11 patients, including 8 adults with spinal cord injuries and 3 children – a 10-year-old boy with a spinal cord injury, a 6-year-old girl with diabetes and a 2-year-old boy with a genetic disorder of the immune system, called congenital hypogammaglobulinemia.[372]

The advance was hailed as a further step toward making therapeutic cloning a practical treatment and was quickly embraced by scientists and journalists alike. Dr. Leonard Zon, a stem cell researcher at Harvard Medical School and the president of the International Society for Stem Cell Research, called it a tremendous advance. Other researchers hope that the breakthrough will galvanize support for similar research.[373]

The third report from the Korean team, published on August 3, 2005, this time in the well respected British journal, *Nature*, announced the successful cloning of the dog, an animal previously resistant to cloning because its reproductive cycles are hard to predict and manipulate:

> Lee B.C., Kim M.K., Jang G., Oh H.J., Yuda F., Kim H.J., Shamim M.H., Kim J.J., Kang S.K., Schatten G., Hwang W.S., "Dogs cloned from adult somatic cells", *Nature*, 436(7051), August 3, 2005, pp. 641.

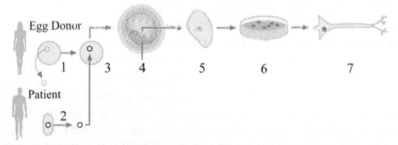

Figure 1. Had the Hwang-Wilmut collaboration been successful, this is how stem cell treatment for Lou Gehrig's disease could have been done: (1) The nucleus from an egg cell is removed; (2) The nucleus from a patient's skin cell is extracted; (3) The skin cell nucleus is implanted into the enucleated egg cell; (4) Early stage embryo develops into a blastocyst, containing about 150 cells and is about 0.08 inch in diameter. It looks like a ball of cells encased in a sphere. The inner cell mass yields stem cells genetically identical to the patient; (6) Stem cells can be cajoled to grow into any of the body cell types; (7) In this case, motor neurons are grown and used to "repair" degenerated motor neurons in the Lou Gehrig's disease patient.

---

[372] Gina Kolata, "Koreans report ease in cloning for stem cells", *The New York Times*, May 20, 2005.
[373] *Science*, 308, June 17, 2005, pp. 1777-1783.

Photo 3. Gerald Schatten (right), professor at the University of Pittsburgh, chatted with Hwang after a news conference in Seoul on October 18, 2005. (Photo: received via e-mail).

The dog clone, an Afghan hound, was named Snuppy (for Soeul National University puppy). This project is separate from the human patient-tailored stem cell research, even though the overall objective is to learn about the root causes of diseases, and the team believe it is possible, if one can responsibly develop the ability to derive stem cells from cloned dog embryos, that our very best friends may turn out to be the first beneficiaries of stem cell medicine.[374]

Shortly after his 2005 "groundbreaking" work, Hwang was appointed to head the new World Stem Cell Hub, a facility, funded by the Korean government, that was to be the world's leading stem cell research center.

The Fall from Grace

Dr. Hwang Woo-suk's fall from grace in December 2005 and professional demise in January 2006 started as a simple issue of ethical concerns. After the August 2005 publication of a report on cloning Snuppy in the prestigious British scientific journal *Nature*, Dr. Hwang's fame skyrocketed, with thousands of South Korean women volunteering to donate their eggs for the "national hero" to further his human stem cell cloning research. But this is not the main issue; the issue is that of earlier human-egg procurements.

The breakthrough reported in the March 2004 *Science* paper was in itself not surprising. Most people assumed that humans would be clonable in a similar way to sheep, mice, cows, pigs and other mammals before them. And the paper used slight variations on very conventional methods, making the familiar ideas easy to digest.

"It was not so amazing to imagine that somatic cell nuclear transfer might well work in the human, if investigators were diligent enough and had access

---

[374] James Brooke, "Korean leaves center in ethics furor", *The New York Times*, November 25, 2005.

to the appropriate clinical material," says Martin Pera of Monash University in Australia.

And Hwang's laboratory had resources that were an envy of others. The 2004 achievement was done with 242 eggs from voluntary donors, a number that other researchers had found impossible to collect. A visit to the previously low-key lab showed why the lab was a natural leader in the field. It was no garage operations, but shiny new facilities and army of capable technicians. Envious researchers who had visited the lab suggested that to stay competitive, other countries would have to redouble their efforts and their funding.

The copious supply of eggs provided a nagging doubt. At first, the main questions raised were how he got his eggs, when other researchers had difficulty getting eggs. A news article of May 2005, in the British journal *Nature*, provided evidence that researchers had been egg donors. Egg donation is an unpleasant procedure that involves a week of daily hormone shots to stimulate ovulation, culminating with the extraction of eggs through a hollow needle. Allowing junior members of a research team to undergo a painful and potentially risky procedure is considered ethically dubious.

In June 2005, just after the publication of the second *Science* paper, Choi Seung-ho, a reporter of "PD Notebook" ("*PD Su-cheop*" for Producer's Notebook), an investigative television program, received an anonymous e-mail on a confidential Internet bulletin board maintained by "PD Notebook." The e-mail, secretly posted a denunciation of Dr. Hwang's work, came from a member of Hwang's lab team. According to the tipster, Hwang faked some of the human stem cell cloning data that had just been published in *Science*.

Another event that led to Dr. Hwang's downfall, after a month of sniping at certain puzzling aspects of his published work, was the posting of a pair of duplicate photos on two Korean Web sites. One of the new duplicate photos appears in the June 2005 *Science* article about the 11 patients and a second in the October 19, 2005 issue of a lesser-known journal, *The Biology of Reproduction*, where it was reported as being of a different kind of cell.

In the *Science* article, the cell colony was labeled as being the fifth of Dr. Hwang's human embryonic cell lines derived from a patient's cells, but in the *Biology of Reproduction* article it was designated as an ordinary embryonic cell line generated in the MizMedi Hospital in Seoul, presumably from surplus embryos created in a fertility clinic.

On November 12, 2005, about three months after the dog clone announcement, widespread murmuring about the ethics of some of the human-egg gathering flared into the public eye when Gerald P. Schatten of the University of Pittsburgh announced that he was severing relations with Dr. Hwang. The abrupt disintegration of the partnership, with Dr. Schatten as Dr. Hwang's American partner, was surprising because the two scientists had collaborated for 20 months and reported important findings together.

Dr. Schatten's stated reasons for the severance seemed comparatively minor. In a written statement, while vouching for Dr. Hwang's work, he said his decision was grounded solely on concerns regarding egg donations in Dr. Hwang's research. Dr. Hwang's source was MizMedi Hospital – the fertility clinic where women donated eggs for Hwang's human embryonic stem cell research. Dr. Schatten's action prompted Roh Sung-il, chairman of the board at MizMedi Hospital in Seoul and a co-author of the second paper in *Science* of June 2005, in which Dr. Hwang claimed to have created stem cells from 11 patients, to disclose at a news conference on Monday, November 21, 2005 that during 2002 and 2003, he made payments of $1,400 to each woman who donated eggs.

On November 22, "PD Notebook," a popular investigative news program of the Korean television station Munhwa Broadcasting Corporation (MBC) network, raised the possibility of unethical conduct in the egg cell procurement process. Despite the factual accuracy of the report, news media as well as people caught up in nationalistic fervor showed their unwavering support for Hwang. They asserted that criticism of Hwang's work was "unpatriotic," so much so that the major companies who were sponsoring the show immediately withdrew their support.

After months of denying rumors that swirled around his laboratory, Dr. Hwang finally admitted on November 24, 2005 that in 2002 and 2003, when his work had little public support, two of his junior researchers donated eggs and a hospital director paid about 20 other women for their eggs. Under international medical ethics standards, researchers are warned against receiving eggs from members of their own research teams who are deemed to be in a dependent relationship. However, by the standard of the Korean law, the egg donations by the junior researchers were not considered a legal or ethical violation; payment for eggs was not illegal either in 2003, but it has been banned since January 2005 by a South Korean law. The murkier point is that in the strict hierarchy of a scientific laboratory in a Confucian society like that of South Korea, junior members often feel great pressure and obligated to please their superiors.

In the interview, Dr. Hwang apologized for his actions. He said, "I was blinded by work and my drive for achievement" – a statement that has an uncanny similarity with the "discovery" of N-rays in the 1900s,[375] or the explanations offered for the Piltdown Man forgery exposed in the 1950s.[376] He added that he had lied about the source of the eggs donated to protect the privacy of his female researchers, and that he was not aware of the Declaration of Helsinki, which clearly views his actions as a breach of ethical conduct.[377]

After the press conference, which was aired on all major South Korean television networks, most of the nation's media outlets, government ministries, and the public gave support to Hwang. Interestingly, sympathy for Hwang outpoured, resulting in an increase in the number of women who wanted to donate their eggs for Hwang's research

Within weeks, things took a tragic turn for science. A report that was greeted with so much elation only six months earlier is now a flashpoint for fraud accusation. PRESSian, a Seoul-based online news site, on December 11, 2005, Sunday reported Hwang might have ordered a subordinate to fabricate photos of nine stem cell batches from just two cell lines that were presented to *Science*. PRESSian said it had acquired transcripts of an interview with Kim

---

[375] The so-called N-rays were a phenomenon described by a distinguished French physicist René-Prosper Blondlot in 1903 but subsequently shown to be illusory. Working at the University of Nancy, Blondlot perceived changes in the brightness of an electric spark in a spark gap which he attributed to a novel form of radiation, naming it the N ray, for the University of Nancy. Following his own failure to detect N-rays, U.S. physicist Robert W. Wood set to investigate further. His thorough investigations, published in the September 29, 1904 issue of *Nature*, showed that these were a purely subjective phenomenon, with the scientists involved having recorded data that matched their expectations. The incident is used as a cautionary tale among scientists on the dangers of error introduced by experimenter's bias. More precisely, patriotism was at the heart of this self-deception. France was defeated by the Germans in 1870, and after the major discovery by Wilhelm Röntgen of the X-ray the race was on for new discoveries.

[376] The so-called Piltdown Man was fragments of a skull and jaw bone collected in the 1910s from a gravel pit at Piltdown, a village near Uckfield, Sussex. The fragments were claimed by experts of the day to be the fossilized remains of a hitherto unknown form of early man. The Latin name *Eoanthropus dawsoni* was given to the specimen. The significance of the specimen remained the subject of controversy until it was exposed in 1953 as a forgery, consisting of a human skull of medieval age, the 500-year-old lower jaw of a Sarawak orangutan and chimpanzee fossil teeth. It has been suggested that the forgery was the work of the person said to be its finder, Charles Dawson, after whom it was named. This view is strongly disputed and many other candidates have been proposed as the true creators of the forgery. It has been suggested that nationalism and racism also played a role in the acceptance of the fossil as genuine, as it satisfied European expectations that the earliest humans would be found in Eurasia. The British, it has been claimed, also wanted a *first Briton* to set against fossil hominids found elsewhere in the world, including France and Germany.

[377] The Declaration of Helsinki, developed by the World Medical Association, is a set of ethical principles for the medical community regarding human experimentation. It was originally adopted in June 1964 and has since been amended multiple times. The Declaration made informed consent a central requirement for ethical research while allowing for surrogate consent when the research participant is incompetent, physically or mentally incapable of giving consent, or a minor. The Declaration also states that research with these groups should be conducted only when the research is necessary to promote the health of the population represented and when this research cannot be performed on legally competent persons. It further states that when the subject is legally incompetent but able to give assent to decisions about participation in research, assent must be obtained in addition to the consent of the legally authorized representative.

Seon-jong, a member of Hwang's team from MizMedi Hospital who was then at the University of Pittsburgh, made by "PD Notebook." MBC delayed the airing of the program containing the interview.[378] "This April Hwang made me create many pictures with two stem cell lines. It was work that should not have been done. But I was not in a position to disobey the order," Kim said in the transcript. Kim added he had received stem cell lines No. 2 and No. 3 from Hwang and made photos for other stem cell batches, possibly from No. 4 to No. 12.[379,380]

By December 15, 2005, Dr. Roh Sung-il disclosed to MBC television that nine of the 11 stem-cell lines Dr. Hwang had said he created did not even exist. Specifically, DNA tests illustrated that those nine lines shared identical DNA, implying that they had come from the same source. Roh stated that "Professor Hwang admitted to fabrication," and that he, Hwang, and another coauthor had asked *Science* to withdraw the paper. Adding fuel to the flame, MBC broadcast the content of the canceled "PD Notebook" show, which substantiated Roh's claim.

Other private sector companies began to waver and show doubt. Maria Biotech Co. Ltd., Seoul, listed on the U.S. Human Embryonic Stem Cell Registry (HESCR) to have three stem cell lines, was an example. The head, Park Se-pill said, "Up until now, I have believed Hwang did derive cloned embryonic stem cells although he admitted to misconduct in his follow-up paper on patient-specific stem cells... Now, I am not sure whether the cloned stem cell really existed."

All these "accusations" prodded the Seoul National University into launching an official investigation by forming an investigative panel – about three weeks after the first exposure of Hwang by "PD Notebook" and one week after a scientific Web site posted the doctored photos that had been used to substantiate Hwang's work. The probe was started on December 17, 2005.

---

[378] In the South Korea's ultra-competitive journalism world, Choi Seung-ho, producer of "PD Notebook," was spurred into using techniques that tarnished his work. In the critical interview, the former co-worker of Hwang, Kim Seon-jong, then at University of Pittsburgh, was led to believe that his former boss was about to be arrested back home for fraud. When the co-worker started to talk about faking photographs for the *Science* article, he could be seen nervously asking if the interview was being filmed. No answer came from the producer, who was holding a bag with a hidden camera. Instead, the producer hinted that if he cooperated with MBC, they would protect him from arrest. To this date, no one has been arrested in the case. The technique used was a flagrant violation of journalism guidelines. As a result, the scheduled broadcast was canceled and the network even made a public apology to the nation, everyone more or less operating under the assumption that the show was at fault and not Hwang.

[379] James Brooke, "'Clone-gate' scoop boomerangs on South Korean news show", *The New York Times*, December 19, 2005.

[380] James Brooke, "A Korean TV show reports, and the network cancels it", *The New York Times*, December 21, 2005.

Photo 4. (Left) Roe Jung-hye, Dean of Research Office, the Seoul National University (SNU), talked to reporters during the first investigative panel press conference held on December 23, 2005. (Right) Hwang waved good-bye to students after announcing his resignation from SNU, on the same day. (Photos: received via e-mail).

The panel sealed off Hwang's laboratory and conducted a thorough investigation, collecting testimonies from Hwang, Roh and other people that were involved with the scandal.

On the black Friday of December 23, Roe Jung-hye, dean of research affairs at the Seoul National University, made a public announcement at an investigative panel news conference. The initial finding was that Hwang had intentionally fabricated stem cell research results creating nine fake cell lines out of eleven, and that the validity of the two remaining cell lines was yet to be confirmed. The panel cited that Hwang's misconduct as "a grave act damaging the foundation of science." Hwang's principal claim of having used only 185 eggs to create 11 stem cell lines, that is, reducing the average requirement from 242 eggs he needed to establish the first cloned human cell line in 2004 to a mere 17 per cell line, was also discredited by the panel. A co-author of the second *Science* paper, Roh Sung-il, had said as many as 1,100 eggs might have been used in research for the paper.

On the same day after the investigative panel report, Hwang apologized and admitted he fabricated results in stem-cell research that had raised hopes of new cures for hard-to-treat diseases. Dr. Hwang tried to resign from his university, but was declined. The university investiagative panel announced additional findings on December 29, 2005. It had determined that all 11 of Hwang's stem cell lines were fabricated.

The days of infamy came on January 10-12, 2006. In its final report published on the first day of infamy, January 10, 2006, the panel reaffirmed its previous findings while announcing additional discoveries. The panel found out that, contrary to Hwang's claim of having used 185 eggs for his team's 2005 paper, at least 273 eggs were shown to have been used according

Photo 5. On August 3, 2005, at a press conference held in Seoul, Hwang Woo-suk (center) and his team (American collaborator Gerald Schatten is at the far right) presented the first successful male dog cloned from adult cells. After the Seoul National University internal investigation, the only claims that Hwang made that were valid are his clones of cows and pig, and the world's first male dog, Snuppy. (Photo: received via e-mail).

to research record kept in Hwang's lab. In addition, the panel discovered that Hwang's team had been supplied with 2,061 eggs in the period of November 28, 2002 to December 8, 2005. Hwang's claim of not having known about the donation of eggs by his own female researchers was also denied by the panel; in fact, it was discovered that Hwang himself had distributed egg donation consent forms to his researchers and personally escorted one to the MizMedi Hospital to take the egg extraction procedure.

The panel stated that Hwang's 2004 *Science* paper was also fabricated and concluded the stem cell discussed in the paper may have been generated by a case of parthenogenetic process. Although Hwang's team did not rule out the possibility of parthenogenetic process in the paper, the panel said, his team did not make any conscientious effort to probe the possibility through the tests available.

The panel cited in conclusion that Hwang's team intentionally fabricated the data in the 2004 and 2005 *Science* papers, and that it was "an act of deception targeted to both scientific community and general public." However, the panel confirmed that Hwang's team actually succeeded in cloning a dog they named Snuppy, an acheivement which he published in *Nature*.[381]

On this day of the final report, Park Ki-young, the presidential Information, Science and Technology advisor, after weeks of silence for her role in the controversy, announced her intent to resign from the advisor post.

---

[381] *Nature*, 439, 2005, pp. 122-123.

The investigative journalism show "PD Notebook" returned to the air on January 3, 2006 and summarized the course of Hwang's scandal until that day. The show had been cancelled in retribution after it aired its momentous show that correctly accused Hwang of oddities in his research on November 22, 2005. Their last show of the year on November 29 covered other topics. It remained off the air for 5 weeks. The second show, aired on January 10, the day of infamy, dealt further with the Hwang affair concentrating on several instances of Hwang's media-spinning tactics and South Koreans' gullibility with it, as was shown by the South Korean public, media and even president Roh.

On the second day of infamy, January 11, 2006, Chung Un-chan, president of the Seoul National University, where most of Hwang's work was conducted, said in an apology to the nation, "Our society has been overwhelmed with the principle of focusing on outcome instead of procedure, and we forgot that ends cannot justify the means... Most of us, in the name of national interests, exaggerated Dr. Hwang's research to make it an aspiration of the nation."

The national post office stopped selling post stamps commemorating Hwang's research on this second day of infamy. On February 12, 2005 commemorative stamps upon Hwang's research were made by Korea Post. The stamps juxtapose an image of stem cells with silhouettes of a man rising from a wheelchair, walking, and embracing another person.

After the investigative panel concluded that Hwang's 2004 and 2005 papers in *Science* were both fabricated, and following on the confirmation of scientific misconduct, *Science* retracted both of Hwang's papers on unconditional terms on this second day of infamy.[382]

Faced with overwhelming evidence, Hwang apologized on January 12, 2006 on national television for having published false research. Nevertheless, he defended portions of his work. The university panel, he noted, effectively backed up his assertions of having succeeded in creating cloned human embryos – he said his team had cloned a total of 101. The panel also verified his claims of having produced Snuppy, the world's first cloned dog, in 2005, in addition to cloned cows and pigs.[383]

---

[382] Nicholas Wade, "Korean scientist said to admit fabrication in a cloning study", *The New York Times*, December 16, 2005.
[383] Anthony Faiola, "Koreans 'blinded' to truth about claims on stem cells", *The Washington Post*, January 13, 2006.

Photo 6. (Left photo) On May 20, 2005, at the zenith of his professional career, journalists surrounded Hwang at Incheon airport. His team had created batches of embryonic stem cells from nine patients. (Right photo) In a very different setting, at the nadir of his professional career, Dr. Hwang (center), flanked by his junior researchers in a press conference held at the National Press Center in Seoul on January 12, 2006. (Photos: received via e-mail).

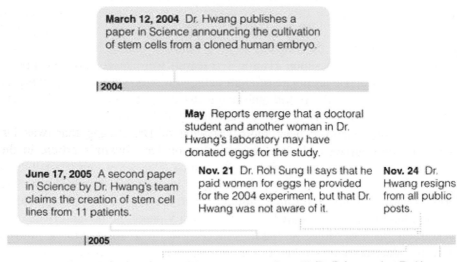

March 12, 2004 Dr. Hwang publishes a paper in Science announcing the cultivation of stem cells from a cloned human embryo.

|2004

May Reports emerge that a doctoral student and another woman in Dr. Hwang's laboratory may have donated eggs for the study.

June 17, 2005 A second paper in Science by Dr. Hwang's team claims the creation of stem cell lines from 11 patients.

Nov. 21 Dr. Roh Sung Il says that he paid women for eggs he provided for the 2004 experiment, but that Dr. Hwang was not aware of it.

Nov. 24 Dr. Hwang resigns from all public posts.

|2005

Nov. 12 An American collaborator, Gerald P. Schatten, severs ties with Dr. Hwang's group, citing "ethical violations" in the way eggs were obtained.

Dec. 15 Dr. Roh says that Dr. Hwang admitted pressuring a former scientist at his lab to fabricate evidence for the June 2005 paper.

Photo 7. The rapid rise and fall of Dr. Hwang Woo-suk. His two papers in *Science* were first called under scrutiny for ethical lapses. By January 2006, the Seoul National University investigative panel declared both papers to be fabricated.

Although as early as on December 23, 2005 Hwang apologized for "creating a shock and a disappointment" and announced that he was resigning his position as professor at the university, the Seoul National University denied Hwang's resignation request, citing a university regulation which dictated "an employee under investigation may not resign from a post." This regulation is effected to prevent premature resignations by investigated employees, thus allowing them to avoid full retributions according to the findings of the investigation (and perhaps avoid involuntary termination), while reaping the benefits of the more honorable and lucrative voluntary resignation.

On February 9, 2006, the Seoul National University suspended Hwang's position as the university's professor, together with six other faculty members who participated in Hwang's team.

The Repercussions

Although the disclosures of Hwang's fraudulent works are a potential blow to Korean science, they can also be seen as a triumph for a cadre of well-trained young Koreans who turned up one flaw after another in Dr. Hwang's work. Almost all the criticisms that eventually brought him down were first posted on Web sites used by young Korean scientists, although vigorous reporting by MBC television and the online newspaper PRESSian also played a leading role.

The young scientists were more skeptical of Dr. Hwang than was Dr. Schatten, who agreed to be senior co-author on Dr. Hwang's article in the June 2005 issue of *Science*, even though all the experiments had been done in Seoul. The referees and editors at *Science* accepted the Schatten-Hwang article without spotting the problems that later came to light, although they did ask for extra tests that may have contributed to the denouement.

The shenanigans is particularly surprising to the many American scientists who had visited Dr. Hwang's lab at the Seoul National University. The scale of the operation, the dedication, and the speed and efficacy of his 65 colleagues were impressive. The lab's production-line organization – specialized into separate units for each aspect of cloning – and the technical skill of those who worked the micromanipulators are the envy of others.

The publication of the two *Science* papers, celebrated by *Science* with great fanfare, has now turned into a debacle. The finding strips any possibility of legitimate achievement in human cell cloning which had propelled the claimant to international celebrity. In his string of splashy papers in prestigious journals, his only one legitimate claim was to have cloned the dog he named Snuppy.[384]

In practical terms, however, the Seoul National University investigative panel's findings are a sharp setback for therapeutic cloning, the much discussed goal of converting a patient's own cells into new tissue to stave off a wide variety of degenerative diseases from diabetes to heart disease. The technique for cloning human cells, which seemed to have been achieved since March 2004, now turns out not to exist at all, forcing cloning researchers back to square one.

Dr. Hwang's escapade may have prevented other researchers from entering the field of human cloning because he seemed to have it all sewn up. Dr. Irving Weissman, a leading stem cell researcher at Stanford University, admitted that he decided not to push the efforts at Stanford because it would have been almost unethical to work with human eggs if Hwang had made the process so efficient. Dr. Robert Lanza, vice president of Advanced Cell Technology, a Massachusetts company that had been active in the human cloning research, said the company's funding for embryonic stem cell research had dried up after Dr. Hwang claimed success in 2004.

The field, however, is still very active. For example, there are two laboratories at Harvard University: Dr. George Daley of Harvard Medical School and Dr. Kevin Eggan's laboratories have been seeking approval to clone human cells. Two groups in England are pursuing the same goal: the International Center for Life in Newcastle upon Tyne University and the other led by Ian Wilmut, the cloner of Dolly the sheep. Advanced Cell Technology of Worcester, Massachusetts, is also back in the game.

As for the field of embryonic stem cells, researchers in the U.S. believe it should not be much affected in the long run, at least on a scientific level, since its theoretical promise is unchanged by one man's misdeeds. One reason Dr. Hwang's results seemed so credible is that he gave lectures in the U.S. in which he explained each step of his process – he would go through the exact method, telling people how to do it. Dr. Hwang also enlisted a leading American expert on attempts to clone monkeys, Dr. Gerald Schatten, as the senior co-author on his 2005 paper, even though Dr. Schatten had done none of the experiments.

Despite Dr. Hwang's failure to do so even with rich funding and copious supplies of human eggs, there is no reason to suppose human cells could not be cloned. Some spectators believe it is only a matter of time before other

---

[384] David Cyranoski, "Verdict: Hwang's human stem cells were all fakes, Korean scientist did not clone a human embryo but did clone a dog", *Nature*, Vol. 439 (7073), January 2006, pp. 122.

groups do what Hwang claimed to have done, although it might be a long time before anyone attains the level of efficiency claimed by Dr. Hwang.

Does the malfeasance surrounding the stem cell research of Hwang Woo-suk make stem cell research work harder? It has to because in the public's mind, the question as to whether there's legitimacy to this kind of science. Fortunately, stem cell research is not dependent on one discovery. Even though Hwang's findings turned out to be fraudulent, nothing he claimed was a fundamental challenge to the principles of embryonic stem cell research.

Hwang claimed to have successfully cloned human embryos, which was only a technical accomplishment. It is something that is already been done in animals. When asked what really holds the stem cell research back, some scientists in the U.S. say that the policies of the Bush administration have affected them more significantly than the Hwang debacle. The lack of federal support keeps many of America's brightest young scientists from working in this area.

How Did It Happen?

How did Dr. Hwang manage to rise so quickly in the scientific firmament and convince so many leading experts that his work was sound? There may be various factors, three of the most important are probably: attracting generous support from the South Korean government, compartmentalizing his laboratory so that few others had any overall view of what was going on, and reporting plausible advances that scientists abroad felt they, too, might have achieved if they had access to as many human eggs as Dr. Hwang obtained. In addition, Dr. Hwang invited well-known American researchers to be co-authors on his papers, which he may have hoped would make his findings more acceptable to leading journals like *Science* and *Nature*. For example, Dr. Jose B. Cibelli, professor of animal biotechnology at Michigan State University collaborated with Hwang and is an author of the 2004 *Science* paper.[385]

The Way to the Top

The starting point of Dr. Hwang's rise to fame was his skill in making the South Korean system work for him. The government had invested about $65 million in his research before the collapse came, and the Ministry of Science and Technology had acclaimed him as an "Outstanding Korean Scientist."

---

[385] Nicholas Wade, "Clone scientist relied on peers and Korean pride", *The New York Times*, December 25, 2005.

The Health and Welfare Ministry committed $15 million in 2005 to set up a World Stem Cell Hub in which Dr. Hwang's technicians would have cloned human cells for scientific customers abroad.

Dr. Hwang had good connections to the government. Dr. Park Ky-young, a former biology professor, is a co-author of his 2004 report on human stem cell cloning. A botanist by training, Dr. Park may not have contributed much scientifically to the task of cloning of human cells; she is, however, the Information, Science and Technology advisor to Roh Moo-hyun, the president of South Korea. Ties with Park yielded a favorable environment for Hwang in the government. A non-official group consisting of high-ranking government officials was created to support Hwang's research that includes not only Hwang and Park, but also Kim Byung-joon, Chief National Policy Secretary, and Jin Dae-jae, Information and Communications minister. The quartet was dubbed as "*Hwang-kum-pak-chui,*" a loose neologism coined from each member's family names which means "*golden bat*" in Korean.

President Roh had been acquainted with Hwang since 2003. He made a number of comments intended to protect Hwang from potential bioethical issues after controversy arose over his human egg procurements. In June 18, 2004, Roh awarded Hwang a medal and said, "it is not possible nor desirable to prohibit research, just because there are concerns that it may lead to a direction that is deemed unethical." In another instance at the opening of World Stem Cell Hub on October 19, 2005, Roh remarked, "politicians have a responsibility to manage bioethical controversies not to get in the way of this outstanding research and progress."

After Hwang's second "groundbreaking" paper was published in *Science* in 2005, support for Hwang came in full swing. In June 2005, the Ministry of Science and Technology selected Hwang as the first recipient of the title "Supreme Scientist," an honor worth US$15 million. Hwang, having already claimed the title of POSCO Chair Professor worth US$1.5 million, secured more than US$27 million worth of support in that year.

Dr. Hwang's rise from peasant's son to international fame has often been attributed to South Koreans' eagerness to embrace new technology and their fiercely nationalistic desire to become No. 1 in the world. Some blame the country's overriding emphasis on quick achievements and its need for a hero who can put the country together at a time of economic uncertainty helped make Dr. Hwang what he is. Over the past four decades, South Korea has grown into the world's 11th-largest economy, while becoming known for its unparalleled respect for higher education. With the government supporting innovation in high-tech industries, South Korea has become a leading player in the fields of semiconductors, automobile manufacturing and shipbuilding. Seeking also to promote its international position in biotechnology research, the government delivered $30 million of funding to Hwang's team with few strings attached. In October 2005, the government launched the World Stem Cell Hub, headed by Hwang and intended to make South Korea a center for the medical cloning industry.

American visitors who had been to Dr. Hwang's growing operation were impressed at the scale and skill of his operation and how he divided his scientists into task forces that specialized in each step of the cloning process. But this compartmentalization may have meant that not all of his co-workers knew what was going on. Few seem to have seen the colonies of embryonic cells Dr. Hwang said he had cloned from patients.

The Way to Demise

The drive to succeed was so strong that many top academics and government officials concede they ignored a series of warning signs. The British journal *Nature*, for example, starting in May 2004 repeatedly raised questions about the ethical standards of Hwang's work. In July 2005, detailed charges of exaggerations and ethical breaches by Hwang's team were posted on a scientific Web site.

One of the first hints that something might not be right came only in November 2005, when a group of young scientists demanded an inquiry into Hwang's work. The people best situated to detect scientific problems are those inside the laboratory who see the raw data being generated and have some practical reason for suspicion. As in many other cases of scientific fraud this was true of Dr. Hwang's, too. It was a whistle-blower in Dr. Hwang's lab who informed the South Korean television network MBC of problems in his work, and that led South Korean journalists to begin to pursue the matter.

The Bits and Pieces

The breakthrough Dr. Hwang said he had achieved would have been a significant advance in therapeutic cloning, in which a wide range of degenerative diseases could be treated with tissues generated from a patient's own cells. On that black Friday – December 23, 2005 when the Dean of Research Office made a public announcement of the investigative panel's unfavorable initial finding – that potential of therapeutic cloning was cited, wistfully, angrily or with renewed determination by scientists and others in South Korea and in the U.S.

The Rev. Kim Je-eun, a Methodist pastor whose 10-year-old son became wheelchair-bound after a car accident and was one of the 11 patients cited in the second *Science* paper, said he vividly remembered the day in April 2004 when the boy first met Dr. Hwang. "My son asked him, 'Doctor, can you make me stand up and walk again?'" Mr. Kim said. "And Dr. Hwang said, 'You will walk again, I promise.'"[386]

The ultimate test of a scientific claim is whether other laboratories independently confirm it. Some scientists have argued that even if Dr. Hwang's errors had remained undetected by the scientific journals and their readers, his work would have fallen under suspicion if no one could repeat it. However, if other scientists had succeeded in cloning human cells before any challenge had emerged to Dr. Hwang's work, it is not so clear whether he would have been exposed. If the procedure indeed worked and other labs had repeated it, the credit could still have gone to Hwang.

The controversy over egg procurement muted any doubts about the science itself. A short cautionary statement in Hwang's 2004 *Science* paper raised the possibility that the cell line may have been created by parthenogenesis. Parthenogenesis is the stimulation of an egg to develop into an embryo. But to give weight that the cell line was cloned, Hwang presented DNA fingerprints as evidence. With such supporting evidence, it was easy to believe that the finding was real.

However, the DNA-fingerprint data and other evidence supporting the existence of a clone were fakes. The cell line, as was hinted in the cautionary statement, was indeed derived from a parthenogenetic embryo. This was the conclusion of the Seoul National University investigative panel on January 10, 2006.

There are various speculations as to why Hwang had faked data. Hwang was under tremendous pressure. The achievements that had brought him celebrity status in Korea, such as the cloning of a cow in 1999, did not have supporting peer-reviewed publication. To have results on human cloning published in a top journal would erase any nagging doubts for good.

Other aspects of the Korean culture may have added to the pressure. Koreans are expected to do everything fast. But like other high-profile fields, in stem cell research, researchers are expected to operate honestly, even when under undue pressure. Perhaps Hwang underestimated the complexity – he assumed that someone, perhaps himself, would reproduce the results more convincingly later. But this confirmation did not come.

---

[386] Choe Sang-Hun, and Nicholas Wade, "Korean cloning scientist quits over report he faked research", *The New York Times*, December 24, 2005.

As others could not replicate Hwang's work, it was thought that this was because other labs did not have access to plentiful eggs, the heavily funded facility or the dexterity of Hwang's technicians. But as time passed, pressure grew for Hwang to reproduce it himself to move forward. According to Han Hak-soo, a producer at "PD Notebook," Hwang was expected to get a patent and to make the stem cell cloning technique medically applicable. By the time of the publication of the 2004 *Science* paper, they had applied for a worldwide Patent Cooperation Treaty (PCT) patent for the technique they had developed and also for the cloned human embryo stem cells. The university would own 60% of the patent; the remaining 40% would go to the other collaborators. Drs. Hwang and Moon Shin-yong, a senior co-author of the 2004 and 2005 *Science* papers, were not participating because they were professors.

And they knew they would have difficulty getting that with the possibility of parthenogenesis still lingering. This could have explained the trickery in the second *Science* paper, particularly when so much money had been invested in him, and when so much was expected out of him. The first *Science* paper in 2004 was a proof-of-concept of stem cell cloning, and the second *Science* paper of 2005 was a giant step forward toward eventual therapeutic cloning.

To make those embryonic stem cells useful for therapy, scientists would then have to turn them into particular types of cells like heart cells or neurons that could be injected into people. Scientists have already learned how to do that for some cells, but not all. Also the cell populations to be implanted must be pure, that is, it is essential that the cells contain no residual embryonic stem cells, because when implanted in the body they tend to form tumors called teratomas that consist of a mix of tissue types including hair, skin and teeth.[387]

The Characters: Accomplices and Foes

Hwang's deception ranks as one of the highest-profile cases of scientific fraud in recent history. Below is a brief description of each main character in this major stem-cell scandal, from the top to the bottom.[388]

Hwang Woo-suk. Hwang, 53 at the time of the unraveling of the sandal, grew up in a poor village in Korea, son of a peasant. He became a veterinarian and later turned to medical science. He, who seemed to have a knack for choosing high-profile projects, was already a known figure before the 2004 and 2005 *Science* papers. As early as 1993, he created the country's first cow born from

---

[387] Andrew Pollack, Cornelia Dean, and Claudia Dreifus, "Cloning and stem cells: Their research, medical and ethical issues cloud cloning for therapy", *The New York Times*, February 13, 2004.

[388] David Cyranoski, "Who's who: a quick guide to the people behind Woo Suk Hwang story", *Nature News*, January 11, 2006.

in vitro fertilization. In 1999, he became something of a national hero after cloning one. Hwang went on to try to clone an endangered tiger, tried his hand at pigs, and even claimed to have developed a cow resistant to mad cow disease in December 2003. The only small criticism in all these early cloning achievements is that they were not reported in peer-reviewed journal publications. Then came the string of splashy articles, two in *Science* and one in *Nature*. The work shot him to international fame, and then infamy within a year and a half.

Moon Shin-yong. Moon, 56 in 2004, is an expert on female infertility and heads the South Korean Stem Cell Research Center that the government formed in 2002 to apply stem cell research to treat diabetes, heart disease and other diseases. Among its goals are to develop technology to raise human embryonic stem cells and to form a genetic database based on the findings. The center is also developing a set of ethics for stem cell research. Moon is a co-author in both the 2004 and 2005 *Science* papers. Eighteen months before the 2004 *Science* paper, Hwang approached Dr. Moon and asked him to join his research team because Dr. Moon would greatly aid his efforts at human cell cloning.

The Egg Donors. Behind Dr. Hwang's work is a copious supply of eggs. Though Hwang had denied using eggs from subordinates or paid donors until late 2005, it is now established that there were 129 known donors, but only 34 were noted in his papers. Many of them were paid donors.

At least two, Koo Ja-min and Park Soon-eul, were workers in his own laboratory. After the fall out, the Seoul National University investigative panel reported that not only did Hwang know of one woman's donation, he accompanied her to the MizMedi Hospital for the procedure.

Koo, 35, is a mother of two who had been working as a Ph.D. student with Hwang since the late 1990s. She had previously worked culturing egg cells from pigs to prepare her for working on human cells. She is a co-author in the 2004 *Science* paper. She said she was a donor in spring 2004 and that she was happy to contribute to the drive to search for cures for spinal-cord injuries.

Park was a research fellow in Hwang's laboratory and a co-author on the 2004 *Science* paper. She helped to develop the method that Hwang's lab uses to remove an intact nucleus from an egg, by squeezing it out rather than sucking it with a needle. Park moved to Gerald Schatten's lab in University of Pittsburgh in 2004. The method was adapted in monkey cloning research at Schatten's lab. It has been questioned whether her egg donation was voluntary, or if she was pressured after hundreds of eggs were ruined in early experiments.

Gerald Schatten. Schatten, 56 at the time of the unraveling of the scandal, is an energetic developmental biologist who has long sought to unravel the secrets of cloning in primates. In January 2000, while still at the Oregon Regional Primate Center in Portland, he led the creation of Tetra, a monkey "cloned" by artificial twinning.[389] Later in 2001, he was recruited to lead a research institute at the University of Pittsburgh Medical School, where he still works today. His investigations into monkey cloning led him to seek advice from Hwang. Their collaboration resulted in co-authorship, with others, of the 2005 *Science* paper. Schatten was also involved with the 2004 *Science* paper. The effusive Schatten often served as Hwang's ambassador in the West. Later, he publicly severed his collaboration with Hwang – an announcement that helped spur investigations into the validity of Hwang's *Science* papers.

Ahn Cu rie. Ahn is a xenotransplant expert at the Seoul National University. Her professional goal is to make pig organs that could solve the organ shortage crisis. As one of Hwang's most ardent supporters throughout 2004 and 2005, she was an unofficial spokeswoman when Hwang was not in contact with the media. She claimed her role in the "scandal" was limited to developing the cell lines produced by Hwang for clinical applications. When Hwang went into seclusion after the November 24, 2005 apology for ethical lapses in human egg procurement, Ahn announced, "Dr. Hwang will not be able to return to the lab, at least, until at the end of this week because he is extremely exhausted, mentally and physically." And when the scandal was unraveling she made a 10-day trip overseas to meet with scientists in the U.S. and Japan, and concluded that the controversy had hurt the lab.

The Whistle-Blowers. In June 2005, only days after the publication of the second *Science* paper, a former member of Hwang's laboratory tipped off MBC producers to the possibility that the data in the *Science* paper were fabricated. Another one gave "PD Notebook" stem-cell line No. 2, one of 11 cells supposedly tailored to patients, and challenged them to test it. A third informer backed up some of MBC's information. According to Soo Han-hak, producer of "PD Notebook," though there had been friction between members of the lab, the friction did not seem to be an issue. Rather, the informers were worried about the future of Korean science.

Roh Sung-il, who comes from a long line of fertility specialists, is a senior fertility expert at the MizMedi Hospital. The hospital copiously supplied

---

[389] Tom Abate, "First monkey born using new method of cloning", *The San Francisco Chronicle*, January 14, 2000.

Hwang with eggs. According to Roh, he secured no less than 1,200 eggs for Hwang's experiments. He turned down co-authorship in the 2004 *Science* paper. He was, however, second author on the 2005 *Science* paper. In November 2005, he publicly admitted that egg donors were paid, later adding a claim that Hwang knew about it. On December 16, 2005 he broke down in tears at a press conference in which he accused Hwang of fabricating data.

Kim Seon-jong is a researcher in his mid 30s, and a co-author on both *Science* papers. He claimed that he doctored photos for the papers at Hwang's behest, telling MBC that he had produced images for eleven lines from two that he was given. He was the very individual interviewed in the "PD Notebook" report that violated journalism guidelines, which led to a 5-week suspension of the program.

Korean Scientists. Korea's young scientist community was very active during the unraveling of the Hwang scandal. Websites such as those of the Biological Research Information Center (BRIC), and Scientists and Engineers' community (SCIENG) are clearinghouses for science-related news; the Science Gallery of DC Inside is a website for digital camera enthusiasts. In early December 2005, the "netizens" and bloggers of these websites started posting doubts about DNA fingerprints and images in Hwang's papers. In a society where it was difficult to criticize a top scientist who had ties in the government, these anonymous portals were havens for freedom of speech. Their contents provided vital leads for investigators.

Journalists. As early as May 2004, the British journal *Nature* in which Hwang published his dog clone paper, questioned the ethical integrity in the way Hwang procured eggs. The query did not amount to much, but instead drew condemnation from Hwang and the Korean press. Then, in 2005, the investigative program from MBC-TV, "PD Notebook," played a vital role exposing Hwang. The program about Hwang's egg procurement on November 22, 2005 led to a confession from Hwang two days later, but the way the reporter of the program had obtained the news was a violation of journalism guidelines. The program was subsequently suspended for 5 weeks. Further evidence gathered by the program concerning Hwang scientific papers was crucial in sparking official investigations.

Park Ky-young. Park is a former plant biologist who later became a consumer-advocate in areas related to genetically modified organisms. She is a co-author on the 2004 *Science* paper. In her professional role as a bioethicist, she said she is limited to surveying people's attitudes on cloning and other biomolecular techniques, but not including issues such as egg donations, in apparent reference to Hwang's egg procurement controversy. Some people suspect Park was added to the paper for political reasons – she later became Information, Science and Technology Advisor to South Korea's president Roh Moo hyun, who was a great supporter of Hwang's research until the unraveling. On January 9, 2006 following the release of the Seoul National University's investigative panel report, which found that the 2004 paper was a fabrication, Park resigned from her advisory position. The Science Minister was also replaced following the scandal.

Investigative Panel. In December 2005, the Seoul National University set up a nine-member committee, led by pharmacologist Chung Myung-hoe from the College of Medicine, to investigate Hwang's lab. The panel quickly filed both preliminary and final reports. The finding was that data in the two *Science* papers were fabricated; but the dog clone claim published in *Nature* stands

Hwang Versus Raelians

On closer examination, there is an uncanny similarity and parallel between the Hwang stem cell embryo scandal and the Raelian claims of having cloned humans.

To begin with, human cloning and human embryo cloning are both very controversial areas. The Raelians refused to disclose the locations of birth of their clones, or claimed that they carried out the cloning in places where the laws did not forbid human cloning. The Raelians obtained their eggs from their cult members. Hwang's confrontation with ethics was his human egg procurement. He procured his eggs in Korea, where the laws against such procurement approach were not well defined, at least before 2005.

In the Raelian claims, the claims covered almost all possible scenarios in which a person or couple would seek cloning: a girl clone of a fertile mother and an infertile father; a girl clone for a lesbian couple; a boy clone of a deceased boy; a clone for a Saudi Arabia parents. The last case is believable because it is known that Muslims are more tolerant of cloning. For Hwang, he had been able to generate embryonic stem cell lines from nine patients. It did not matter whether the patient whose cells were being cloned was young or middle-aged, male or female, sick or well – the process worked.

The Raelians were right on target to have five clones by February 5, 2003. The rate of success was unbelievable. For Hwang, a success rate of one in 17 made the technique too efficient to be true. The process was so efficient that Dr. Davor Solter, the director of the Max Planck Institute for Immunobiology in Freiburg, Germany commented, "You almost have no reason not to do it."

He added that it seemed more efficient to clone and obtain human stem cells than to do the same experiment in animals, although no one knows why.[390]

The Raelians were facing keen competitions from various groups. In November 2002, Severino Antinori announced that he had successfully implanted clones in three women, with the first birth expected in January 2003. The Raelians, to be first, rushed to their announcement of Eve on December 26, 2002. Hwang was also under steep competitions from a number of stem cell experts. One of the biggest competitors to Dr. Hwang's Korean Stem Cell Hub is the U.K. Stem Cell Bank, a government lab near London established in 2002, receiving the first deposits in 2004, almost two years ahead of the Korean Stem Cell Hub.

The Raelians' first clone, Eve, the seven-pound baby girl was cloned from the DNA of a skin cell of a 31-year-old American woman. The developing embryo was implanted in her own womb to grow to term, that is, the surrogate mother is also the donor. In Hwang's first paper, he and his team started with 242 human eggs donated by 16 Korean women. Of these, they got about 20 embryos from which they tried to extract stem cells and managed to produce only one stem cell line – one that was derived with eggs and somatic cells from a single female donor in her 30s. The scientists managed to produce embryos only when the person to be cloned was also the donor of the egg used.

Then there are also differences between the Raelians and Hwang. The Raelians, besides press conferences to announce the clones, showed no clones in public settings, and declined DNA validations of the clones. Hwang, however, had two papers in the prestigious *Science*, one in 2004 and one in 2005. Hwang also got well known Western scientists to vouch for his work. An additional reason Hwang's results seemed so credible is that he gave lectures in the U.S. in which he explained each step of his process – he would go through the exact method, telling people how to do it. Another difference is that the Raelians, in the minds of many, are cultists; Hwang's lab was a legitimate research center, handsomely funded by the Korean government, and one that was frequented by visitors and experts.

Whatever the cause that led Hwang to such an extraordinary tale of delusion and deception, it is fair to assume that both the Raelians and Hwang were lured by the fact that fame, scientific prizes and, often, lucrative patents await those who finish first. But only Hwang had been more cautious when he embarked on the escapade, he would have learned from the Raelian hoaxes, which took place in December 2002 and January 2003, merely a year

---

[390] Gina Kolata, "Koreans report ease in cloning for stem cells", *The New York Times*, May 20, 2005.

and a half before his first *Science* paper, that under the scrutiny of science's microscope, it does not pay to play any trickery.[391]

Also, had these fraud maestros been less ambitious and been content with a great breakthrough, they would probably have pulled a blind over many eyes. The greatest mistake they made is never know when to stop: the Raelians came up with five clones, when Eve would have suffice to prove that they were the first to clone a human; Hwang came up with 11 cell lines and an efficiency that was questionable, when his 2004 paper was already earth-shattering.

Most interestingly, when the Hwang scandal was unraveled down to the last thread, the Raelians extended their friendly hands to collaborate with Hwang. Hwang did not take up the offer.

Synopsis of the Dogged Reports

We have been crediting the media for unraveling the Hwang clone-gate. In actual fact, the media also had a lot to do in spinning Hwang's "breakthroughs" out of proportion. As a starter, the controversy over egg procurement muted any doubts about how the embryo in the 2004 paper was actually obtained. In the paper, there was a short statement raising the possibility that the cell line might have been created by parthenogenesis. Such possibility was hardly or never raised in the midst of egg procurement media frenzy.

After this paper, both *Science* and *Nature* were vying for Hwang's papers. Even though the review process was still as rigorous as for any other contributions, anything coming from Hwang was treated as a high-profile contribution to be looked at seriously and rushed to print.

For example, when the second paper came into *Science* office on Tuesday, March 15, 2005, via e-mail, it was treated as a high-profile paper. The lead author, Hwang Woo-suk of Seoul National University, was a known entity at the journal. He had previously published a paper in *Science*, on February 12, 2004.

As in its normal practice, *Science* put this latest paper from Dr. Hwang through the same review process it put the nearly 12,000 other papers received in an average year. Each paper is sent to one or two outside experts on the journal's board of reviewing editors. The editors advise on whether the paper is appropriate for *Science* magazine. On average, 70% of submitted

---

[391] Elisabeth Rosenthal, "Under a microscope: High profile cases bring new scrutiny to science's superstars", *The New York Times*, December 24, 2005.

papers are rejected outright. Each of the others is sent to at least two additional scientists in the pertinent expertise area for in-depth reviews. These reviewers normally comment on the paper and assess its quality. They check off boxes ranging from "reject" to "publish without delay." About 25% of those reviewed end up being published.

The roles of reviewers are not science police, however. The normal review practice is to assume the integrity of the contributors and that the data are real. The question is, "whether the data support the conclusions."

In the case of Hwang's second paper, after having passed scrutinies of three outside reviewers, the paper was accepted for publication on May 12, faster than the journal's average time from submission to acceptance, which is about three months. The paper appeared in May 19, 2005 issue of *Science*.[392]

Other media and organizations also play critical roles.

Besides the Korean media, Western media also interviewed Hwang and reported Hwang's "breakthroughs." The reports and interviews were soon published all over the electronic media.[393] There is even a book about him (in Korean),

Hwang Woo-suk: The scientist who is changing the world,

written by the science/tech department of Publisher Mae-kyung, Seoul.[394]

Korea Post issued on February 12, 2005 commemorative stamps on Hwang's research. The stamps, later discontinued on January 11, 2006, juxtaposed an image of stem cells with a man rising from a wheelchair to embrace another person. In February 2005, Hwang had just published his first paper. The report was a proof-of-concept of stem cell cloning. Even though the second paper was a giant step toward eventual therapeutic cloning, it was still far away from it.

In June 2005, the Korean Ministry of Science and Technology endowed the accolade of "Supreme Scientist" on Hwang, an honor worth US$15 million. At that point, Hwang had claimed the POSCO chair professorship worth US$1.5 million, and secured more than US$27 million worth of support for his research.

Though it is always easier to exonerate suspects than to convict them, in this case, it is fair to speculate that with all these media coverage and financial

---

[392] Gina Kolata, "A cloning scandal rocks a pillar of science publishing", *The New York Times*, December 18, 2005.
[393] Claudia Dreifus, "A conversation with Woo Suk Hwang and Shin Yong Moon, 2 friends, 242 eggs, and a breakthrough", *The New York Times*, February 17, 2004.
[394] Mae-kyung Science/Tech Department, *Sesang-eul Bakkuneun Gwahakja Hwang Woo-suk*, (Mae-kyung, Seoul, Korea, 2005), 272 pages, in Korean.

support, Hwang had to deliver to live up to his reputation. But he just went astray. At the end of the day, what the now disgraced Hwang can still claim, besides some pigs and cows, is a cloned dog named Snuppy and his method of extracting nuclear material from an ovum. What Hwang's team did differently was to squeeze the nuclear material from the ovum, which caused minimum damage to the egg.

●●●●●   ●●●●●   ●●●●●

Most scientists support cloning to make embryonic stem cells, but at the same time, they are nearly unanimous in denouncing so-called reproductive cloning to create babies. Religious groups also oppose reproductive cloning, but they split on therapeutic or research cloning for stem cells. Roman Catholics, evangelicals and many mainline Protestant churches oppose cloning for a variety of reasons. Ethicists note that some Christian churches and most Jewish groups back therapeutic cloning.[395]

Stem cell research is always going to raise a hue and cry for people who feel that the embryo is the moral equivalent of an adult human life. As we have seen, the questions multiply and research is supplying answers at a record pace, stirring up ethical and moral issues along the way. As we learn more and more about how to understand and to control life we will need to work just as hard to gain the wisdom needed to make good choices about life.

---

[395] Laurie Goodstein, Denise Grady, John Files, and Cornelia Dean, "Cloning and stem cells: The debate; Split on clones of embryos: research vs. reproduction", *The New York Times*, February 13, 2004.

13 Legal and Ethical Issues

"And just as science constantly changes with new developments, we come to the startling discovery that ethics are not rigid but must also alter in response to our changing in understanding of the world and the new technology *milieu exterieur* that we continually create."

- Harold Morowitz, in *Shaping the Future*, 1989

► How? What? Why?

Biology is the most intimate of the sciences. It deals with some of life's most wondrous occurrences, with reproduction and birth, with human abilities and limitations, with diseases, and with death. It seeks scientific explanations of these things, an understanding of how they occur and often why they occur. But explanations do not eliminate the wonder.

With the new repertoire of tools of genetic engineering and cloning, biology is not just the science of what we are and of how we came to be who we are, it has become the science of what we can become. Still, no matter how much we know about being human, we will always be human, guided by societal rules and customs.

Conventional wisdom holds that science and technology speed at a dizzying pace headlong into the future, leaving law, ethics and social policy laboring to catch up. While this wisdom may be true in many of the cases, the reverse can also be true when society at large has a lot of warning of what is coming. But policy makers might move too fast toward rigid social policies in an effort to avoid being caught off guard.[396]

---

[396] Tabita M. Powledge, "Ethical and legal implications of genetic testing: A synthesis", *The Genomes, Ethics and the Law*, (American Association for the Advancement of Science, Washington, DC, 1992), pp. 1-21.

Cloning is here, and society at large has ample warning of the eventual arrival of therapeutic cloning and possibly the first human clone. In order not to enact overly restrictive guidelines, we should understand cloning a little better. In cloning, the nucleus of a human body cell is transplanted into an egg whose nucleus has been previously removed, and the zygote is stimulated to divide to produce a blastocyst embryo. From this point of embryo development, we should differentiate the two forms of cloning, depending on the end purpose. In the first form, reproductive cloning, the blastocyst is placed into a uterus with the intent of creating a newborn. In a related second form but a different procedure, cells are isolated from the blastocyst and the cells are used to create stem cell lines. In what is sometimes called therapeutic cloning, the donor of the nucleus for transplantation to produce stem cells can be the person (patient) in whom the stem cells will be used to regenerate damaged tissues.

Using stem cell for regenerative medicine is, but one medical use. There are other medical uses of stem cells produced using nuclear transplantation. For example, stem cells derived from a body cell or a disease cell of a patient who has inherited the risk for that disease could be powerful tools for medical research and could lead to improved therapies.

So we are on the verge of regenerative medicine, and on the verge of making humans. But in many aspects, as described above, the two forms of cloning overlap strongly, making it very difficult to draw up effective guidelines. Are both forms of cloning unacceptable to society? Are they bad because it is premature to talk about them? If therapeutic cloning is beneficial, how should this kind of information be made available to the public? How does society go on from truly groundbreaking research to delivering safe and effective technologies? How can technologies derived from such groundbreaking research be developed so as to minimize their potential for abuse? If sex is so good, is cloning necessary at all? If sex is so good, should cloning be banned? And many other questions, too numerous to enumerate one by one.

In June 2001, the U.S. National Academy of Sciences' Panel on Scientific and Medical Aspects of Human Reproductive Cloning (PSMAHRC) was charged with the responsibility of examining the scientific and medical issues relevant to human reproductive cloning, including protection of human subjects, and to clarify how human reproductive cloning differ from stem cell research.

The Panel determined whether the current state of the art for reproductive cloning is scientifically feasible, reproducible and medically safe, and whether human participants in the process could be adequately advised and protected. The Panel did not extend to examinations of the ethical issues related to human reproductive cloning.

To achieve its goal, the Panel studied scientific and medical literature; held workshops with world leaders in relevant technologies and pioneers who plan to clone human beings. The Panel found that animal studies of reproductive cloning demonstrate that only a small percentage of the attempts are successful. Many of the resulting clones die during all stages of gestation, or newborn clones often are abnormal or die, and that the procedures carry serious risks for the mother. However, the data on nuclear transplantation to produce human embryonic stem cells (hESCs) show that these cells are functional.

In a January 24, 2002 testimony before the 107th Congress of the Senate Committee of Labor, Health and Human Services and Education Subcommittee, Committee on Appropriations, U.S. Senate, Irving L. Weissman of Stanford University, as chairman of the National Academy of Sciences' PSMAHRC Policy, recommends that[397]

- ❑ Human reproductive cloning should not be practiced at this point in time. It is dangerous and likely to fail. The Panel therefore unanimously proposes that there should be a legally enforceable ban on the practice of human reproductive cloning.
- ❑ The scientific and medical considerations related to this ban should be reviewed within five years in the form of a sunset clause. Then the ban itself should be reconsidered only if at least two conditions are met:[398]
  - o A new scientific and medical review indicates that the cloning procedures have been perfected and are likely to be safe and effective; and
  - o A broad national dialogue on the societal, religious, and ethical issues suggests that a reconsideration of the ban is warranted.
- ❑ The scientific and medical considerations that justify a ban on human reproductive cloning at this time are not applicable to nuclear transplantation to produce stem cells. Because of the considerable potential for developing new medical therapies for life-threatening diseases and advancing medical fundamental knowledge, the Panel supports the conclusion of a previous National Academy of Sciences' report that recommends biomedical research using nuclear

---

[397] *Scientific and Medical Aspects of Human Reproductive Cloning*, (National Academy Press, Washington, DC, 2002).
[398] A sunset law is a law requiring periodic reevaluation to justify its continuing existence.

transplantation to produce stem cells be permitted. However, a broad national dialogue on the societal, religious, and ethical issues is encouraged on the matter.

The Panel also recognizes that scientists place high value on the freedom of inquiry – a freedom that underlies all forms of scientific and medical research. Thus recommending restriction of any research is a serious matter, and the reasons for such a restriction must be compelling. In recommending a ban on human reproductive cloning, the Panel cites the potential dangers to the implanted fetus, to the newborn, and to the mother carrying the fetus as constituting such compelling reasons. In contrast, the Panel believes there are no scientific or medical reasons to ban nuclear transplantation to produce stem cells, and such a ban would certainly close avenues of promising scientific and medical research.

Therapeutic and Reproductive Cloning

Cloning can be used for therapeutic purposes and therapeutic cloning holds out the promise of new medical interventions and cures. The damage done by degenerative disorders such as Parkinson's disease or Alzheimer's disease might be reversed, that is, therapeutic cloning is looking for new ways to provide regenerative cures for degenerative disorders.

There are supporters and opponents for therapeutic cloning. Most of the supporters are in the camp of biomedical and healthcare research while most of the opponents are in the religious camp. In fact, many in the community of therapeutic cloning are feeling the stigma of being associated with cloning. Its proponents fear that the two kind of cloning – therapeutic and reproductive, have merged in the public's mind. At least some leaders in therapeutic cloning have voiced not to call the field "therapeutic cloning," but anything else such as nuclear transplantation or stem cell research.

Cloning can also be used for reproductive purposes. This is where there is a lot of confusion. When people say they are against reproductive cloning, they fail to draw a clear distinction between purposefully cloning of a whole individual and cloning used in assisted reproduction. Cloning of whole individuals is a distasteful idea to most people, so is the idea of purposefully propagating a family of clones. But cloning may be justifiable in cases where a woman may not be able to reproduce successfully on her own. Eventually research may prove that human cloning used as a form of assisted reproduction can be done at no greater risk to the child than *in vitro* fertilization (IVF). In fact, reproductive cloning can be a part of IVF: cloning technique such as artificial twinning can be used to increase the number of embryos for implantation, thus increasing the success rate of IVF.

Supporters of cloning, therapeutic or reproductive, hope that such research would be done openly in the U.S., Europe or Asia, where established government agencies exist to provide guidelines and careful oversight of the studies. Otherwise the research may lead to the bad aspects of cloning.

A scientific group, the International Society for Stem Cell Research of Northbrook, Illinois, USA, called for a complete ban on human cloning, while strongly supported cloning experiments to derive stem cells. The president of the group, Dr. Leonard I. Zon, who is a professor of pediatrics at Harvard, explained that too many cloned animals have been unhealthy with problems like arthritis and obesity. But he doubts any serious researcher would even try to clone an entire human being, adding, "You never know about outliers. But all responsible scientists and doctors would not do this."

Dr. Lee M. Silver, a geneticist who is a professor of molecular biology at Princeton University, agrees that few researchers would want to duplicate people. "We know that the recipe for creating cloned embryos could in theory be used by somebody for reproductive purposes," Dr. Silver said. "I think it is going to happen. I'm not saying it's good, but I think it's going to happen."[399]

It is clear from the above few comments that, as with other technologies, cloning has its good, bad and ugly aspects.[400]

Secret Projects

Unreasonable government restrictions may lead undesirably to clandestine activities in some offshore laboratory where a researcher or a group of researchers are seeking instant fame. Whether the claims of Clonaid of having produced five "human clones" are valid or not, the location of the activity has never been disclosed. Like it or not, Clonaid has certainly received a lot of media coverage. The nondisclosure of the location of cloning can only add another dimension of mystique to the Raelian claim.

Even the funded U.S.-Italy-Israel effort is still looking for a nation willing to host their work. The collaboration believes that the attitude of Judaism is more accepting about creating human life than Christianity, or Buddhism or Islam since "the first men were created by God and God also gave humans the intelligence that at the end of the cycle human clone themselves." As voiced

---

[399] Laurie Goodstein, Denise Grady, John Files, and Cornelia Dean, "Cloning and stem cells: The debate; Split on clones of embryos: research vs. reproduction", *The New York Times*, February 13, 2004.
[400] For an excellent review, see Ronald M. Green, "I, clone", *Scientific American*, September 1999.

by a Rabbi, in principle, Judaism takes a positive view on technological developments and medical progress, but also puts limits on medical science.[401]

After her experience with Nazi Germany's gruesome experiments, Israel has a natural uneasiness with anything that has something to do with genetic engineering. The fact notwithstanding, Israel is a leader in the use of *in vitro* fertilization to treat infertility. But following the lead from Europe, in 1999 Israel enacted a law that put a 5-year moratorium on human cloning.

So far, Cyprus appears to be no more receptive to the idea of hosting a human cloning project. Zavos, who was born on the Mediterranean island, met in March 2001 with Greek Cypriot president Glafcos Clerides, as well as with the health minister. A Cypriot government spokeswoman said that the country is unlikely to host human cloning because it could damage its bid to join the European Union.

Given the pace of events, it is possible that there are other researchers at work. Besides looking for a nation to host the cloning effort, a current technical limiting factor is the availability of a sufficient number of ripe human eggs since hundreds of eggs might be needed to produce only a few viable cloned embryos. Though hormone-induce egg maturation produces at best only a few eggs during each female menstrual cycle, scientists might soon resolve this problem by improving ways to store frozen eggs and by developing methods for inducing the maturation of eggs in egg follicles *in vitro*.

Stigma of Cloning

Instead of educating the public, the press and Hollywood often times do a disservice to the public. The nightmarish scenarios much talked about in the press and in Hollywood include a dictator using cloning to amass an army of "perfect soldiers" or a wealthy egotist seeking to produce multiple copies of himself; popular films such as *Multiplicity* and *Blade Runner* feed these nightmares by glossing over the fact that cloning cannot instantaneously yield a copy of a live or deceased person, but produces embryos for implantation in surrogate mothers. To make an adult, cloning first needs a lot of eggs, and then has to go through the little understood process of development in the womb, the laborious and time-intensive child rearing to reach adulthood. Saddam Hussein, who was a constant eye prick for the U.S. until the overthrow of his government by Coalition Forces on April 15, 2003, would have to wait at least 20 years to realize his dream of a perfect Republican

---

[401] "Scientists look for nation willing to host human cloning project", *Tribune Business News*, March 16, 2001.

army, and the Donald Trump of the world would have to enlist thousands of women to be the egg donors and to be surrogate mothers of his clones.

During development and growth, environmental factors play a significant role. Genes contribute to the abilities and shortcomings of a person, but from conception forward their expression is constantly shaped by environmental factors, by the unique experiences of each individual and by purely chance factors. Even identical twins – natural human clones, possess different physical and mental characteristics to some degree.

Thus there is a strong interplay between nature and nurture: cloning Adolph Hitler might instead produce a modestly talented dancesport Hitler II who can do the Waltz but not rhythm dances. Thus those seeking to mass-produce children or to make an exact copy of someone would almost certainly be disappointed in the end.

▶ Ten AD (After Dolly)

As we think about "The Good, the Bad and the Ugly" of human cloning, it is useful to keep a few things in proper perspective. Otherwise, cloning Clint Eastwood will produce "A Fistful of Dollars" for many, "For a Few Dollars More" for some, and those who think Clint Eastwood II would be an exact replica of Clint Eastwood will be "The Beguiled."[402]

Cloning as an Assisted Reproductive Technology

First, when cloning is finally perfected, however many years AD (After Dolly), it will probably not be a commonly used reproductive technology (reprotech). For many reasons, sex is so good that the majority of heterosexuals who can reproduce unaided will still prefer the old-fashioned, sexual (that is, coital) way of producing children. No other method better expresses the loving union of two individuals seeking to make a baby. Sex is so good, why clone?

Second, as we think about those who would use cloning because they are infertile, we should remember that an important factor affecting the quality of a child's life is the love, affection and devotion the child receives from parents, not the method of the child's birth. Children produced by cloning will probably be extremely wanted children, and thus with good counseling

---

[402] Clint Eastwood is the movie star who acted in "The good, the Bad and the Ugly", "For a Few Dollars More", "For a Fistful of Dollars", and "The Beguiled".

support for their parents, there is no reason to question these children will not be showered with the love and care they deserve.

The life of the first generation of cloned children, being at the center of scientific and media attention, will not be easy for them. They also face the risk of being swept into the hassle of potential lawsuits. Florida attorney Bernard Siegel subpoenaed Clonaid soon after the announcement of the birth of Eve on December 26, 2002. There is as yet no written law forbidding human cloning, Siegel could only file a petition that a guardian be appointed for Eve. His concern was the alleged cloned girl might need extensive medical treatment which Clonaid might not be able to provide. The case was later dropped because the court had no jurisdiction over Eve.

In the future, if human cloning would be legal in some countries while illegal in others, would clones legally delivered in their countries of birth be forbidden to enter countries which forbid cloning? Would cloning laws override immigration laws in these cases? Would clones be able to adopt citizenship of their parents if the parents come from a country forbidding cloning and the cloned child was born in a country not forbidding cloning?

In addition to media attention and potential legal actions, clones and their parents will also have to negotiate the worrisome problems created by genetic identity and unavoidable expectations. But all the above issues are not without precedent. The issues are reminiscent of early days of *in vitro* fertilization (IVF): IVF has not become a commonly used reproductive technology (reprotech) except as an assisted reprotech; the children of IVF have not received less love than naturally born children. Louise Joy Brown, being the first IVF baby born on July 25, 1978, received a lot of media attention. Subsequent IVF babies are almost unheard of. Indeed there are more than 380 infertility clinics in the U.S. in which about 100,000 assisted reproductive procedures were carried out in 2002 with a success rate of about 25%. By the 25th birthday of Louise Brown, there were about one million IVF children born worldwide. Births of these babies were hardly "newsworthy" and they represent only a small fraction of all births.

All is Fine Unless You are the Clone

It is now 10 AD (After Dolly), evidence is mounting that creating healthy animal clones is more challenging than scientists had expected. While barnyard animals and mice have been cloned with regularity, no pets, with the exception of Cc:, the copycat, commercially cloned cats (Tabouli and Baba Ganoush, Peaches, Little Nicky and Little Gizmo) and Snuppy (not counting Idaho Gem, a mule, and Prometea, a horse), have been cloned. The cloning of primates (monkeys and humans), with the exception of the claims by Clonaid, has resulted in frustration.

The clones that have been produced often suffer from large offspring syndromes (LOSS), including developmental delays, heart defects, lung problems, malfunctioning immune system, and growth abnormalities analogous to acromegaly in which the bones become excessively enlarged. Indeed Dolly was euthanized in 2003 at age of six and half years after being diagnosed with arthritis and lung problems, possibly related to her corpulence. Since the life expectancy of Dolly's breed, the Finn Dorset, is eleven to twelve years, and Dolly was cloned from the mammary cell of a six-year old ewe, was Dolly six years old when she was born? Was Dolly genetically twelve years old when she was only chronologically six years old?[403]

At issue here is not that one particular thing goes wrong or one specific aspect of development goes awry, but rather, the cloning process seems to create random errors during nuclear reprogramming. These errors could arise from the enucleation process (as Gerald Schatten has shown), the culture (as Richard Schultz has shown), and many yet unknown factors. These errors can produce any number of unpredictable problems, at any time in life.

We still do not understand how an egg reprograms an adult cell's genes, and that can be a great source of cloning calamities. The problem seems to be that an egg must do the programming task in minutes or hours in cloning that in normal fertilization takes days. The difference in time scales in reprogramming is a cause for concern. Rudolph Jaenisch at the Whitehead Institute of the Massachusetts Institute of Technology believes that breathtakingly rapid reprogramming can introduce random errors into the clone's DNA, subtly altering individual genes with consequences that can halt embryo development and kill the clone *in vivo* or *in utero*, accounting for the low success rate. Dolly was the result of more than 200 tries; with cattle, 100 attempts to create a clone typically result in a single live calf. Cloning mice is more efficient, but even then, only 2% to 3% of the attempts succeed.

Or the gene alterations during reprogramming can lead to problems that prove fatal soon after birth. Ryuzo Yanagimachi affirmed that cloned mouse embryos have serious developmental problems, which usually kill them before birth, and more die right after birth, suffering usually from lung problems. In addition, genetics also play a huge role: inbred strains are much harder to clone than hybrid strains of mice. Presumably inbred animals have much less genetic diversity to overcome genetic errors than hybrid animals.

---

[403] Gina Kolata, "Researchers find big risk of defect in cloning animals", *The New York Times*, March 15, 2001.

Figure 1. Genetically identical mice, one naturally born and one cloned. Guess which is the clone?

Or the genetic alterations during reprogramming can lead to major problems that do not show until later in life. According to Yanagimachi, some mouse clones grow fat, sometimes enormously obese, even though they are on the same diet as otherwise identical mice that are not clones. The mouse clones seem fine until an age that is the equivalent of 30 human years, after which their weight soars. Mouse clones also tend to have developmental abnormalities such as taking longer to eye opening and ear twitching after birth. Mark Westhusin reported cow clones are often born with enlarged hearts or lungs that cannot develop properly. Dolly the sheep clone, while apparently healthy, grew fat and had to be separated from the other sheep and put on a diet. She was also diagnosed to suffer from lung problems, terminating her life at six AD.

We should also emphasize that in cloning by nuclear transplantation, when the DNA is transferred from the donor cell, the mitochondria DNA (a fraction of 1% of the total DNA) is left behind in the donor cell. The donor DNA is then implanted into an enucleated egg whose mitochondria DNA is left in the cell cytoplasm of the egg cell. There could be a conflict between the donor DNA and the egg mitochondria, something that we do not as yet have much knowledge about.

▶ Where Science Meets the Public

The advent of human cloning has been foreseen for more than a decade and there have been substantial on-and-off discussions of its legal, social, and ethical issues. The first heated debate in the U.S. was probably the one in the U.S. Senate right after Dolly the sheep and the announcements by various individuals, including Richard Seed, Severino Antinori, Panayiotis Zavos, and Brigitte Boisselier, that they were planning to clone humans. President Bill Clinton proposed a 5-year moratorium on cloning in 1997. The debate was then interrupted by a politically motivated investigation of President Clinton's affair with Monica Lewinsky in 1998. The debate resumed in 2001, under a new presidency, but it did not get very far. It resumed again in 2003 when Boisselier of Clonaid announced the births of five "human clones," which are now believed to be hoaxes. Unfortunately, the debate was interrupted again by the U.S.-led invasion of Iraq in March that year.

Often, the various interest groups and interested individuals have not talked to each other. Then again, critics postulate nightmarish visions of the human cloning future while proponents paint a rosier picture. To ensure the identification of real problems based on science, it is necessary to bring scientists and nonscientists together. This is why there is still an ongoing debate in the U.S. Senate on the issue, involving policy makers, scientists, ethicists, and nonscientists.

To scientists, acquisition of new knowledge is both wondrous and good. Scientists reinforce the belief by the conviction that original research is a creative endeavor. The quest for better comprehension is a fundamental human trait, one that sets *Homo sapiens* apart from other living species.[404]

The public's perception of science can be ambivalent and sometimes less optimistic. Nonscientists find the continuing quest for knowledge somewhat frightening but the attitude is numerous, diverse, and at times contradictory to characterize. People are eager to make use of fruits of scientific research, and many are avid followers of scientific breakthroughs.

The scientists, biologists in particular, have also devoted considerable time in recent years to explain their work to the public. As the public acquires detailed knowledge about their genetic makeup, human nature begins to seem more biological and less psychological. This idea contrasts strongly with the reverse notion that most of us grew up with, the primacy of an almost purely psychological persona has shaped for many decades our culture's idea about human nature, personality, and individuality.

Once the concept of genetics caught on, the public has gone so far as to blame everything undesirable on the genes. If, after gaining pounds, one is unsuccessful in losing weight, one puts the blame on obesity genes; if one is caught for violent behavior, one blames the violence gene; there is a gene for alcoholism; there is a gene for chain smoking... What they have not taken into account is environmental factors, upbringing and self-discipline.

Some could feel constrained by this new understanding of the importance of human biology, which they may see as implying a rigidly programmed destiny. Then enter genetic engineering and cloning. Others, using genetic engineering or cloning to avoid or delay disease and disability, will achieve a

---

[404] Attributed to Maxine Singer, president of the Carnegie Institution, Washington, DC.

feeling of mastery over their futures in a way never before possible. In any case, these new technologies seem, at a minimum, likely to recast people's notions about human individuality.

But the tension between the acquisition of new knowledge and the fear of that knowledge remains widespread in society. This is a troubling tension to scientists because they rely on the public for support. On a purely financial level, the majority of scientific research is paid for with public funds. And more broadly in a democracy, scientific work, whether public-funded or private-funded, on controversial subjects can be slowed or halted by public opposition, even if engendered by unwarranted fears.[405]

The public's ambivalence towards science also emerges in other ways. The media covers new scientific discoveries and their implications extensively. The press also devotes considerable time to scientific controversies that most scientists consider relatively minor or beside the point. Transgression of scientific standards, whether substantial or insignificant, becomes front-page news. The views of a small minority may be presented as a counterpoint to widely held scientific outlooks, giving the minority viewpoints a credence that they would otherwise not deserve.

Much of the public's unease over scientific advances can arise because new scientific discovery can conflict with long-standing premises, explanations and authorities. In many cases, scientific explanations of natural phenomenon are becoming available where mythological explanations have traditionally held sway. These conflicting viewpoints influence public debate in a number of ways.

An excellent example in the U.S. is the continuing debate over whether evolution should be a part of school curriculum? Modern genetics fully supports the conclusion of evolution that human beings evolved from earlier forms of life. Facts notwithstanding, surveys show that over half of the Americans think that biblical creation myths should be taught in science curricula in schools, even though mainstream religious leaders do not support the view of creation.

How the question is posed can also have effects on the outcome of surveys. For example, in a poll, when gene therapy was presented as a means of curing fatal diseases and preventing inherited birth defects, 84% of the respondents were in favor of it. Posed as a way to alter the genetic code, many of the same group of respondents also inconsistently said it was morally wrong to tamper with nature.[406]

---

[405] Steve Olson (ed.), *Shaping the Future*, (National Academy Press, Washington, DC, 1989).
[406] Steve Olson (ed.), *Shaping the Future*, (National Academy Press, Washington, DC, 1989).

Understanding of the origin of diseases and birth defects also reveals the gap between scientific and mythological explanations of events. There are still some people in the U.S. who believe that AIDS is a form of divine retribution against homosexuals and drug abusers. To add weight to their claim, they echo the views of John Woolman (1720-1772), a prominent American thinker, who in 1759 wrote,

> "The more fully our lives are conformable to the will of God, the better it is for us; I have looked on the smallpox as a messenger from the Almighty, to be an assistant in the cause of virtue, and to incite us to consider whether we employ our time only in such things as are consistent with perfect wisdom and goodness..."[407]

If such were the case, what would Woolman say about the eradication of smallpox? Donald Ainslie Henderson was the one who led a World Health Organization's (WHO's) 11-year (1966-1977) effort that successfully wiped out the scourge of smallpox from the face of the globe.[408]

Replacing mythological explanations with scientific ones is a difficult task. Many myths are associated with accepted authorities, such as religion, or with unquestioned assumptions, such as the inviolability of nature. Interestingly, myths tend to be seen as human, and science is regarded as less human. This is probably because myths have been passed from generation to generation, while scientific explanations are relatively newcomers, and at times, harder to comprehend.

To overcome this, there are a few things scientists may do. First, widespread ignorance about science needs to be tackled. Scientists need to teach the substance of science more broadly and deeply, starting with the youngest newcomers at schools, freshmen and sophomores at universities. Such education effort requires time from scientists to be at meetings, hearings, talking to the media and dealing with the legislators. Scientists should come out of their ivory towers, instead of writing only scientific papers understandable to very few, they should also write for the laypeople. Some TV educational programs are already doing a great job in this respect.

Second, scientists must convince nonscientists the human nature of science and scientists. Scientists must be prepared to criticize when the results of their work are misapplied. Unfortunately, the current climate is that Hollywood is doing a better job making science and science fiction movies, projecting scientists as white-coat, bespectacled characters. Misconceptions

---

[407] Andrew Melrose, *The Journal of John Woolman*, (Morrison and Gibb Limited, Edinburgh, 1898).

[408] Donald Henderson obtained an MD from the University of Rochester School of Medicine (1950-1954). HAL, the author of this book, is an alumnus of The University of Rochester (1981-1986).

and unfounded nightmarish scenarios of science projected by the media and Hollywood should be minimized.

Third, scientists must also be vigilant about the possible misuse of their work in warfare: biological warfare and genetic warfare. The military contends that current work on biological warfare is purely for defensive purposes, but there is no sharp distinction between research for defensive and offensive ends. The U.S., on the one hand, is very demanding when it comes to stopping others from developing such weapons; on the other hand, it itself is systematically developing such weapons.

Scientists must guard against mythologizing their own work by looking to science for answers to every human problem. Scientists should realize their ignorance far exceeds their knowledge and they are far from being know-it-alls![409]

▶ Issues with Human Cloning

It would seem from the work of Schatten and colleagues with rhesus monkeys that the cloning of humans and other primates might not be possible using current cloning technology. Schatten transferred a variety of rhesus monkey cell types into 724 eggs from female monkeys. Thirty-three embryos developed well enough to be transferred into surrogate mothers after initial cell divisions; however, no pregnancies were established. Imaging of DNA and basic cell structure of the embryos revealed chromosomal abnormalities, despite the apparent normal division of the cloned cells.

The genetic similarity between monkeys and humans means the findings of the monkey studies can raise serious questions about the feasibility of human reproductive cloning with existing technology. Although there have been claims by the Raelians that they successfully cloned humans, most scientists dismiss their claims.

If Schatten is right and if the hindrance is only technical, given enough time and materials, a way would be worked out to overcome the technical barrier. Also Schatten's work does not prove that cloning will never work. Someone else using a different methodology could potentially solve this problem.

Also from the experience of Cibelli's successful cloning of hybrid human-cow embryos and not knowing about it for another three years, or from Wilmut's successful cloning of Dolly and not knowing about her existence for seven months, there is no telling that there is already something in existence

---

[409] Attributed to Maxine Singer, at a National Academy of Sciences Symposium to commemorate the opening of the Arnold and Mabel Beckman Center, Irvine, California, 1988.

that we should have known, but not know! We may just as well be better prepared if a clone (besides those of the Raelian Group) should be announced.

Technical issues aside, there are other issues in human reproductive cloning. In many ways human cloning techniques are being adopted from animal cloning techniques, *in vitro* fertilization (IVF), and genetic screening, therefore many of the issues arising from the deployments of these techniques also exist in human reproductive cloning.

Just as reproduction in cattle and barnyard animals have been transformed into a quasi-industrial process, with the goal to turn out whole herds of animals having the same weight, speed of maturity, and overall characteristics, women would be reduced to living laboratories, mere vehicles for the creation of a race of desired characteristics.

Then what are desirable characteristics? What is perfection? What is beauty? And if some African people regard being obese as beautiful and most Western nations regard slim as more desirable, which definition of beauty is the better one? Who will decide desirable characteristics? By engaging the wholesales of cloning, which has the look and feel of an industrial process, it is turning babies into another commodity, like cars or luxury items.

This is not unprecedented. In earlier times, during Renaissance Europe, people deigned to consider human only those children who survived the rigors of wet-nursing and parental neglect. In the modern era, we have commodified children to a far greater extent than we dare to admit. We want them to be delivered like an automobile, in the model, shape and color of our choice. A slick-looking baby, presumably a nice-looking one; a bright color baby, presumably a tanned skin one; an aerodynamic baby, presumably a slimmer-looking one; a heavy-duty baby, presumably a strong one; not a lemon, presumably a healthy one; not a clunker, presumably an agile one... Now we want offspring whose genotypes have been declared free of defects.[410]

The trend is rather clear, as noted by Bentley Glass (1906-), a distinguished *Drosophila* geneticist, that bearing a child who is not perfect might one day become tantamount to a crime. In the present climate, we already have wrongful birth lawsuits in which the children sue their parents for endowing on them physical inconveniences. It is not unreasonable to extend the suits to wrongful identity lawsuits in which clones sue their parents for endowing on them an identity that they would rather not be identified with.

In the commercial sector, we have implied warranty of merchantability – a warranty that the goods shall be merchantable is implied in the contract for

---

[410] Rita Arditti, Renate Duelli Klein, and Shelley Minden, (eds.), *Test-Tube Women*, (Pandora Press, Winchester, Massachusetts, 1989).

their sale; and implied warranty of fitness – a warranty that the goods will be fit for the buyer's particular purpose.[411] If babies were viewed as commodities, would we not have similar warranties in human reproduction such as the warranty of clonability and warranty of nurturability, or variations thereof? Implied warranty of clonability is a warranty that the babies shall be clonable is implied in the contract for the service, whereas implied warranty of nurturability is a warranty that the babies will be fit to be nurtured in particular environments.[412]   And in legal rhetoric, each word has a very precise meaning: "vitality" is the capacity to live, grow or develop; "vigor" is the capacity for natural growth and survival...[413] Then what will the legal definition of perfection be? What will the legal definition of beauty be?...

Attempts to create a race of desired characteristics would have immediate adverse impacts on others. Those in the disability rights movement may construe such attempts as a retroactive death sentence on them, a declaration that does not belong in the world and that the world of the future would be better off without people like them.

Ruth Hubbard (1924-), whose reputation for lucid analysis of the sociology and politics of science extends far beyond feminist circles, argues the effect of genetic and fetal screening would be to shift responsibility for children's disabilities squarely onto the shoulders of the mother. Rather than blaming fate, society would blame women; to avoid censure, women and their families would begin implementing the society prejudices by choice.[414] Such arguments can be extrapolated to cloning for desirable characteristics, except that in cloning, it could be the donor, surrogate mother, or the medical expert effecting the cloning who could be blamed.

Others, including biologist Clifford Grobstein (1916-1998), hold that humanity is on the verge of a revolutionary transition, from chance to purpose, from genetic roulette to genetic determinism. These people argue that humans can no longer shift responsibility, whether to divinity, chance, or unkind fate. Rather, we humans will have to become the creator of ourselves.[415] Some regard these arguments invariably contrapuntal, pitting helpless against control, unintentional actions against forethoughtful actions. An assumption is

[411] Michael Metzger, Jane Mallor, James Barnes, Thomas Bowers, Michael Phillips, and Arlen Langvardt, *Business Law and the Regulatory Environment*, (Irwin, Boston, 1995).
[412] Hwa A. Lim, *Genetically Yours: Bioinforming, biopharming and biofarming*, (World Scientific Publishing Co., New Jersey, 2002).
[413] Martin Rispens & Son v. Hall Farms, Inc, *601 N.E. 2d 429 (Ind. Ct. App. 1992)*.
[414] Rita Arditti, Renate Duelli Klein, and Shelley Minden, (eds.), *Test-Tube Women*, (Pandora Press, Winchester, Massachusetts, 1989).
[415] Clifford Grobstein, *From Chance to Purpose: An appraisal of external human fertilization*, (Addison-Wesley Publishing Co., 1981).

made that we can now grasp evolution, which proceeded previously not in quite the most desirable manner. Some argue that even if we could grasp evolution, should we want to?[416]

Still others argue that what children need is not "good genes." Besides what good genes are has yet to be defined. Rather it is – love, affection, food, clothing, and shelter, rules, disciplines, moral instruction, acceptance, mental stimulation, together with a sense that in their lives justice, fair play, humor, friendship, and a valued place in their families and communities – that are critical to their well being. Neither genetic tinkering nor cloning is likely to provide these; only a concerted commitment by parents and families, schools and other societal organizations will. There is no doubt that the government too has to be responsible for its subjects: some propose greater government expenditures in parental care, day care, health care, and education for children, and greater corporate and government support for childbearing and child-raising.

▶ Cloning as Reprotech

When we think of a disease, we think of seeing a doctor. In this interpretation, then, gynecology as a discipline tends to pathologize pregnancy, treating it more as a disease state rather than a normal manifestation of female physiology. Similarly, fertility clinics will treat infertile women as patients. To help these infertile patients, fertility doctors use assisted reproductive techniques such as *in vitro* fertilization (IVF). But there is no reason why cloning cannot be used as an alternative form of assisted reproductive technology (reprotech). If such is the case, cloning clinics (clonics) and supporting services such as third-party arrangements will begin to emerge.

Egg donation and cell donation, two forms of third-party arrangement, carry their own set of complications. The psychological risks to offspring due to familial secrecy immediately come to mind. Most IVF programs have been scrupulously careful to avoid giving critics any grounds for accusing them of being involved in the purchase of oocytes (eggs).

There is another ethical issue: donor oocytes are increasingly being seen as the solution to the infertility of older would-be mothers, those over forty years old, whose chances of conceiving through IVF with their own eggs are virtually zero. Jean Benward, chair of the American Society of Reproductive Medicine's Mental Health Professional Group had observed that there is a trend afoot in which increasingly younger donors are being roped in for the

---

[416] Gina Maranto, *Quest for Perfection: The drive to breed better human beings*, (Lisa Drew Book, New York, 1996).

sake of increasingly older mothers. The astonishing births to women who might be grandmothers, achieved primarily by Mark Sauer of Columbia University and by Severino Antinori in Rome, have only been accomplished thanks to the vigorous eggs of younger women.

Despite the legal imbroglios, in the U.S. there are services offering to enlist surrogate mothers. Like artificial inseminations by donors, such third-party arrangements do not challenge social norms, but at the same time they do not sit well with some fertility specialists, psychologists, ethicists, and legal professionals.

Surrogacy or gestating youngsters for a fee represents a radical departure from the way in which our society understands and values pregnancy. It substitutes a monetary value for the web of social, affective and moral meanings associated with human reproduction. In the interest of couples' desire to have genetically related offspring, young women serve essentially as incubators, and may do so primarily for mercenary reasons. Any possible bond they may develop with the infant they have borne must be torn asunder. But this may not always be the case. There have been cases in which the surrogate mothers refuse to give up the babies, and lengthy legal battles ensure, at the expense of the welfare of the innocent newborns.[417]

In a human cloning experiment, after the nuclear transplantation, the embryo has to be implanted into a mother's womb. Given that the embryo will develop for months in the womb, interacting with the surrogate mother's body, it is not unreasonable to expect "parents" to hire a young, healthy mother to be the surrogate.

Conversely, should surrogates drink, smoke or otherwise expose fetuses to harm during a pregnancy, "parents" may be put in the uncomfortable position of taking action against the surrogates. If as a result of a surrogate mother's actions, the child is born impaired, or if it has a physical or mental defect due simply to the vagaries of development, who bears responsibility for the child's care?

And if defects or diseases begin to show later in the life of the clone, who will be responsible? Not to mention that the defects or diseases could have rooted in any of the sources: the cloning technique, the surrogate mother's action, the surrogate mother's bodily substance exchanges with the fetus, the birth process, the donor cell nucleus, spontaneous mutation or environmental factors!

So cloning, as a form of reproductive technology, takes us from genetic roulette to genetic determinism only to a certain extent. We only know for

---

[417] Johnson v. Calvert, May 1993, 5 Cal. 4th 84; 19 Cal. Rptr.2d 494; 851 P.2d 776.

sure the nucleus used in transplantation, beyond that, we still have little control over development.

▶ Ethical Issues with Cloning

Cloning raises concomitant ethical questions. Therapeutic cloning is dimmed somewhat by the fear of eugenics revival. If we can manipulate embryos for therapeutic ends, why not go further to enhance human abilities? Disease resistance, intelligence and even looks, size (both length and width) all beckon alluringly. And then who will be the judge of the alleged enhancements? In such a scenario, will we be cloning for attributes we like? Or will we be cloning what is best for the welfare of the clone? In the process, will we tamper with the mainstay of human survival in the past – human diversity?[418]

My Father is My Twin

Even if human cloning is physically safe for the clones, it raises its own share of ethical dilemmas.

Intra family relationship is disrupted. To clone a boy, it is necessary to have a man as a DNA donor, a woman as an egg donor, and may be another woman as a surrogate mother. For the boy clone, who is the father? The doctor (if the doctor is a man), the donor, or the donor's father, that is, the grandfather? If it is the grandfather, what is it like to be born his father's identical serial twin? Who is the mother? The doctor (if the doctor is a she), the surrogate mother, the egg donor, the donor's wife (if he has one), or the donor's mother, that is, the grandmother?

To clone a girl, it is necessary to have a woman as a DNA donor, may be an egg donor, and may be a surrogate mother. For the girl clone, who is the mother? The doctor (if the doctor is a she), the DNA donor, the egg donor (if there is one), the surrogate mother (if there is one), or the DNA donor's mother, that is the grandmother? The girl clone may not have a father (if the DNA donor is unmarried), unless she accepts her "mother" as identical twin so that the grandfather is the father, or that the doctor as the father (if the doctor is a he). If you think this is confusing enough, think of the case in which the clone is born with her identical twin as her doctor, mother, egg donor, as well as surrogate mother?

If grandparents are parents, is there sort of a missing generation?

---

[418] Andrew Kimbrell, *The Human Body Shop*, (Harper, San Francisco, 1993).

Douglas R. Hofstadter, cognitive science and computer science professor at Indiana University, once posed in *Metamagical Themas* the following riddle:[419]

> A father was driving his son to a ballroom dance practice when their car stalled on the railroad tracks. In the distance whistle blew of an approaching train. Frantically, the father tried to ignite the engine, but in his panic, he could not turn the key, and the onrushing train bulldozed the car to the ground. An ambulance sped to the scene and picked them up. On the way to the hospital, the critically injured father died. The son was still alive but his condition was very serious, and he needed immediate surgery. The moment they arrived at the hospital, he was wheeled into an emergency operating room, and the surgeon came in, expecting a routine case. However, on seeing the boy, the surgeon blanched and muttered, "I can't operate on this boy. He's my son."

How could it be? Was the surgeon lying or mistaken? No. Did the dead father's soul somehow get reincarnated in the surgeon's body? No. Were the surgeon the boy's biological father and the dead man the boy's adopted father? No. What, then, is the explanation? Think it through and you will see there is a simple solution in the usual intra family structure. But in a clone family, there are multiple trivial solutions.

Cloning can lead to more acute problem if parents seek to clone a deceased child. Is there a right to one's unique genetic code? Will cloning lead to violations of human dignity? The psychological well being of a clone may be adversely affected. What pressures will the clone experience if, from birth onward, the clone is constantly compared to the esteemed or beloved person who has already lived? Under such pressure, Einstein II will almost definitely not grow into a genius.

In wrongful birth lawsuits, plaintiffs sue their parents for having been born physically challenged. It is likely that we will have more wrongful birth lawsuits if cloning techniques are not reasonably reliable. On top of that, it is also likely that we will have a new class of lawsuits – wrongful identity lawsuits in which clones sue their parent for "endowing" on them identities they would rather not be associated with. Surely not too many will be thrilled to be born Hitler II. In this case, which parent is the defendant: the DNA donor, the egg donor, the surrogate mother, or the doctor?

---

[419] Douglas R. Hofstadter, *Metamagical Themas: Questing for the essence of mind and pattern*, (Basic Books, New York, 1985).

Here we recall the interplay between nature (genetics) and nurture (environment) in human development and growth: a clone is likely to be nurtured into an individual genetically identical (barring mutations), but socially different from its donors in a different environment. A clone of Osama bin Laden, an elusive prick in the eyes of the U.S. Whitehouse, in today's society could very well produce a walking embodiment of the U.S. democracy system.

In addition to existing implied warranty of merchantability and implied warranty of fitness for merchandise, implied warranty of clonability and implied warranty of nurturability will have to be enforced. But the law is still lagging behind cloning research, and the warranties will probably not come about until after a precedent-setting court case.

Some of the fears, that cloning may be used to produce a subordinate class of humans as tissue or organ donors, are less substantial than others. Existing and new laws and institutions should protect clones from being exploited as a two-legged biofactory because clones, just like IVF babies, should have the same legal rights as any natural humans.

Psychological and familial harms are a worry, but these harms are not unique to cloning. Many parents have imposed unrealistic expectations on their children to excel at schools and in sports, for example. In the wake of rising rates of divorce, remarriage, adoption, wedlock birth and assisted birth, we have grown accustomed to unusual intra family structures and relationships. Clearly, the initial efforts at human cloning will require good counseling for the "parents" and careful follow-up of the clones.

▶ Can a Clone Run for U.S. Presidency?

Whether it is ethical to clone a human or not, someone may already be on the way to do it. Having a criminal record may bar one to a public office, but having been conceived by a convicted felon is not necessarily so. After all, the surrogate mother is only a medium, and that two separate individuals enjoy their respective, distinct rights. And after all, the U.S. First Family, the Bush Family, constantly has run-ins with the law, yet these troubles do not forbid their parents serving as the U.S. president and the governor of Florida![420]

The U.S. Constitution, as it currently stands, does not have a clone clause. To be eligible to run for the top office, U.S. presidency, the candidate must be born on U.S soil, 35 years or older. This age requirement means that if a

---

[420] Michael D. Lemonick, "Could a clone ever run for president?" *CNN News*, November 1, 1999.

clone were born today, it will not throw the country into political dilemma for another thirty-five years, unless someone has already succeeded in cloning humans secretly somewhere in the U.S., obtained the clones U.S. birth certificates without the authority knowing about their true identities. The Raelian clones (if they were granted U.S. citizenship) would not be able to run for U.S. presidency until the year 2037, the earliest.

Some may argue that a clone should not be eligible for citizenship, but the argument may not be watertight. According to the law, a human born on U.S. soil is by default a U.S. citizen, whether the mother just wanders across the U.S. border purposely for delivering the baby on U.S. soil.

Cloning is illegal in the U.S. A likely scenario is the cloning is performed offshore in a country where cloning is allowed or in a clandestine effort.[421] The clone is then implanted into a surrogate mother, best if she is a U.S. citizen so that she would not have problem returning to the country. If not a citizen, the surrogate could obtain a visa to enter the country. The clone is then delivered like any other babies, with a U.S. birth certificate.

All would be fine, unless, like in all recent U.S. presidential races, the media begins to dig into the candidate's past; or that the surrogate mother, the donor or the clone expert now wants his or her share of fifteen-minute fame. Imagine the fiasco if the clone is now president-elect, or more interesting still, already a sitting president.

A victorious President Clone is not unlikely. From the way the political election has been going, voters are not voting for the best candidate but plutocrat, the candidate who can best project himself, or the candidate with some physique.[422] This was probably how Arnold Schwarzenegger was elected the governator of the State of California.[423] A clone with the desired external appearance is not difficult to come by. Just look for a donor with the right look. The only slight complication is that this type of external appearance should still be fashionable more than thirty-five years afterwards.

A potential obstacle to President Clone will come if President Clone carries serious side effects. Dolly the sheep suffered from some health problems before she was euthanized on Valentine's Day (in the U.S.) of 2003. In this case, the external appearance may or may not help. In historical retrospect, Abraham Lincoln had genotype linked to Marfan's syndrome, a

---

[421] As of this writing, there is no country that has publicly announced approving human reproductive cloning.

[422] We say "himself" because as of date, there has not been a female president in the U.S. yet. May be one day we will really elect a truly smart female president, like Hillary Rodham Clinton.

[423] Erin Hallissy, Simone Sebastian, and Peter Fimrite, "Shock and awe at outcome of recall election", *The San Francisco Chronicle*, October 9, 2003.

disorder of the connective tissue that is heritable and affects the skeleton, lungs, eyes, heart and blood vessels. The successes and failures of Lincoln's great life seem not to have been affected by the disorder at all. But in modern days, many a presidential candidate with health problems cannot even be victorious in primaries, let alone making it all the way to the presidential election.

Biologists are already talking about cloning the woolly mammoth and other extinct animals, may be the Elephant (Republicans) and the Donkey (Democrats) would want to try to do something similar. After all, political candidates are always trying to associate themselves with great leaders of the past.

If this were possible, more than thirty-five years from now, we may have Albert Gore clone running against George Bush clone one more time, but this time without Jeb Bush clone as the governor of the State of Florida; or Jimmy Carter clone against Ronald Reagan clone, but this time without William Casey clone meddling in the Iran hostage crisis; or William Clinton clone as president without Monica Lewinsky clone to tar his otherwise unblemished presidency... This could provide a form of historical fact check, and a means to unearth the dirt of political maneuverings and deceits at the expense of all the taxpayers and voters.

Having said this, we must not forget that in the prevailing environment, John F. Kennedy II or Abraham Lincoln II may turn out not to be politicians. They might grow up to be dotcom guys. And thirty-five years from now, they would likely turn out to be biotechnologists or nanotechnologists who loathe politicians. Assuming, of course, by then we will still have human presidents and not nanobots (nano robots) as presidents?

We use the top office of the U.S. as an example for the sake of illustration. It is conceivable that the same scenario can play out for the top jobs in other countries, barring differences in local customs and cultures. Imagine, for example, the scenario being played out in a country where the checks and balances of power are not well established, or where dictatorship still reigns.

Other possibilities are also conceivable. Imagine the scenarios where leaders and bosses are clones? Or when the president has a clone who is not too much younger than Mr. President, and Brother clone had a back scene coup. We can keep extending the possibilities, but then we will have become Hollywood, which tends to sensationalize cloning than educating the public about it.

We are not saying that only clones will abuse powers, human presidents can do the same. The U.S. presidency of 2001-2008 is a good example.

Conversely, if we could have a very good President Clone who is better than President Human, why not have a clone? But for Hollywood, a movie about a clone of abusive traits makes a better box office.

▶ The Genie is Out of the Bottle

Should there be legal guidelines to human cloning, or a total ban on human cloning? The fact with human cloning is this: the technology is here and is being perfected. May be the techniques are still riddled with problems, as yet not reliable and controllable enough for the community to plunge ahead. Improvements on the techniques and better understanding of human development are currently underway.

Undeterred by public calls and the imposed moratorium with a sunset clause [424] to reevaluate the situation in five years (2004), several groups publicly announced in 1998 that they would charge ahead. The first group consists of Panayiotis Zavos, Severino Antinori, and Avi Ben Abraham; the second group, led by Claude Vorilhon and Brigitte Boisselier, has actually made claims they had produced the first generation of five clones, claims that are now believed to be hoaxes. [425] Other groups, such as the one led by Richard Seed, are almost inactive. These are groups which have publicly announced that they have been experimenting with human cloning. Conceivably, there are other clandestine efforts, such as those in the Commonwealth of Independent States (CIS) and in an Islamic country. [426] So watch out, the guy sitting next to you could be a clone!

The genie is out of the bottle and the focus should be redirected: it is time for us to define civilization not by the things that the cloning technology makes possible, but by which of these possibilities we choose to undertake. Cloning technology can be used without the end result being a clone, such as in developing stem cells for transplant in the so-called therapeutic cloning.

Some experts may balk at the thought of the cloning technology is being perfected by nonacademic, nontraditional entities, but *in vitro* fertilization will provide an exemplary precedent. The history goes back to when Joseph Califano was directing the Department of Health, Education, and Welfare (HEW). The Department made the decision that the government would not fund fertility research but it would not stop the research either. With a lack of

---

[424] A sunset law is a law requiring periodic reevaluation to justify its continuing existence.

[425] Tom Abate, "Leader emerges for disciples of human cloning", *The San Francisco Chronicle*, October 12, 2000.

[426] Timothy Bancroft-Hinchey, "Human cloning: The race is on", *Pravda*, May 21, 2001.

funding, federally funded research came to a stall, and as a result many advances in reproductive technology have come from efforts in the private sector, often financed by the patients themselves.

In Vitro Fertilization Debate

In the U.S., the public debate over assisted reproduction became very heated in the early 1970s. Vocal opposition to IVF includes: James Watson who claimed that IVF is not therapeutic; medical ethicist Leon Kass who questioned the acceptability of IVF;[427] and theologian Paul Ramsey who argued that it was not the proper goal of medicine to enable women to have children and marriages to be fertile by any means and thus IVF and embryo transfer were misbegotten and unethical.[428] The disparate reticence crystallized in 1974 when the Congress forbade the HEW to fund any research on human fetuses, except that which might improve the survival rate of fetuses. A year later, HEW promulgated the regulations in the work since late 1973 – regulations which required suspension of government funding of IVF research until such time as an advisory had made a thorough investigation of the field and advised the department's secretary Califano whether such research was ethically acceptable. Four years would pass before the Ethical Advisory Board (later renamed Ethics Advisory Board) was appointed. Though legislators had not banned IVF outright, cutting off federal funding pending the board's review served as a *de facto* moratorium.[429]

However, cutting of federal funding did not take air out of IVF research; many IVF researchers received support from private philanthropies such as the Ford Foundation and the Rockefeller Foundation, the Population Council, the Planned Parenthood Federation, and the Carnegie Corporation. They also had grants from the National Institutes of Health, the U.S. Public Health Service, the National Research Council and other entities as well as monies channeled to them indirectly from the federal government via their own institutions. Congress' action forced IVF researchers to rely solely on private sources, or to turn their IVF efforts into lucrative enterprises and the IVF community responded.

---

[427] Leon R. Kass, "Babies by means of *in vitro* fertilization: Unethical experiments on the unborn", *New England Journal of Medicine*, 285, (1971), pp. 1177.

[428] Paul Ramsey, "Shall we 'reproduce'? I. The medical ethics of *in vitro* fertilization", *JAMA*, 220, (1972), pp. 1348.

[429] Sherman Elias, and George J. Annas, *Reproductive Genetics and the Law*, (Year Book Medical Publishers, Inc. Chicago, 1987).

Scientific inquiry freedom is really a potent myth, trotted out only whenever scientists feel their privilege is under attack. But researchers are obsequious to the requirements of funding sources, including federal agencies, thus the federal granting process can exert a tremendous influence over how science should proceed, determine which fields of research should flourish and which should flounder. By sitting on the fence, the government lost its opportunity and authority to guide and regulate the course of IVF research.

As it turned out, the high demand for IVF services provided a ready flow of cash to fund the research: the segment of the American population who was desperate and could afford not only were willing to allow physicians to use them as experimental subjects, but also to foot the bill for IVF. These private sector dollars eliminated the need for future federal involvement and gave researchers far more latitude than they would otherwise have had they relied on federal funding.

Reproductive Cloning Dé JàVu

The advisory committee to debate on human cloning was convened by Bruce Alberts, president of the National Academy of Sciences (NAS) and a highly accomplished cell biologist. The committee included a number of experts who are actively involved in exploring the pros and cons of stem cell therapies. The committee's findings, announced in January of 2002, advocated therapeutic cloning in parallel with alternative strategies. Incidentally, the German committee, convened by Ernst-Ludwig Winnacker, head of German National Science Foundation, reached a similar conclusion.

President George W. Bush, apparently anticipating the NAS panel's conclusion, appointed an advisory committee all but guaranteed to produce a report much more to his liking. The presidential committee chairman was none other than Leon Kass, who has gone on record as being against all forms of cloning, and was against IVF in the 1970s.[430]

Human cloning is currently privately financed either by private organizations or by "patients." Charities are lining up to bankroll stem cell experiments as well. But as has been noted, the recipe for creating cloned embryos could in theory be used for reproductive purposes.[431]

It is only a matter of time before the first human clone becomes a reality. When that happens, it is a *dé jàvu* all over again. When that happens, the

---

[430] Robert A. Weinberg, "Of clones and clowns", *The Atlantic Monthly*, June 2002.
[431] Laurie Goodstein, Denise Grady, John Files, and Cornelia Dean, "Cloning and stem cells: The debate; Split on clones of embryos: research vs. reproduction", *The New York Times*, February 13, 2004.

impact will be much greater than IVF. The basic notions of sociality and equality could be radically changed: instead of meritocracy we would have genetocracy, with individuals and ethnic groups increasingly categorized and stereotyped by genotype – a form of genetic caste system.[432]

Humanity is the product of millions of years of evolution. Julian Huxley extended that notion to imply "humanity is now obligated to continue the creative process by becoming the architect for the future development of life."[433] Darwin's "survival of the fittest" suggests humanity compete for their own life; in the new cosmology, humanity will be the creator of life. The new cosmology consists of Mother Nature, Father Time, and Author Children, and Children Author is trying to change Mother Nature by shortening Father Time, to accelerate the evolutionary process.

▶ Overcoming Legal and Ethical Impasse

When dealing with a complicated subject, such as ethical issues, it is easy to see the forest as a bunch of trees, and to suffocate in the minutiae, while missing the point. But we do not have to know everything before we do anything. If we feel that way, we are doomed to paralysis and irrelevancy, and would never be able to make progress.

In much of ancient history, questions of guilt or innocence, and what to do or not to do, were relatively easy to resolve – the gods knew the right answers. In this situation, the only problem in finding the real truth was to understand the language of the gods.

In other societies gods were not always that communicative. It was necessary to either trick them or cajole them, to get them to reveal the facts of the matter. In many ancient societies (and some not so ancient), the priests and other potentates were supposed to be in direct contact with the gods, so they could serve as intermediaries in these judgments, or could read the signals sent by gods. Since the gods knew (and know), it was (and is) only a matter of getting them to tell. A vestige of that view persists in our present practice of requiring witnesses to swear an oath, now often secularized into an affirmation. Presumably because there is somewhere a higher authority who knows the truth, and who will be offended by lies. And presumably, witnesses believe that and therefore they are more motivated to tell the truth under oath. Tell this to national leaders and politicians who lie from both sides of their mouths.

---

[432] Jeremy Rifkin, *The Biotech Century*, (Jeremy P. Tarcher/Putnam, 1998).

[433] Julian Huxley, *Evolution in Action*, (New American Library, New York, 1953).

Add to this one more dimension – the law – and the whole situation gets even more complicated. In the common-law countries, an important ingredient of the law is the body of customs that have grown up with the society, and the laws are codified expression of accepted common experience. In such countries, championed by the U.K. and the U.S., precedents – earlier decisions by other courts, and just plain traditions – are vital in any legal argument. Firm reliance on precedents in interpreting ambiguous laws is important to minimize the potential for internal contradiction.

In common-law countries like the U.S., the law is cumulative for two reasons: judges issue decisions that add to the body of precedents (case law is the term used to describe this process), and legislative bodies pass many laws, but repeal few. In addition, in the U.S., there was an explosive proliferation of regulatory agencies in the second half of the 20th century. These agencies issue realms of regulation in their areas of responsibility, regulations that have the force of law. In less certain situations, since we cannot talk about probability in laws, we handle these subjects through circumlocutions and portentous and polysyllabic but meaningless or amorphous words.[434]

Another way to resolve many of the uncertainties is to write laws itself with enforcement in mind. We can build guilt into the law by writing it so simply that proof of guilt is as easy as possible. We can do that by, in effect, sacrificing the law's social purpose to its enforceability. All through the law books we can find laws and rules written to facilitate enforcement, providing sharp boundaries for what is in fact the very fuzzy line that divides right from wrong. This is fundamentally bad for a democracy because it hands to the law enforcement agencies the decision making process that ought to belong to the judicial system and this promotes discrimination. An excellent example is the 9:00pm EDT, August 9, 2001, President Bush's announcement of the list of accepted stem cell lines eligible for federal funding.[435] It is very simplistically loud and clear: before that time, your stem cells are eligible for federal supports; after that time, you are on your own. This may be good for a politician trying to make his camp happy, but terrible for science and progress.

The clear, but scientifically and economically rather silly, announcement of Bush, does not forbid private funds. To overcome the restrictions, scientists and researchers have found ways to go around the law: 1) using private funds,

---

[434] H.W. Lewis, *Why Flip a Coin? The art and science of good decisions*, (Barnes & Noble Books, New York, 1997).

[435] Another excellent example is speed limits on highways. Most speed limits on highways are supposed to be surrogate for safety. Obviously, there is no magically safe speed on any highway for all drivers, for all vehicles, and for all traffic, and road and weather conditions. Nor is there any agreed standard for how safe the highway ought to be.

2) using model organisms, and 3) extracting stem cells without creating viable embryos. We shall discuss each of these ways in turn and see how scientists come up with innovative ways to overcome these legal and ethical impasses.

Private Funding

This way of overcoming legal and ethical impasses is rather self-explanatory. We shall just discuss briefly some of the difficulties.

The U.S. federal government will pay only for research with human embryonic stem cells that were created before August 9, 2001. It will not pay for the creation of any new human embryonic stem cell lines, per Dickey-Wicker amendment, which prohibits federal research where human embryos are destroyed, discarded or subjected to substantial risk. Scientists are free to use private funds.

It is incredibly difficult to raise private money to sustain a reasonable research program. Companies and venture capitalists have been reluctant to invest in the field, partly because of the ethics debate, but also because investors perceive it will take a long time for such therapies to reach the market and provide a return. Moreover, injecting cells, particularly if they are customized to each patient, is perceived as a less attractive business proposition than mass-producing a pill that everyone can take.[436]

Many companies pursuing cloning and cell replacement therapies, not all from embryonic stem cells, have gone out of business. PPL Therapeutics, the Scottish company that helped clone Dolly the sheep, was dismantled in 2004. Infigen, a pioneering cloning company in Wisconsin, laid off all of its employees, also in 2004.

The federal government funds 95% of basic research. So if the federal government will not fund embryonic stem cell research, the research community will have to use ingenuity.

Model Organisms

Scientists use a number of model organisms for investigating biological phenomena. These model organisms include vertebrates, invertebrates, plants and microorganisms. The ethical implications of using animal models, particularly mammals, in experiments are as important, but much less controversial than using humans or human parts, and not as hotly debated.

---

[436] Andrew Pollack, Cornelia Dean, and Claudia Dreifus, "Cloning and stem cells: their research, medical and ethical issues cloud cloning for therapy", *The New York Times*, February 13, 2004.

Vertebrates

Vertebrates are particularly useful model organisms because they are relatively closely related to the human evolutionarily and genetically. Hence what happens in a mouse, for example, in terms of physiology, can be similar to that which happens in a human being. There are clear ethical concerns with using vertebrates (and indeed all animals) in experiments since to varying degrees they are able to suffer fear, pain and trauma in scientific studies.

Studying humans (*Homo sapiens*) has obvious advantages. The complete human genome is now available and the research is directly applicable. The disadvantage is that it is fraught with ethical implications in controlled medical trials. Chimpanzee (*Pan troglodytes*), whose genome is also available, and the other great apes are also closely related to us, but using them as models are obvious moral minefields as well.

Pigs (*Sus scrofa*) are described as "horizontal humans" because their organs are of similar sizes to our own. They are used in some medical research, such as organ factory. Dogs (*Canis familiaris*) and cats (*Felis catus*) are much more rarely used. The dog genome is available. Dogs have long been used to study human diseases. Rabies, in fact, was first discovered in dogs, insulin was discovered in dogs, and the first open heart surgery was in dogs.[437]

Mice (*Mus musculus*) are quick breeders, and are very similar to the human in their biochemistry and physiology. The complete genome is available. Rats (*Rattus norvegicus*), whose genome has been sequenced, are a favorite in medical research. For nearly 200 years now, scientists have used animals as models on which to test ideas about human biology. Today, the rat, along with its rodent cousin the mouse, account for more than 80% of all laboratory experiments. Lab rats are used to practice surgery; to study cancer, diabetes, and cardiovascular disease; to investigate psychiatric disorders, neural regeneration, and space motion sickness. In drug development, the rat is employed to demonstrate therapeutic efficacy and assess toxicity of novel drug compounds before human clinical trials. Scientists have hundreds of strains of mice and rats that mimic human illnesses.[438]

Clawed toads (*Xenopus laevis*) are easy to keep, with very large and easy to manipulate eggs. It has been a mainstay of embryological research, but unfortunately, this toad has an enormous genome that is more polluted with junk DNA than ours. The genome sequence is not likely to be forthcoming any time soon.

---

[437] Nicholas Wade, "Researchers decode dog genome", *The New York Times*, December 7, 2005.

[438] Jonathan Amos, "Rat's 'life code' read by science", *BBC News*, March 31, 2004.

Zebra fish (*Brachydanio rerio*) have the advantage of coming in transparent varieties. Used in water pollution tests to determine the probable effect of toxins on aquatic life, and recently in studies of heart formation.

Invertebrates

Invertebrates have the advantages of being relatively smaller and easier to handle than vertebrate model organisms. There are fewer ethical problems experimenting on these organisms.

The fruit fly (*Drosophila melanogaster*) genome is available. They are extremely easy to breed in enormous numbers, and their salivary glands in the maggots of these flies have giant *polytene* chromosomes to allow the study of individual gene expression.

The roundworm (*Caenorhabditis elegans*) is the best-understood multi-cellular organism on Earth. Not only is its genome available, so is the entirety of its embryology. We know when and where every cell in its body develops from the egg.

Plants

Plants are economically important, and easy to study. There are normally no ethical issues to worry about in studies using plants, except in cases where genetically modified crops and genetically modified foods are involved.

Thale cress (*Arabidopsis thaliana*) is a tiny cress having very few chromosomes. Its genome has been completely sequenced. In the plant world, it is the equivalent of the botanist's fruit fly in the animal world.

Maize (*Zea mays*) is an important crop species. It has an extremely convenient mechanism for doing genetics: all a researcher needs to do is just count the colors and shapes of the kernels on a cob. Rice (*Oryza sativa*) and wheat (*Triticum aestivum*) are similarly useful, and the rice genome has been sequenced.

Microorganisms

Microorganisms are even easier to grow than plants. A vast majority of all biological research is carried out in microorganisms, such as using yeasts and bacteria.

Brewers' yeast (*Saccharomyces cerevisiae*) is one of the simplest eukaryotic single-celled organisms. It is economically very important in brewing and baking. Other economically important fungi include *Penicillium notatum* and *Aspergillus niger*, which are, respectively, the sources of penicillin and most commercial citric acid.

Fission yeast (*Schizosaccharomyces pombe*) is similar to brewers' yeast. A difference is the reproduction process – by fission rather than by budding.

Escherichia coli is the best-understood bacterium. It is found in the human gut, normally harmless but an occasional pathogen (such as *E. coli* O157:H7, first isolated in 1982, and which causes food-poisoning). It is commonly used for the manipulation of genes from various other organisms.

Bacillus subtilis is a spore-former. It is a eukaryote that does complex cellular differentiation, and is closely related to the causative agent of anthrax (*Bacillus anthracis*).

Phage lambda ($\varphi \lambda$) is an *E. coli* virus. HIV (Human immunodeficiency virus) is an unusual virus replicating *via* RNA. It was the first human retrovirus to be well characterized. SV40 (Simian virus 40) is a well-known oncovirus (cancer-causing virus), much studied in the realm of DNA replication.

Circumventing Ethical Impasse

Stem cells can in theory grow into any of the body's tissues and organs. But embryonic stem cells are extracted from human embryos after they have grown for about five days in the lab, and obtaining those cells requires that the embryos be destroyed. The moral objection has been that that is destroying human life.

On one side are those who say human life is a continuum that begins with a fertilized egg. A human embryo, however early, is human life. They find it unacceptable to destroy human embryos to extract their stem cells, and that the end cannot justify the means. For example, Carrie Gordon Earll, bioethics analyst for Focus on the Family, a conservative ministry in Colorado Springs, called this type of research "nothing short of cannibalism." "We don't sacrifice one human life in order to possibly help another human life," Ms. Earll said. "This really is discrimination against the most vulnerable human being."[439]

In the middle are those who view human embryos have "a unique moral status" that should be respected. There's a significant weight to the decision to use human embryos. But using human embryo stem cells to find ways to

---

[439] Laurie Goodstein, Denise Grady, John Files, and Cornelia Dean, "Cloning and stem cells: The debate; Split on clones of embryos: research vs. reproduction", *The New York Times*, February 13, 2004.

relieve human suffering "pays respect to their unique moral status." And these people fully accept the ethical tradeoff.

Yet another group says that for them there is no means-end calculus. Early embryos, they say, are simply microscopic balls of cells with no particular moral status. These balls have no body parts, they look nothing like a fetus, and most die anyway when they are implanted in women. For this group of people, embryonic stem cell research poses no ethical issue.

And that impasse has led to a search for other ways of getting these precious cells. So while most stem cell scientists focus on obtaining stem cells from early embryos, Dr. Rudolf Jaenisch of M.I.T. and Dr. George Daley of Harvard Medical School have begun asking if they can get stem cells another way, perhaps by creating aberrant cell clusters that contain stem cells but could never survive more than a week or so. The idea is to produce embryonic cells without the embryos and make nearly everyone happy.

For now, Dr. Jaenisch and Dr. Daley are testing the idea in mice. In particular, Alexander Meissner and Rudolf Jaenisch of the Whitehead Institute in Cambridge, Massachusetts, have created mouse nuclear transfer embryos that are inherently incapable of implanting in the uterus. They did so by switching off a gene in the donor nucleus that is needed for the implantation process. The gene was switched back on later because it is needed to form the intestinal tissues. Dr. Jaenisch says he is convinced the method can work.[440]

The research has caught the attention of some members of the U.S. Congress, who have proposed bills to allow federal funding for such methods. The idea also has attracted scientists, like Dr. Markus Grompe, director of the Oregon Stem Cell Center in Portland, who says he is about to start human embryonic stem cell work for the first time because the new method offers him a way to do so without violating his moral principles.

Dr. Daley's interest in the new methods "is being driven by the realities of federal funding and the political climate in the U.S." The U.S. federal government will pay only for research with human embryonic stem cells that were created before August 9, 2001, and not pay for the creation of any new human embryonic stem cell lines.

Dr. Jaenisch's motivation is pragmatic. At issue is the question of who decides what research should be pursued, and why. And the players include not just scientists but also a group of fervid observers who are looking for compromise solutions. They include, most prominently, Dr. William Hurlbut,

---

[440] Alexander Meissner, and Rudolf Jaenisch, "Generation of nuclear transfer–derived pluripotent ES cells from cloned *Cdx2*-deficient blastocysts", *Nature Online*, October 16, 2005.

a physician by training who teaches ethics courses at Stanford University and a member of the President's Council on Bioethics.

For the last three years, Dr. Hurlbut has been trying to get a consensus on alternative methods of obtaining stem cells, after deliberating on the moral status of the human embryo for a president's council report on cloning. He personally finds it morally unacceptable to destroy a human embryo, but he also understood the immense promise of stem cells.

Then he came up with an idea to obtain embryonic stem cells in the following way: Instead of fertilizing an egg, you start the cloning process but in an altered way so that no embryo is produced. Ordinarily, with cloning, scientists slip an adult cell into an enucleated egg. The egg reprograms the adult cell's genes. These reprogrammed genes then direct the development of an embryo, a fetus, a newborn, and finally, an adult that is genetically the same as the donor adult.

Dr. Hurlbut proposed something slightly different. First remove, silence or alter the pattern of expression of genes from the donor adult cell that are responsible for the full development of an embryo. Then start the cloning process by adding the altered cell to an egg.

"What I'm suggesting is creating something that never rises to the level of a living being," Dr. Hurlbut said. "No embryo is ever formed. It's not a human embryo if it doesn't have the potential to develop into the human form." He called it a "biological artifact."[441]

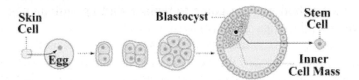

Figure 2A. In using the current nuclear transfer method to create embryonic stem cells, an adult skin cell, for example, is implanted into an enucleated egg. The egg reprograms the adult cell into an embryonic state and the embryo begins to divide. Four or five days later, a blastocyst forms. It has an inner cell mass and an outer shell. Stem cells are extracted from the inner cell mass, destroying the embryo in the process.

---

[441] Gina Kolata, "Embryonic cells, no embryo needed: hunting for ways out of an impasse", *The New York Times*, October 11, 2005.

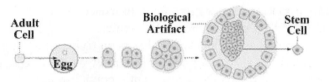

Figure 2B. In altered nuclear transfer method, genes necessary for embryo formation are altered in an adult cell before it is implanted into an enucleated egg. Because of the altered genes, when the cell divides, no embryo is formed. Four or five days later, an equivalent of blastocyst in this case, called a biological artifact, is formed. The outer-shell cells are loosely connected, but an inner cell mass develops fully. Stem cells can be extracted from the fully developed inner cell mass. (Figure adapted from that of Dr. William B. Hurlbut, Stanford University).

Some people wonder whether "biological artifacts" are really a solution. After all, there is no way to assure that the genes researchers think are deleted are really gone, or they may still be partly active, producing a human embryo although the scientists may not realize when extracting the cells.

An alternative idea, a fancy of scientists, is to bypass human eggs entirely. The crux of this idea is to figure out how an egg reprograms a cell's genes. Scientists could then recreate the process and turn adult cells into stem cells without the egg, and embryo, as an intermediary. Kevin Eggan recently reported that human embryonic stem cells could reprogram cells. When Eggan and colleagues slipped an adult cell into an embryonic stem cell, the adult cell became a stem cell with two sets of genes – those of the original stem cell and those of the adult cell. As long as the double set of genes remained, the method was not going to replace starting with an egg. Finding a way to remove the extra genes could be years away.

Robert Lanza and colleagues at Advanced Cell Technology (ACT), a biotechnology company in Worcester, Massachusetts, have developed yet another alternative way of generating embryonic stem cells that leaves the embryo viable. They let a fertilized mouse egg divide three times until it contained eight cells. This is a stage just before the embryo becomes a blastocyst. They extracted one of these cells, then coaxed it into growing in

Figure 3. In yet another way to create embryonic stem cells, scientists first learn how an egg reprograms an implanted adult cell. They then reconstitute the conditions *in vitro* by adding proteins to the adult cell, eliminating the egg completely. The cell reprograms and then divides into different kinds of cells, including stem cells.

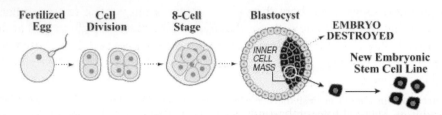

Figure 4A. In the current method to create embryonic stem cells, an egg is fertilized. Four or five days later, the fertilized egg develops into a mass of cells called a blastocyst. Stem cells are extracted from the inner cell mass of the blastocyst, destroying the embryo in the process.

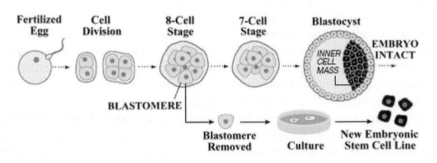

Figure 4B. In a new way to create embryonic stem cells, an egg is fertilized. After the third division at the 8-cell stage, a single cell (a blastomere) is removed. The blastomere is cultured to form a new cell line, while the 7-cell stage develops intact. Note that with the exception of culturing the extracted blastomere to form a new cell line, this process is very similar to well-established preimplantation genetic diagnosis. (Figure adapted from *Nature*).

glassware and forming cells that have all the same essential properties as embryonic stem cells derived from the inner cell mass. The remaining seven-cell embryo was implanted in a mouse uterus and grew successfully to term.[442]

This procedure is known to work with humans too, because it is the basis of a well-established test known as preimplantation genetic diagnosis. In such a test, one cell is removed from each of a set of embryos to be tested for any of 150 genetic defects. Hopefully, the parents would thus have the choice of implanting an embryo that is as disease free as possible.

The technique has other positive sides. If it proves to work in humans, it could do more than just provide researchers with a new source of cells. It

---

[442] Young Chung, Irina Klimanskaya, Sandy Becker, Joel Marh, Shi-Jiang Lu, Julie Johnson, Lorraine Meisner, and Robert Lanza, "Embryonic and extraembryonic stem cell lines derived from single mouse blastomeres", *Nature Online*, October 16, 2005.

might allow every child born through preimplantation genetic testing to bank its own line of embryonic cells for the future. The blastomere removed at the eight-cell stage could be allowed to divide, with one cell being used for genetic testing and the remaining for growing a culture of perfectly matching embryonic stem cells.

With the parents' consent these cells could also be used for research, providing many new embryonic stem cell lines for laboratories. The procedure might even be offered for all embryos generated in fertility clinics when it is perfected.

There may be concerns of birth defects. So far, after some 10 years of experience, there is no indication that preimplantation diagnosis causes health problems in humans, said Andrew R. La Barbera, scientific director of the American Society for Reproductive Medicine. Children born after the preimplantation diagnosis procedure have the same incidence of birth defects as those who did not undergo the procedure. Some people see the evidence as the beginning of a day when every fertility clinic embryo has a cell removed and banked for future tissue use or organ replacement.[443]

If ACT technique does work in people, which could take some time to find out, the technique might divide the anti-abortion movement into those who accept or reject *in vitro* fertilization, because the objection to deriving human embryonic stem cells would come to rest on creating the embryos in the first place, not on their destruction.

This gets around all of the ethical arguments, except for that small minority of the pro-life community that does not even support *in vitro* fertilization. Dr. Lanza's technique is likely to be welcomed by many in the middle of the stem cell research debate, although it has not won over the United States Conference of Catholic Bishops. Richard M. Doerflinger, its deputy director for pro-life activities, dismissed the technique, saying that preimplantation genetic diagnosis itself is unethical. The technique "is done chiefly to select out genetically imperfect embryos for discarding, and poses unknown risks of future harm even to the child allowed to be born."

In the meantime, Douglas Melton, a Harvard stem cell researcher, says he takes umbrage at ethicists and lawmakers telling scientists what to do. Dr. Leonard Zon, a stem cell researcher at Harvard Medical School, says some of the ethicists' ideas sound good, but he adds: "Are they practical? And if they are practical, are they necessary?"

Scientists who succeed have an instinct for the best experiments to do, the ones most likely to work out, Dr. Zon said. "Some successful people are called

---

[443] Nicholas Wade, "Stem cell test tried on mice saves embryo", *The New York Times*, October 17, 2005.

lucky, but they seem to be lucky over and over again. I think it's an intrinsic quality," he says. "What it is is somehow choosing a path that is likely to be very successful and having faith enough to choose bold paths and things that will work."

Procuring human eggs is still a complication. Dr. Hurlbut and others do not encourage asking young women to take drugs to produce copious amounts of eggs for use in research. Instead, they may be able to use eggs that are normally discarded by fertility clinics, or they may eventually be able to remove eggs from ovaries of women who were having their ovaries surgically removed, or from the bodies of women who had just died. Though for now researchers are not yet able to prod immature eggs in ovaries to mature, scientists are confident that process can and will be accomplished soon.

Egg procurement complication brings us to yet another alternative to overcome ethical impasse – hybrid cloning. Most people are very uneasy about dismembering human embryos to obtain stem cells for therapeutic purposes. These people might feel more comfortable with a hybrid solution. For example, if hybrid cow-human stem cells were viable for producing tissues and organs, then hybrid clones might kill two birds with one stone: first, cow eggs are used, and second, no human embryos, but hybrid embryos, are destroyed.

However, Maisam Mitalipova, a pioneer of hybrid cloning at the University of Wisconsin, claims that hybrid embryos do not give good quality stem cells. Researchers are working hard to improve the quality. A disturbing issue is how many human genes does a cow have to have before we give it human rights?

▶ Message in a Bottle from the Future

Human cloning technology is within grasp and a clone may be born soon. Once the territory of a handful of "marginalized" scientists, the cloning of human embryos is becoming increasingly common around the world. Many unaccounted for cloning clinics (clonics) have surfaced from indirect sources. The Russian Federation, USA, China, France, Italy and England are in the race to clone the first human being. An unverified source reported that Severino Antinori declared that women in the Commonwealth of Independent States countries and another in an Islamic country were carrying human fetuses he had cloned – a claim made in 2001, but so far no news has come out of it.[444]

---

[444] Timothy Bancroft-Hinchey, "Human cloning: The race is on", *Pravda*, May 21, 2001.

Many researchers believe multiple U.S. clonics are quietly preparing to create human embryos. Thus the best measure is to prepare for such an eventuality.

Polls in the U.S. show the American public opposed human cloning, but the opponent unsuccessfully pushed for a ban on therapeutic cloning in the U.S. Senate in May 2002. Banned or not in the U.S., U.S. law has no jurisdiction overseas and there is no practical way to ban worldwide. In fact, cloning experiments (for therapeutic purposes) have already proliferated abroad, notably in Sweden, U.K., Singapore, Australia, China and others, and many scientists in the U.S. are simply waiting for some political calm to embark their own human cloning work.

Xiangzhong (Jerry) Yang said he had been briefed on human cloning in "half a dozen labs" in China. At least one of the labs successfully cloned a human embryo before Advanced Cell Technology. Of note is Guangxiu Lu of the Xiangya Medical College, located some 800 miles from Beijing in Changsa of Hunan Province. Lu has been cloning human embryos regularly for years. Yang is working hard to get these Chinese colleagues to publish in Western peer-reviewed journals and to lure some of them to the U.S. to present their work so that the world will know of their achievements.

In England, embryo-related work has won government approval and according to the Human Fertilization and Embryology Authority, the British government agency that monitors embryo research, some institutes have obtained licenses to conduct experiments.[445]

In the U.S. ethical qualms have complicated research. The controversy has kept many interested researchers away, at least for now. In May 2002, the University of California at San Francisco admitted to conducting a cloning experiment that ended in failure. It is not unreasonable to assume that other U.S. labs and companies are pursuing the same experiment in secrecy, particularly that Hwang Woo-suk's claims of having cloned human embryos have now been nullified.

To Clone or Not to Clone?

Despite George W. Bush's decision to restrict funding for therapeutic cloning and impose a total ban on reproductive cloning, it is best that therapeutic cloning, such as stem cell research, should be promoted. Reproductive cloning should be carefully curtailed and monitored until existing physical risks have been overcome and until scientists become more proficient handling cloning. Past instances will indicate that a total ban or prohibition will not serve the

---

[445] "Human cloning researchers quietly building worldwide labs", *Boston Globe*, June 21, 2002.

purpose, but will only stifle related research and encourage clandestine activities. Regulation with caution and careful oversight will be the most ideal policy.

The social and ethics communities should also do their parts: they should ready the public for the day; instead of inundating the general public with unfounded fact and nightmarish scenarios, news media should educate and make the pubic better informed.

However, the worry is whether the law could ever keep pace with biotechnology. Clones differ from inanimate products (electronic gizmos) in many aspects: our legal system does not permit us to terminate a life, just like we can destroy a gizmo when it is found defective. If a clone should turn out to have serious health problems, the clone would become a social liability. The clone is alive so it can migrate and reproduce.[446] Clones are thus inherently more unpredictable in the ways they interact with other living things in the environment, making it almost impossible to assess all of the potential impacts.

It may seem that once a clear distinction between therapeutic cloning and reproductive cloning can be drawn, the issue of banning or not banning will become easier. May be not. The initial stages of therapeutic and reproductive cloning are exactly the same and there is no telling that a researcher may do one form of cloning under the disguise of the other! Nuclear research and chemical research can lead to good ends, but they can also lead to nuclear bombs and chemical weapons. Implementation of different legal guidelines for therapeutic cloning and reproductive cloning may be as difficult as trying to make sure that nuclear power plants are not secretly making nuclear bombs. Unless we can have equable laws for both forms of cloning, the difficulty in implementation will remain.

▶ Why? Why Not!

George Bernard Shaw once said, "Some men see things as they are and say 'Why?' I dream of things that never were and say 'Why not?'" Probably more so in the West than in the East, our modern culture and society have an implicit faith in technology. Advances in technology have repeatedly improved life expectancy, prosperity and provided more conveniences. Given this track record, it is understandable that our culture has tacitly let technology

---

[446] We use "it" to refer to the clone to show that the clone could be a male or a female, not that the clone is not a human.

evolve at its own pace, trusting that new technologies would yield more blessings.[447]

This is not to say that technology has been left entirely unfettered, otherwise we would be under a barrage of chemical attacks by now. But two emerging trends are forcing us to question our implicit faith in technology and our position. The first trend is the suspicion that our watchdog institutions are unable to keep abreast with change. Such cynicism is not new for the accelerating pace of innovations would overstretch even a most perfect regulatory system, and we know our system is no way near perfect. The second and newer trend is that technologies that seemed like science fiction a few years ago are now possible or are within the reach of small teams of experts. When we talk about this trend, we have in mind genetic engineering and cloning of barnyard animals and humans.

Clone the Law

In 1 AD (After Dolly), California, Michigan, Louisiana and Rhode Island banned human cloning; the U.S. federal government, on the other hand, passed laws against the use of federal funds for human cloning experiments, but placed no restrictions on what private entities could do. Twelve nations worldwide also banned human cloning.[448]

If the law cannot ride herd with the pace of change of technology, it will lumber along like an unwieldy dinosaur wending its way to extinction. The law is by definition conservative: it attempts to bring order out of chaos only as fast as consensus can be reached among social groups who are willing to conform to norms they each believe are fair and equitable. When the law falls behind, conflicts arise to present new questions that are not easily answered by reference to established precedents – sort of cloning a precedent case to apply in the present case. And human cloning has no precedents because no verifiable human clone has existed yet.

The law becomes particularly hairy when a fair and equitable ground does not exist. Such is the case when a (religious) group alleges that no one owns human life, and indeed, life is nature's creation too inherently divine for us humans to interfere; in a second (human rights) group, the argument that the commercialization of human bodies is in conflict with established notions about the right of individuals.

---

[447] Tom Abate, "Biotech is pushing the possibilities past the breaking point", *The San Francisco Chronicle*, February 5, 2001, pp. B1.

[448] Mariam Falco, and Matt Smith, "House members opens hearing on human cloning", *CNN News*, March 28, 2001.

But these critics are laying the groundwork for a much needed culture change to ride herd with modern technologies. May be we must learn to control the inexorable growth of technology, biotechnology included. But this is not likely to be the case. So biotechnology will be central in this change for science will make it possible to do things with life that our moral sensibilities will not allow.

Let us keep these issues in mind as we march into "The Brave New World" of Aldous Huxley.[449]

••••• ••••• •••••

When considering the morality and ethical implications of using animals in experiments, it is worth considering the fact that most people object to experiments on cute fluffy kittens more than they do to those on verminous rats, and although they are horrified by cultures where puppies with big doleful eyes are considered food, they may still be happy to keep the aesthetically challenged chicken in batteries for their eggs, and have them killed when they stop laying.

It is also interesting to observe that those in position of power who are most vocal against this and that, they are also the very individuals who are beneficiaries of scientific and technological advances that common people can only yearn for.

---

[449] Aldous Huxley, *The Brave New World*, (Penguin Modern Classic, 1932).

14 At the Crossroad

"Living bodies are even in the smallest of their parts machines *ad infinitum*."

- *Monadologie* (1714), Gottfried Wilhelm von Liebniz (1646-1716)

▶ Relatively and Genetically Speaking

When early Taoist alchemists seeking an immortality pill played around with heating mixtures of saltpeter and sulphur, they left the world sooner than they had anticipated. The inflammable concoction, described in a Taoist book in the eighth century BC as an elixir too dangerous to experiment with, turned out to be gunpowder![450]

Innovation can serve up very unpredictable results, but the spirit of questioning why things are the way they are and how they can be improved lies at the very heart of progress. From tool-making experiments of the earliest hunter-gatherers of the Paleolithic era to the controversial cloning of Dolly the sheep, Cc: the cat, Snuppy the dog; stem cell research and regenerative medicine... innovations have been pushing the boundaries of science, technology and medicine. What was science fiction only yesterday is quickly becoming science fact.

---

[450] Trish Saywell, "Yesterday's dream, tomorrow's reality", *Far Eastern Economic Review*, October 17, 2002.

In a 1996 work, HAL[451] wrote, in comparing biology with the physical sciences:[452,453]

> In these remaining centennial years, if we look back on the twentieth century, we can conclude that the first half was shaped by the physical sciences but its second by biology. In the first half were revolutions in transportation, communications, mass production technology and the beginning of the computer age. The first half also, pleasantly or unpleasantly enough, brought in nuclear weapons and the irreversible change in the nature of warfare and in the environment, and pinnacled with the moon shot. All of these changes and many more rested on physics and chemistry. Biology was also stirring over those decades. The development of vaccines and antibiotics, discovery of the structure of the DNA, early harbingers of the green revolution are all proud achievements. Yet the public's preoccupation with the physical sciences and technologies, and the immense upheavals in the human condition which these brought, meant that biology and medicine could only move to the center stage somewhat later. Moreover, the intricacies of living structures are such that their inner secrets could only be revealed after the physical sciences had produced the tools – electron microscopes, radioisotope techniques, chemical analyzers, laser technology, nuclear magnetic resonance, ultrasound technique, PCR, X-ray crystallography, and rather importantly, the computer – required for probing studies. Accordingly it is only now that the fruits of biology have jostled their way to the front pages…

In the opening statement of a lecture by one of the greatest physicists of the twentieth century and a Nobel laureate, Chen Ning Yang, said:[454]

> It has been said that the 20th century was the century of physics. There are ample reasons to support this statement. It was in that century that man discovered, for the first time since our ancestors discovered fire, the second and the vastly stronger source of energy –

---

[451] HAL are the initials of the author, and coincidentally, also the name of the supercomputer HAL9000 in the flick Odyssey 2001. Write hal_lim@yahoo.com for comments.

[452] Hwa A. Lim, "Bioinformatics and cheminformatics in the drug discovery cycle", in: *Lecture Notes in Computer Science, German Conference on Bioinformatics*, Leipzig, Germany, September/October 1996, Ralf Hofestädt, Thomas Lengauer, Markus Löffler, and Dietmer Schomburg (eds.), (Springer, Berlin, 1997), pp. 30-43.

[453] G.J.V. Nossal, and Ross L. Coppel, *Reshaping Life: Key issues in genetic engineering*, (Cambridge University Press, Melbourne, 1989).

[454] Chen Ning Yang, "Thematic melodies of twentieth century theoretical physics: quantization, symmetry, and phase factor", lecture at UNESCO Conference, Paris, July 27, 2002.

nuclear power. It was in that century that man learned to manipulate electrons to create the transistor and the modern computer, transforming thereby human productivity and human lives. It was in that century that man learned how to probe into structures of atomic dimensions, discovering thereby the double-helix, a key to the secrets of life. It was in that century that man ceased to be earth-bound, taking first steps on the moon. In short, it was a century in which man made unprecedented progress on many fronts of human activities. And these progresses were largely ushered in by breathtaking advances in the science of physics...

The two statements, by Yang and HAL, seem to agree on a broad view and seem to corroborate each other. In fact, another form of "fire" was harnessed in biology. Stanley Cohen of Stanford University and Herbert Boyer of the University of California performed in 1973 a feat in the living world that some biotechnology analysts believe rivals the importance of harnessing fire.[455] The two researchers took two unrelated organisms that could not mate in nature, isolated a piece of the DNA from each and then recombined the two pieces of genetic material. The so-called recombinant DNA is a kind of biological sewing machine that could be used to stitch together the genetic fabric of unrelated organisms. Of course, Herbert Boyer, funded by Bob Swanson, founded Genentech, South San Francisco, on April 7, 1976, and in so doing, launched the biotechnology industry.

Retrospectively, in the closing two decades of the twentieth century, genetic data had become the gold standard of biological investigation. Popular magazines frequently discussed the prospects and potential problems of genetics. Whether the proclamations then of a genetic era represented a journalistic hyperbole or truly marked a new epoch in the relationship of genetic to human existence is now a historical fact.

In these intervening years, though the public has become more cognizant of genetics and can understand more of the vocabulary of molecular heredity, the public still fails to discern what the new genetics can and cannot do. The public also fails, as fully informed consumers, to participate in decisions about how the genetic information will be applied.

While the public is working hard to catch up, science and technology are accelerating at full throttle ahead: the human genome has been sequenced, barnyard animals have been cloned, some pets have been cloned, and the human is next...

---

[455] Peter King, "Genetic engineering: The biggest thing since fire", *e.nz Magazine*, July/August 2000, pp. 24-33.

The similarities and differences between physics and genetics are much more subtle. Both fields attempt to reduce and explain complex observable phenomena with basic components, atoms or quarks in the former and genes in the latter. Both disciplines have benefited greatly from innovations produced in the twentieth century and have been profoundly influenced over the past 150 years or so by great imaginative intellects. Both fields have produced methods that have altered human experiences (lasers, magnetic resonance imagers; computers, prenatal diagnostics, recombinant DNA technique, cloning technique…) and posed serious human dilemmas (nuclear bomb, eugenics, bioweapons, genetic weapons…).

One significant difference is apparent. Though the laws of physics constraint human existence, physicists rarely claim comprehensive explanations for the variety of human life. Despite the predictive power of the theory of relativity and the great intellect of Albert Einstein, an estimate says 99% of those people who have lived and are living do not know what relativity is and who Einstein is or who Einstein was. However, many individuals within and without the scientific community believe that the study of human genetics will reveal answers to why and how human beings are what they are; why they get sick and die; and even why they behave the way they do.[456]

The belief in the explanatory power of genetic information arises in part from the scientific evidence produced in genetic investigations, in part from the fact that we all come from a family and resemble our relatives in many ways. From before conception till after death, we are constantly influenced by individuals around us, many of whom may be related to us genetically, as well as experientially, culturally, socially, economically and politically. This reductionist biological science of human heredity – including its great explanatory power and inescapable determinism – differs in impact from that arising from we live in a world governed by the laws of physics.[457]

▶ Divinity Versus Greed

Past civilizations addressed the questions of great public concern – whether waging a war, planting crops, having a child, or deciding to move or stay – by having those with occult powers to pray and meditate for divine guidance. Priests, oracles and shamans, and men and women believed capable of receiving divinely inspired answers or revelations, were called upon.

---

[456] Not a very good or fair comparison. But it serves the purpose to highlight the fact that biology affects us intimately.

[457] Paul R. Billings, "The scientific basic of the genetic revolution", in: *The Genome, Ethics and the Law*, (AAAS, Washington, DC, 1992), pp. 23-45.

The Greeks were no exception. Dating back to 1200 BC, the Oracle of Delphi was the most important shrine in all of Greece. Built around a sacred spring, Delphi was considered to be the *omphalos* – the navel, that is, the center of the world. People came from all over Greece and beyond to have their questions about the future answered by the Pythia, the priestess of Apollo. The Pythia, a role filled by different women from about 1400 BC to 381 AD, was the medium through which the god Apollo spoke. And her answers, usually cryptic, could determine the course of everything from when a farmer should plant his seedlings to when an empire should declare war. For fourteen centuries it helped determine the course of empires: Sophocles, Alexander the Great, and Croesus of Lydia all consulted the Oracle at one time or another. The power of the Oracle of Delphi fluctuated and eventually lost favor as Christianity became the dominant religion of the land, and in the 4th century AD Christian Rome silenced the prophesying.[458]

In today's world, similar practices are still exercised, except that some of the practices have evolved in sophistication. Interestingly, this is how modern government of the U.S. works, sometimes by consulting astrologers. It is known that, for example, Ronald and Nancy Reagan relied on astrologers including Jeane L. Dixon and Joan Quigley to help plan the president's schedule.[459]

But there is another dimension to today's world. If we extend the view of economist Robert Nelson, when today's policy-makers encounter public policy dilemmas – whether on employment policies, environmental protection, healthcare, stem cell research, cloning – they turn to the modern day oracles, the "statisticians," including surveyors and pollsters. These specialists construct their own prophecies on vital questions of the day. They no longer rely on religious revelation or creed, but on reams of arcane statistical data and the interpretation of behavioral principles. They then use the inference as litmus of the popularity of intended actions. Nelson said, "If the priests of old usually asked whether an action was consistent with God's design for the world, in the message of contemporary economics the laws of economic efficiency and of economic growth have replaced the divine plan."[460]

The Invisible Hand

In another arena, humans have bartered and traded with one another since the last Ice Age about 18,000 years ago. Though trading was an important adjunct to early societies, the market was never an autonomous entity. The

---

[458] John Roach, "Delphi Oracle's lip may have been loosen by gas vapors", *National Geographic News*, August 14, 2001.
[459] "Psychic Jeane Dixon dies", *CNN News*, January 26, 1997.
[460] Robert H. Nelson, *Reaching for Heaven on Earth*, (Rowman & Littlefield, Savage, Maryland, 1991).

means by which these societies solved their basic economic problems, such as allocating community resources for different uses and distributing goods within a community, obeyed the religious dogmas and taboos of society.

Unlike goods, land was not a commodity nor was it manufactured, but in history it was successfully invaded by the market and commercialization of land has had an enormous impact.

John Locke (1632-1704), the most influential English thinker of the seventeenth century, looked on nature as a vast, unproductive wasteland. His view of value meant that the natural world was only of worth when human labor and technology transformed it into useful commodities. The more quickly that land could be transformed into a store of material goods, the more secure a society would become and the more progress civilization would have achieved. As Locke states it, "Land that is left wholly to nature... is called as indeed it is, waste... on the other hand, he who appropriates land to himself by his labor, does not lessen but increases the common stock of mankind."[461]

Locke's view was revolutionary. For centuries, British land had not been marketed, but was handed down through families or kept by the church. The market doctrine provided the rationale for the unchecked exploitation of nature and the view that land is a commodity like any other. Locke had transformed nature from a good into goods, radically changing our relationship with nature. Suddenly, the natural world, including humans, was not a community of subjects, but rather a collection of exploitable objects.

The market ideology rose in importance during the scientific and technological upheavals of the seventeenth and eighteenth centuries. The Enlightenment thinkers, imbued with optimism arising from the new age of inventions and consequent productivity, were committed to discovering the rules of human behavior that were scientific and not based on religious dogmas or taboos of society. Analogous to the mathematical laws discovered for the physical universe, these thinkers created the market philosophy in a self-conscious attempt to codify principles for human conduct as efficient and predictable.

A stereotypical Enlightenment thinker is philosopher Francis Hutcheson (1694-1746), an early proponent of a physical law of society. Hutcheson was aware of the enormous influence of the traditional Judeo-Christian concept of benevolence as a principal mover of people, but he posed another social force of even greater influence – self-interest. According to Hutcheson, self-interest was to social life what gravity was to the physical universe. He is

---

[461] Andrew Kimbrell, and Kirk B. Smith, (eds.), *The Green Lifestyle Handbook*, (Henry Holt, New York, 1990).

reported to have said "Wisdom denotes the pursuing of the best ends by the best means."[462]

Nineteen years after the death of John Locke, Adam Smith (1723-1790) was born. A student of Francis Hutcheson, Smith, profoundly influenced by Locke, would take the whole concept of self-interest one step further and transform self interest into a revolutionary new social doctrine. For Smith, self-interest was a principle of human behavior for a new economic order, for self-interest and market principles were an invisible hand that led always to the good of all. Smith's economic principles were circumstantial in that when he published his famous treaties on economics, *The Wealth of Nations*, in 1776, England was in the midst of the Industrial Revolution (~1700-1900): the invention of the steam engine, advances in the use of chemicals and in metallurgy, and a huge surge in productivity.[463] Immersed in the economic growth, Smith felt that there would be little limit to the wealth humankind could create if left free to do so. This laissez-faire philosophy asserts that as each pursues his or her self-interest, the law of supply and demand will be applied to all commodities and will govern price and production; and ideally, if the government keeps its hands out of economic affairs, the natural selfish order of each person will lead to economic well-being for everyone, since competition and self-interest in the market will lead to a cornucopia of economic wealth.

Fictitious Commodities

Few of us know John Locke, Francis Hutcheson or Adam Smith, and still fewer of us have read their works, yet self-interest has become our second nature: greater productivity, individual self-interest, upward mobility, glorification of competition, greater consumption of goods and services, suspicion of government intervention... The rich and famous are idols of our society. These are the tenets of the U.S. capitalism, the world leader in consumerism economy.

The laissez-faire market system is not without problems. In fact, the social history of the past two centuries has in many respects been the result of the contradiction and tensions inherent in the market system's creation of fictitious commodities. Over time, treating non-commodities as commodities becomes a double-edged sword: on the one hand, it leads directly to massive increase in wealth, technological development, and consumption, as modern market

---

[462] Robert H. Nelson, *Reaching for Heaven on Earth*, (Rowman & Littlefield, Savage, Maryland, 1991).
[463] James Watt's improved steam engine was patented in 1769.

economy spreads over the face of the earth; on the other hand, it also leads to the downfall of pure laissez-faire market system. The selling of human labor, nature, and other fictitious commodities (such as money, stocks) precipitated innumerable social abuses, dislocations, unfair practices including child labor, inhuman working conditions, impairment of health, destruction of families and communities, uncontrolled exploitation of nature, extinction of species, air and water pollution, wildly fluctuating economic conditions, deflation and inflation, spiraling governmental and individual borrowing.

The contradistinction of the invisible hand is best exemplified by Golden Rules of different sectors. The Golden Rule of the new economic seers is provided by the ideology of the market forces: would a proposed activity or policy inhibit or interfere with market forces? Would it impede an individual's ability to contract for goods and services? If the answer is affirmative, the policy is not desirable. At a federal level, the Golden Rule is: does the proposed action or policy serve to advance the overall economic efficiency and the long-term national and international productivity? If the answer is negative, the policy is undesirable.

The Golden Rule of political candidacy seekers is provided by the ideology of popularity: would the proposed activity or policy lead to election or reelection of the individual? If the answer is yes, the policy is good. For interest groups, the Golden Rule is a very simple rule of thumb: Do unto others, in the self-interest of group members. The selfish gene character is now recast as the invisible hand character.

Now it is easy to see why there are so many different opinions about the human body shop, stem cell research, cloning, regenerative medicine and the biology of life? Depending on which Golden Rule one subscribes to, the human body shop, stem cell research, cloning, regenerative medicine and the biology of life can be good or can be bad.

Tapping into Human Resources

At the outset of the Industrial Revolution, the sale of labor was the most important and visible noncommodity treated as a commodity. The commodification of labor was made possible by the growth of the factory system. Before mechanization, skilled workers bartered or sold many of their products, but not their labor. The late eighteenth century textile industry was the first to witness the transformation of labor, through the displacement of human weavers by machines. The new textile factories replaced individual and family workshops with factories and wage work; mass-produced products began to replace those fashioned by individuals.

The early advances in industrializing the manufacturing of textiles were a threat to many local communities that had survived on the pre-mechanical production of cotton. In 1870, the story goes, Ned Ludd, a boy from Leicester, England, expressed his anger toward mechanization by demolishing stocking frames. Thenceforth those who took up machine sabotage were known as "Luddites."

At the time Smith was propounding his market theory, a revolution in the demographics of England had created the massive pool of laborers required to make the commodification of labor a reality. For peasants desperately seeking to sell their labor in an industrial setting, the transfer requires a drastic adjustment: no longer did they work at their own pace, but at the pace of the machine; no longer were slack seasons determined by the weather, but by the market. It was a grim age during which long hours of work, the general dirt and clangor of the factories, and the lack of even the most elementary safety precautions all combined to give early industrial capitalism an ill-reputation from which it would never recover.

Labor became the technical market term for human beings, in so far as they sold their work as employees. It should be more aptly called human resources, just like natural resources, which can be extracted and exploited. They in turn no longer worked in the interest of the community, or for family or religious duty, but rather for bare subsistence wages. Laborers' lives and the lives of their family had become an accessory of the economic system. The structure of the family, its mobility, and the relationships therein were determined by how and where each member could sell his or her labor!

Mendel and Darwin

Gregor Mendel (1822-1884) and Charles Darwin (1809-1882) were about a century behind the invisible hand pioneers such as Francis Hutcheson and Adam Smith. Though they were contemporaries, Darwin and Mendel led very different lives. Mendel was a monk whose laboratory was the garden of the monastery; Darwin came from the British leisure class and sailed over the world observing and collecting. They were both gifted with visionary minds and could see basic biological principles in data, which to others may have been unclear or incorrect. In the period 1850 to 1870, Mendel and Darwin changed biology and the view on human life forever. Mendel conceptualized the gene and demonstrated the principles of its transmission while Darwin described variation and selection.

There was another important difference between the careers of these two great founders of modern genetics. Darwin's book, *The Origin of Species by*

Natural Selection, made him a celebrity. In fact, the book has never gone out of print since its debut. Mendel's work was forgotten for decades.

When genetics scholars rediscovered Mendel's work some three and a half decades later, they linked it to Darwin's work and showed that it was applicable to humans as well to plants: the hemophilia which afflicted Queen Victoria's family was transmitted like a gender-linked Mendelian trait; a shortening of the fingers, not associated with a disease, was an easily detected dominantly inherited variation... Familial disorders, which affected the chemical nature of the blood or urine, usually inflicted siblings but not their parents. They were more common in children produced by incestuous or consanguineous mating. These observations suggested these disorders were Mendelian recessive traits.[464]

But human genetic variation cannot be understood outside of the specific environment in which mutations arise. The gene for sickle cell anemia, an illness that is common in the U.S., confers resistance to malaria, a benefit for those living in areas where the infection is epidemic. When the gene for human muscular dystrophy is put into some transgenic animals, it may not cause the disorder at all. The genes involved in nearsightedness may have predisposed our ancestors to be the prey of predators, but now only increase our risk of wearing designer spectacles, wearing contact lenses, or being a patient of lasik operation to correct vision. In modern days of foraging at local restaurants and supermarkets, and high-speed and smart weapons, the genes responsible for tallness will only make an individual a more conspicuous and bigger target!

Because genes mutate spontaneously, genetic screening programs will never eliminate genetic variation or genetically linked disorders. In fact, the science of human genetics will never explain the disease burden caused by bias or ignorance, nor why some individuals with a gene suffer and die prematurely while others adapt and excel. Abraham Lincoln had genotype linked to Marfan's syndrome, a dominant genetic disorder affecting the connective tissues that affects both men and women of any race or ethnic group. The disease neither demonstrates that Lincoln had this illness nor speaks for the successes and failures of his life.

Then when does an inherited trait become a disability? In a culture in which being tall confers certain advantages, are parents justified in requesting growth hormone for their otherwise normal children, and are physicians justified in refusing? Are clone specialists justified in refusing to clone a tall person?

---

[464] Paul R. Billings, "The scientific basis of the genetic revolution", in: *The Genome, Ethics and the Law*, (AAAS, Washington, DC, 1992), pp. 23-45.

Despite its overall complexity, developmental biology is built on a solid foundation of biological principles. Most important, it is clear that development is choreographed by the messages carried in an organism's genes. Nongenetic factors, including environmental factors, can also play a key role in development. But the genes establish development's basic pattern and timing.

Judged from the level of genes, because virtually every cell in an organism contains the same collection of genes, an organism is essentially a clone of a fertilized egg! Human cloning is transferring whole genetics to the offspring with the intention to have an offspring of certain characteristics; genetics manipulation is editing and modifying the genetics of an offspring who has been born, or of modification of genetic attributes to create a founder herd.

► The Invisible Body

After industrialization, the invisible hand has led us from natural resources and financial resources to include human resources.

Demographics and resources are a ticking time bomb. Two thousand years ago, the world population reached a mere 250 million. In the Middle Ages, the population doubled to 500 million, and after World War II, the numbered rose to 2 billion. Within the life span of this current generation, the population is expected to grow to more than 11 billion by 2050.[465]

With 95% of the growth taking place in the least developed countries, the gap between the rich and poor will continue to widen. As a consequence, a new wave of migration will wash into wealthy countries and urban areas. By the year 2005, more than half of the populations of Asia and Latin America lived in mega-cities.

The tremendous need for new natural resources to support the explosion will force industries to exploit new revolutionary possibilities. Genetically engineered food will appear on each family's table. But this is an issue beyond the scope of our discussion here. The issue of pertinence is now the invisible hand has led us to further encompass body resources. Today we are at the dramatic crossroad in our treatment and thinking about the human body and human reproduction.

It took only a short time after the dawn of *in vitro* fertilization (IVF) in the late 1970s for business to recognize that the field has commercial potential. It took only a short time longer for scientists to see that business was a new source of both research funds and personal incomes. This marriage of science and commerce has wrought changes in the traditional research culture of the

---

[465] Guenter Schoenborn, *Entering Emerging Markets*, (Motorola University Press, 1999).

biological sciences, a process spawning considerable worry to many in the scientific community.

Of major concern has been increasing secrecy, where in the past scientists routinely shared ideas, many now tend to be close-mouthed and tight-fisted. Objections to this new ethos embody more than just nostalgia. Science traditionally has thrived on free exchange, and there is fear that, if scientists stop talking to each other, science will progress more slowly and sometimes not at all.

We can choose to continue in our blind faith in technology, the invisible hand of the market force, and the economics of self-interest; we can also choose to adhere, in near religious manner, to the centuries-old dogma of mechanism and the market; we can continue to view our bodies as machines and commodities, we can continue to remake our bodies with surgery and genetic engineering, we can continue to manipulate the reproductive process by eliminating the birth of children with undesirable traits; we can continue to alter with drugs and genetic therapies those with abnormal traits; we can continue to clone life forms and human body parts; we can continue to clone humans; we can continue to permit international sale of organs, the commercialization of fetal parts, the sale of sperm and eggs, the patenting of animals and human genes and cells... This is the course, with a few exceptions, that our governments, regulatory agencies, scientific review boards and courts have decided on.

We can also decide to do nothing, hoping that experts will come to rational consensus about the human body shop and cloning, and implement sound limits to the technology and the marketing of life. However, history has shown time and again that some marginalized technologists and scientists have virtually never limited a technology to its beneficial purposes: nuclear revolution brings in X-rays, but also brings us to the brink of global destruction; petrochemical revolution brings in increase in agricultural production, industrial output and transportation mobility, but also leads to species extinction, top-soil erosion, acid rain, ozone depletion, global warming, and myriad other environmental threats. The lessons of these technological instances are clear: Unless human choices control technology, technology will control human choices.

The new medical, genetic engineering, reproductive and regenerative technologies have and will save lives, and perhaps will provide new cures for humanity; but without appropriate limits these technologies and the market ideology behind them will also lead to the devaluation and commercial exploitation of the technologies. Just because a technology can be implemented does not mean that it should be implemented. It will require the realization that in a civilized society something is just not for sale.

To put control over unfettered technology advances will require a counterrevolution against the current march. The revolution in holistic health, natural childbirth, natural foods, and environmental movements help to develop respect for human life and the environment.

Our society is such that every revolution is in part a revival. The current legal guidelines are to see how new areas of technology can be inserted into the legal framework with the least disruption to existing legal interpretation.[466] The human body revolution needs to revive the understandings of numerous moral and ethical traditions that have always understood the body as meaningful and sacred. The Judeo-Christian argument is straightforward in its general rule – human body should not be used as mere means for ends. But not all religions share the same view.

A caution is in order here. Legislative enactment, particularly enactment regarding new technologies, often assumes a certainty that is lacking. Seizing on one or two reliable applications, statutory language tends to expand to the limits of its logic, thereby creating a legal structure too rigid or misdirected for the more slippery reality it tries to capture.[467]

While well established legal doctrines and concepts of informed consent, autonomy, procreative liberty and confidentiality will answer some questions, many situations do not fit neatly into existing categories. The human body shop, regeneration, and cloning are three such situations. Resolving these dilemmas will require more basic normative judgments about values of autonomy, privacy, discrimination, fairness and justice.

Yet each of these important values is itself a contested matter, over which individuals and groups often fiercely struggle. This is where the Golden Rules of different sectors tend to conflict each other. Rather than searching for broad, all encompassing solutions, we must analyze and reconcile the conflicts as they arise in particular areas of genetic testing, the body shop, regeneration and cloning, with all the certainty that entails.

The current behavior toward the body, as toward most of the natural world, is governed by the principle of efficiency. Whether in labor or in medicine, the code is clear: minimum input for maximum output in minimum time; hasten evolutionary processes via improving Mother Nature by shortening Father Time. This is a form of directed evolution.

Efficiency makes sense. Who after all, opposes efficiency? But if we treated life forms solely on the basis of efficiency, the consequences could be

---

[466] Hwa A. Lim, *Change: Business, corporate governance, education, scandals, technology and warfare*, (EN Publishing, Santa Clara, 2003).
[467] John A. Robertson, "Legal issues in genetic testing", in: *The Genome, Ethics and the Law*, (AAAS, Washington, DC, 1992), pp. 79-110.

very dire. Imagine giving children minimum food and affection for maximum loyalty and performance? Such utilitarian behavior will be viewed as pathological and illegal. Social welfare department will have taken the children away and put the parents in jail. Imagine treating friends in an efficient manner, those close to us will immediately recommend psychotherapy. Imagine if we treat our pets in an efficient manner, our neighbors will have called animal cruelty centers. Whether it is a child, a spouse, a friend or a pet, love and not efficiency is the primary mover. Yet this realization has not quite made its way into our public policies on the human body, and the natural world.

The French scientist Julien Offray de La Mettrie (1709-1751) called humans "perpendicularly crawling machines."[468] But we are more than machines: we regenerate our blood and our skin, and we now believe that moderate exercises will regenerate neurons. Parts of our internal organs, liver for example, have regenerative capability. Machines, on the other hand, are incapable of regenerating. Biochemist Rupert Sheldrake (1942-) notes:

> "One of the most striking ways in which living organisms differ from machines is their capacity to regenerate. No man-made device has this capacity. If a computer, for example, is cut into pieces, these cannot, of course, form new computers. They remain pieces of a broken computer. The same goes for cars, telephones, and any other kind of machinery."[469]

As we transcend market forces, we encounter a different view of the technologies of the human body shop. Empathetic consciousness about the human body recognizes the value of the techniques of transplantation, reprotech, genetic screening, genetic engineering, but does not idolize them. The majority of liver transplant patients are individuals who are suffering from liver disease due to alcohol abuse. Of all the liver disease patients, transplants can save at most a fraction of them. The real solution to liver disease is not and never will be transplantation, but rather lies in prevention. We should probably change the stressful, efficiency-oriented work and life habits that lead to alcoholism. The same argument applies to other preventable man-made diseases, including smoking. Curing cancer resulting from smoking is not the issue; stop smoking, which imparts harm to one's own health and the health of secondary smokers, is a much better prevention.[470]

---

[468] Julien Offray de La Mettrie, "*L'homme Machine*", 1748.

[469] Rupert Sheldrake, *The Presence of the Past; Morphic Resonance and the Habits of Nature*, (Random House, New York, 1988).

[470] Andrew Kimbrell, *The Human Body Shop: The engineering and marketing of life*, (HarperCollins Publishers, New York, 1993).

Biology is not useful if we cannot lead a meaningful life. Biology can do only so much. The rest lies in our own hand. We should probably espouse the following ideas:

> Prevention is better than cure,
> Regeneration via prevention,
> ...

▶ Sexual Reproduction Versus Cloning

Science and technology are advancing simultaneously in various fronts, not only in reprotech and cloning. These advances are great. For example, a person's outlook and behavior is a function of his or her past: a bitter past is likely to make one more pessimistic; while a pleasant past is likely to make one more optimistic. It is likely that in the future, in brain research, we will be able to "erase" and "load" memories to change a person's personality!

Science and technology are also being applied in many fronts, with the invisible hand of the market force – the catalyst. There is an abundance of products and treatments, surgery and drugs available today. They promise slimmer stomach, larger biceps, straighter and whiter teeth, youthful hair, quicker minds, and firmer and perhaps more voluptuous breasts... Not all of these treatments, surgery or drugs, work, of course. Not that we suggest everyone should do so, but if image is all that we want, these are potential alternatives. The diet industry alone is worth $50 billion annually in the U.S.!

But if passing on our genetic legacy is what we want, consider this: With sexual (coital) reproduction, only one half of our genome is passed onto our offspring at a time. Therefore a particular trait (gene) is often passed from a parent to the offspring with a probability smaller than unity. Thus, in order to transmit as many of the genes onto the next generation, we have to produce as many offspring as possible to maximize the probability. This might constitute a difficult task for all competitors are trying to do the same and resources are getting more and more scarce. There are other sources of our genes available: in relatives. Hamilton's kin selection tells us that we should be nice to our relatives. Otherwise we may have to opt for cloning...

But sexual reproduction sustains diversity; cloning cannot sustain, let alone increase, diversity...

Consider also: In the late 1970s, Japanese robotics expert Masahiro Mori published what would become a highly influential insight into the interplay between robotic design and human psychology. Mori's central concept is that in a plot of similarity to humans on the x-axis against emotional reaction on the y-axis, a strange thing happens as the android becomes more lifelike.

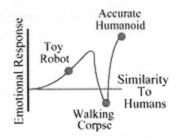

Figure 1. Masahiro Mori's Uncanny Valley.

Predictably, the curve rises steadily when emotional embrace grows as the android becomes more human-like. But at a certain point, close to true verisimilitude, the curve plunges down, through neutrality, indicating emotional embrace turns into real revulsion. The plot then rises again to a second peak of acceptance that corresponds with 100% human-likeness. This chasm is the well-known Mori's Uncanny Valley. The Uncanny Valley represents the notion that something that's like a human but slightly off will make people recoil.

In robotics, lifelikeness is approached from the left, getting increasingly more lifelike. Most robotics experts and designers take clues from Mori's Uncanny Valley and build robots that are not too lifelike. Otherwise onlookers' response may dip into the valley, indicating repulsion. But recently David Hanson of The University of Texas at Dallas built K-Bot, a replica of his girlfriend Kristen Nelson's face. But this is a different story of crossing the valley.[471]

Consider a hypothetical situation that we are sitting next to someone. We strike up a lively discussion. A few moments later, we begin to feel extremely comfortable, to the point of liking the individual. Then just at that point, the individual discloses to us that "it" is a clone. If Mori's Uncanny Valley is applicable in this case, we are approaching the Uncanny Valley from the right because we just realize "it" is less than "human" in the sense of what we know human. Would we recoil?

There are other possible situations. But this will suffice.

So would it be sexual (coital) reproduction:[472]

A sperm and an egg cell unite. This seemingly straightforward event sets in motion one of the most awe-inspiring process in all of biology. Within a short time the fertilized egg begins to divide,

---

[471] Dan Ferber, "The man who mistook his girlfriend as a robot", *Popular Science*, September 2003.
[472] Steve Olson (ed.), *Shaping the Future*, (National Academy Press, Washington, DC, 1989).

producing two cells, four cells, eight cells... The dividing cells form a ball, which then becomes hollow. Parts of the ball form dimples and ridges, with layers of cells moving inside other layers. Soon these layers thicken, forming different kinds of tissues. Other cells leave their points of origin and migrate through the developing embryos, eventually to become vertebrae, muscles, nerves... Tissues fold and protrude to form organs and limbs. The outline of a living creature takes shape. And already within the growing embryo, the sex cells that someday will repeat the entire process are growing...

Or would it be reproductive cloning:[473]

"Bokanovsky's Process," repeated the Director, and the students underlined the words in their little notebooks.
One egg, one embryo, one adult-normality But a bokanovskified egg will bud, will proliferate, will divide. From eight to ninety six buds, and every bud will grow into a perfectly formed embryo, and every embryo into a full-sized adult. Making ninety-six human beings grow where only one grew before. Progress.
"Essentially," the D.H.C. concluded, "bokanovskification consists of a series of arrests of development. We check the normal growth and, paradoxically enough, the egg responds by budding."[474]

●●●●●   ●●●●●   ●●●●●

Judging from the current state of the art of reproductive technology,

Sex sells,

Sex cells (eggs and sperm) sell

and the efficiency of reproductive cloning technology, we can hardly deny it,

Sex is so good, why clone?

And whatever it is,

Prevention is better than cure,

Regeneration via prevention

---

[473] Excerpt from Aldous Huxley, *The Brave New World*, (Penguin Modern Classic, 1932).
[474] We know that the description of the cloning process in *The Brave New World* is not quite the way cloning is being done. However, it is interesting to cite paragraphs from a fantasy from as early as 1932!

are better options. In the future when the cloning technique has been perfected, then we might talk about cloning on a case-by-case basis. And in unpreventable cases, we might talk about regenerative medicine.

Excellent References

"Better to be good than to be original."

- Ludwig Mies van der Rohe

1. Stan Davis, and Bill Davidson, *2020 Vision: Transform your business today to succeed in tomorrow's economy*, (Simon & Schuster, New York, 1991).
2. Gina Kolata, *Clone: The road to Dolly and the path ahead*, (Penguin Press, New York, 1997).
3. Andrew Kimbrell, *The Human Body Shop: The engineering and marketing of life*, (HarperCollins Publishers, New York, 1993).
4. James Le Fanu, *The Rise & Fall of Modern Medicine*, (Little, Brown and Company, London, 1999).
5. Hwa A. Lim, *Genetically Yours: Bioinforming, biopharming and biofarming*, (World Scientific Publishing Co., New Jersey, 2002).
6. Hwa A. Lim, *Change: Business, corporate governance, education, scandals, technology and warfare*, (EN Publishing Inc., California, 2003).
7. Gina Maranto, *Quest for Perfection: The drive to breed better human beings*, (Lisa Drew Book, New York, 1996).
8. Benjamin A. Pierce, *The Family Genetic Sourcebook*, (John Wiley & Sons, Inc., New York, 1990).
9. Jeremy Rifkin, *The Biotech Century: Harnessing the gene and remaking the world*, (Jeremy P. Tarcher/Putnam, New York, 1999).

Food for Thoughts

Chapter One

1. What is the difference of the definitions of sex in science and in everyday usage?
2. What are "survivals" in the sense of Sir Edward Tylor?
3. Why do politicians long for one-handed scientists?
4. What are some of the inherent aspects of science?
5. Compare decision-makings in science, business and the law
6. What should journalists try to do in accurate reporting, particularly of scientific breakthroughs?
7. What do scientists mean when they say the Y chromosome is decaying?
8. Cite some of the differences between the X-chromosome and the Y-chromosome.
9. Give a possible reason at the genomics level why different women behave differently?
10. Give a possible reason at the genomics level why men and women are different?

Chapter Two

1. Why do we call ourselves the Children Author of Mother Nature? How are we shortening Father Time?
2. What are the characteristics of living things? Give examples in which a living thing may lie dormant as if it is a nonliving thing.
3. Are lifestyle drugs good? Can you provide some examples, their uses and potential harm, if any?
4. What is the significance of Charles Darwin's "survival of the fittest"?
5. What are the differences between classical genetics and modern genetics?
6. Discuss historical observations of genetics in different parts of the world.
7. Can you give other examples of how genetic diseases changed the course of history other than hemophilia?
8. How are Gregory Mendel's observations of plant breeding important? How can Mendelian observations be related to Charles Darwin's "survival of the fittest"?

9. What is PCR? Why is it an important genetic tool?
10. Discuss the significance of the human genome project and its potential ethical issues.

Chapter Three

1. Discuss how some fish spawn.
2. What is parthenogenesis? Can you provide examples of animals reproducing by this means?
3. Why is sex an evolutionary disaster? If sex is an evolutionary disaster, why does it persist?
4. What are the two dominant hypotheses of evolution? Compare and contrast them?
5. Do you know of any supporting evidence of the Red Queen hypothesis?
6. What are selfish genes?
7. What is kin selection? How does this explain some of the problems faced by "survival of the fittest" of Charles Darwin?
8. Explain, using hymenopterans as an example, how eusociality works.
9. What is reciprocal altruism?
10. Can you explain why a drunkard would gladly die for two brothers, four cousins and eight second cousins? (Hint: Mendel's inheritance).

Chapter Four

1. What are survivals?
2. What are the stages of falling in love? What are Cupid's chemicals and their effects? Why are VNO important in the search for a mate?
3. What are the key differences in the sexual dances of the human and the chimpanzee?
4. Provide a plausible explanation of human male-female bond.
5. What is polygamy? Polygyny? Polyamory? Polyandry?
6. For the purpose of passing on genetic legacy, what is the advantage of taking on a lover?
7. Why is facial resemblance such a reassurance for a new father?
8. Can you explain the high divorce rate in modern society?
9. Why do aborigines farm out their children to other countries? Compare this with what the people in the West would normally do in similar situations.
10. Discuss the tug-of-war *in utero*.

Chapter Five

1. Discuss some of the definitions of beauty in history. What is the golden section? Discuss some of the recent research on the search for beauty.
2. What are the differences in the beauty of a woman and that of a man? Why is beauty not as important for the man?
3. Compare and contrast the concept of beauty in different cultures.
4. Discuss examples from the animal kingdom in which mothers cull weaker offspring to ensure survival of those most likely to survive to ensure the genetic legacy is passed on.
5. Discuss infanticides in ancient time. From a certain anthropological point of view, how was female infanticide practiced as a form of population control?
6. Compare and contrast selective paternal and maternal selection reproductions.
7. How may we say that the founders of Rome were exposure survivors?
8. Discuss how sex hormone levels can be controlled as a form of birth control.
9. What is *in vitro* fertilization? What is ICSI? What are the potential risks of IVF?
10. Explain the statement "what reprotech produces is parental choice, not the baby's welfare."

Chapter Six

1. What are the key differences between meiosis and mitosis? Does mitosis occur in germ line cell division?
2. What is ploidy? Haploidy? Diploidy? Why are they important in germ line and somatic cells?
3. What are the differences between the germ line cell production in the male and the female?
4. How can the food intake of expecting mothers affect the development of babies *in utero*?
5. What role does the invisible hand play in the commercialization of reproduction?
6. Compare and contrast alchemy and algeny.
7. Briefly discuss the stages of eugenics.
8. What are cacogenics and aristogenics?
9. Explain why eugenics eventually fell out of fashion?
10. How does modern genetics raise the possibility of a revival of eugenics?

Chapter Seven

1. What do scientists do to effect parthenogenesis?
2. What are the different meanings of cloning?
3. Why is IVF sometimes called sophisticated plumbing problem?
4. Discuss how Hollywood movies can do a disservice to science.
5. Discuss how the birth of Dolly the sheep made such a news splash even though a number of barnyard animals had been cloned before her.
6. What are the problems with the current cloning technologies? Why is the first attempt to clone the gaur so newsworthy?
7. Why is the donor's age important in the study of cloning?
8. What are the potential uses of artificial twinning for scientific research?
9. What can barnyard animal clones be used for? How is cloning more efficient in the pharming of therapeutic proteins? Why is one pharm animal preferred to another pharm animal for a specific purpose?
10. What are the potential uses of pig clones? Why are they the ideal choice for such purposes? What are the potential problems of using pig clone parts?

Chapter Eight

1. What is gene banking? Is gene banking restricted to only pets?
2. What are the differences in cloning a dog? A sheep? A cat? A rabbit?
3. Cc: the cloned cat does not quite look like the donor mother. Why is this so?
4. What is a key success factor in the French rabbit cloning effort?
5. Why are mules sterile? If so, is cloning the only way to produce a new mule? (Think over the second part of the question carefully).
6. Why is cloning a horse so difficult? What is so unique about Prometea as a clone?
7. Why are most champion horses gelded? If so, is cloning the only way to produce a new champion horse?
8. Argue for and against "if cloning of champion horses is possible, can a polo team of only identical champion horses be created?"
9. "A pet clone will not resurrect a pet." How do you explain this statement?
10. Discuss the efficiency of cloning. Provide plausible explanations for the efficiency.

Chapter Nine

1. What are the differences between identical and nonidentical twins?
2. Explain why identical twins are natural clones.
3. If identical twins are natural clones, why are they different when they grow up?
4. Why are identical twins more identical than clones?
5. In concordance studies of twins, why is it the difference in concordances that is a more reliable indicator than high concordances?
6. What are the methods of cloning? What are the key differences?
7. Why is synchronization of the donor cell with the egg cell critical in cloning?
8. Do cloning methods always involve no fertilization of an egg by a sperm?
9. Can you think of how cloning by nuclear transfer can be combined with twinning to obtain multiple clones?
10. Why is Fibro the cloned mouse so noteworthy?

Chapter Ten

1. Why is cloning a supermodel an attractive idea?
2. How do embryos develop into a boy or a girl?
3. How does sexual reproduction sustain genetic diversity?
4. What is nuclear reprogramming in cloning?
5. Why is it believed that monkeys cannot be cloned? Does this have any impact on the possibility of cloning humans?
6. What is the physiology of human cloning that is potentially making it difficult to clone a human?
7. Discuss the backgrounds of some of the human cloning mavericks.
8. Why is it disturbing that some scientific results are disclosed only years afterwards?
9. Why are cloning experiments aborted sometimes?
10. Why clone hybrids?

Chapter Eleven

1. Provide a reason why early transplantations ended in failure.
2. Do you know of a reason why the lung is the hardest organ to transplant?
3. What are stem cells and what are their medical uses?

4. How are stem cells derived? Why is there so much controversy over stem cell research?
5. What is the newfound use of cord blood? Of fat?
6. In cancer treatment, why do we say "first kill the patient, then save the individual?"
7. Why are multipotent adult stem cells not as useful as embryonic stem cells?
8. What is regeneration? Regenerative medicine?
9. Discuss the lifetimes of some of the cells in the human body. Why are some longer-lasting than the others?
10. If regeneration is a part of life, why do humans still die?

Chapter Twelve

1. Discuss the attitudes of the people in the east and that of the people in the west towards stem cell research.
2. Discuss some of the legal guidelines of stem cell research in some of the stem cell nations.
3. Discuss why most of the stem cell nations are also those which have been active in *in vitro* fertilization.
4. What are the criteria for a stem cell research to be eligible for federal funding in the U.S.?
5. What are the three phases of stem cell lines? And explain why the 78 cell lines in the U.S. stem cell registry is an over count.
6. To keep cells undifferentiated, stem cells are currently grown on mouse tissue, which acts as soil or feeder, and nourished with a liquid derived from cows and pigs. Discuss some of the contamination problems.
7. Why is the California Institute of Regenerative Medicine critical to the U.S. stem cell research preeminence?
8. What is the Declaration of Helsinki?
9. How do you think the Hwang clone-gate will affect stem cell research?
10. Give two examples, other than the Hwang clone-gate, of deliberate fraudulent scientific findings.

Chapter Thirteen

1. Why does biology have more ethical and moral implications than other sciences?
2. What are the different aspects of cloning?

3. Why is it that new sciences sometimes find it difficult to go mainstream?
4. Do scientists have responsibilities in educating the public? If so, how?
5. How do we say that if we clone, we are transitioning from "genetic roulette" to "genetic determinism?" Do you agree?
6. Why is third-party arrangement plagued with ethical dilemmas? What are the potential risks?
7. What sort of disruptions to the intra-family relationship will cloning cause?
8. Argue if a clone can run for the top office in your country.
9. Why is IVF a good precedent to the current cloning debate?
10. What is the west attitude towards technology in general? Towards cloning technology in particular?

Chapter Fourteen

1. Why was the early 20th century the century of the physical sciences?
2. What are the major differences between the restrictions of physics and biology on life?
3. What is the invisible hand? What do you think is a factor that prompted the birth of the concept?
4. What are the Golden Rules of different sectors: economists, political candidates, businesspeople, and a country?
5. Compare and contrast the lives of Charles Darwin and Gregory Mendel.
6. Do you agree that biological resources, like capital resources and natural resources, should be tapped?
7. If the invisible hand is so strong in the market, why is the commercialization of science undesirable in some cases?
8. Discuss how if a technology is left unfettered, it may lead to social problems.
9. Why is efficiency not practical when it comes to life?
10. Argue if sexual reproduction or cloning is more desirable.

Index

G

K

L

M

N

O

Q

S